Essentials of Molecular Biology

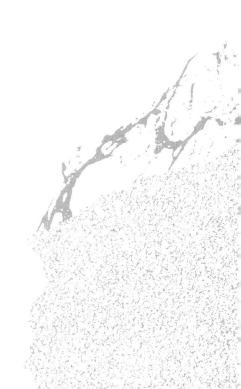

————The Jones and Bartlett Series in Biology————

SECOND EDITION

Essentials of
Molecular Biology

DAVID FREIFELDER
UNIVERSITY OF CALIFORNIA,
SAN DIEGO

EDITED BY
GEORGE M. MALACINSKI
INDIANA UNIVERSITY

JONES AND BARTLETT PUBLISHERS
Boston   London

Editorial, Sales, and Customer Service Offices
Jones and Bartlett Publishers
One Exeter Plaza
Boston, MA 02116

Jones and Bartlett Publishers International
PO Box 1498
London W6 7RS
England

Library of Congress Cataloging-in-Publication Data

Freifelder, David. **1935-1987.**
    Essentials of molecular biology / David Freifelder; edited by
George M. Malacinski. —2nd ed.
        p.   cm.
    Includes bibliographical references and index.
    ISBN 0-86720-137-1
    1. Molecular biology.   I. Malacinski, George M.   II. Title.
QH506.E87 1992   91-42634
574.8'8—dc20   CIP

PRINTED IN THE UNITED STATES OF AMERICA

96 95 94 93 92   10 9 8 7 6 5 4 3 2 1

It is better to understand a little than to misunderstand a lot.

Anatole France

Contents

4.

The Physical Structure of Protein Molecules *Alexandra Newton*

60

5.

Macromolecular Interactions and the Structure of Complex Aggregates
George M. Malacinski

80

6.

The Genetic Material
Robert L. Sinsheimer

98

Prologue

It is safe to say that no discipline in biology has ever experienced the explosion in growth and popularity that molecular biology is enjoying. Most scholars acknowledge that the discipline began its rapid growth in the early 1950s with the simple yet elegant descriptions of the structure of the DNA double helix. Intense intellectual curiosity provided the initial driving force for progress. Then a series of technological advances propelled the discipline to the point where its horizons now appear to be virtually unlimited.

During the early years, the focus was on the purification and characterization of the key macromolecules involved in DNA, RNA, and protein synthesis. From information on the biochemical and physical properties of the key macromolecules, functional roles were deduced. Aided by discoveries in microbial genetics, the general blueprint for gene structure and expression was elucidated in a remarkably short time. More recently, success with gene isolation (recombinant DNA technology or genetic engineering) has provided all the confidence molecular biologists needed to believe that a complete molecular understanding of virtually any biological phenomenon is within reach. That attitude, right or wrong, has in turn spawned an unprecedented interest in the commercial applications of biology. Molecular biology has led to the establishment of novel ties between universities and industries, and has powered the development of new disciplines such as molecular medicine.

Because of the broad extent to which the concepts of molecular biology have been integrated into most of the other disciplines of biology (including cell biology, developmental biology, genetics, and even evolutionary biology), the faculty at Indiana University, for example, revised its basic curriculum to position Molecular Biology as the first course beyond the freshman level.

In order to provide an effective learning experience for entry-level biology majors, an appropriate textbook was required—one that would not intimidate students by its depth, or overpower them with its extent of detail. Freifelder's *Essentials of Molecular Biology* was selected because it emphasizes the fundamental features of various aspects of DNA, RNA, and protein structure, function, and expression without delving into an inordinate amount of detail of the relevant molecular processes.

The first edition of *Essentials of Molecular Biology* has been hailed by students as a very successful text, both at Indiana University and at a variety of other institutions. One of the key features of this text is the "layering" approach to knowledge construction. Complexity is developed chapter by chapter, rather than presented all at once as a series of detail-laden descriptions. Students appreciate this approach. Recently, the introduction of a "critical thinking" approach to the undergraduate biology learning experience at Indiana University made a somewhat more suitable text necessary. Acknowledging the strengths of *Essentials of Molecular Biology* and its proven effectiveness, this second edition is particularly well suited to fit a "critical thinking" curriculum.

The following preface explains how this edition differs from the original book and, I hope, improves on it.

Preface

This second edition of *Essentials of Molecular Biology* represents a revision of Freifelder's highly successful original edition. It is intended to serve as the textbook for a short first course. Initially, a survey questionnaire was sent out to a few dozen faculty who use the textbook routinely in their molecular biology classes. The result emphasized that the chapter sequence should be left intact, and that by all means the brevity of presentation and emphasis on fundamentals that characterize the first edition should be maintained.

Because David Freifelder is deceased, the following strategy for updating and revising the text was undertaken. Experts in the areas included in individual chapters were invited to edit single chapters. Because molecular biology is indeed a human endeavor, a deliberate attempt was made to engage editors who represented all levels of career development, and who would therefore be able to provide in their personal profiles a glimpse of the diverse backgrounds, motivations, and perspectives that individual scientists have brought to the discipline.

Accordingly, senior researchers who participated in the early days of the development of molecular biology, as well as graduate students aspiring to careers in molecular biology, have contributed to this revision of *Essentials of Molecular Biology*. They were given the following instructions: review this chapter to ensure that it is accurate and complete; update where needed; and revise the prose, if necessary, to better explain a phenomenon. Experts were encouraged to be "anti-encyclopedic" in their approach. Simple two-color illustrations were to be employed, and tables of data were to be added only infrequently. Adhering to those instructions was probably not easy and required the exercise of considerable discipline, for contemporary molecular biology is in a data-collection mode, encouraged by such rapid technological advances as protein characterization and DNA sequencing.

I would like to thank the expert editors for succeeding in adhering to those guidelines. They responded enthusiastically, and all chapters have been revised to one extent or another. In a few cases chapters were completely rewritten, while in other cases they were only slightly modified. The editors avoided the temptation to cram too much information into a single sentence. They also held closely to the use of "stick diagrams" as illustrations. Those illustrations are the type a professor would draw on the blackboard, and are therefore easy to comprehend, for they communicate only the essential points. That is, of course, what this book is all about.

Essentials of Molecular Biology differs from the "mega-books" both in purpose, and in scope and depth. The big books contain more background information and substantially more detail. They are, by nature, not as ideally suited for beginning undergraduate courses, which emphasize intellectual development, as this book does.

Among the features that have been added to the revised edition are the following:

- ☐ Expanded lists of study questions, which have been graded in difficulty and conceptual complexity

- ☐ Personal profiles of the expert editors

- ☐ Chapter summaries

- ☐ *Scientific American* references

- ☐ An extensive glossary at the end of the book

- ☐ An Appendix that explains some chemical principles of molecular biology

ACKNOWLEDGMENTS

Herbert Nolan provided superb editorial and production services. His painstaking efforts in dealing with the manuscripts and his limitless patience with the editor and the contributing authors greatly enhanced the quality of this textbook.

Various Indiana University students, including graduate students Gregg Ashcroft, Wayne Weaver, Linda Martin, Dave Gilley, and Alex Burgin, and a team of undergraduates who were familiar with the original edition of *Essentials of Molecular Biology* read and re-read the manuscripts for clarity of expression and depth of coverage. They also assisted the editor in the preparation of chapter summaries and revised and expanded drill questions, problems, and conceptual questions, and assisted in reading the galley proofs.

Susan Duhon of the I.U. Axolotl Colony guided the preparation of the index and participated in editing several chapters.

Finally, several additional graduate students and undergraduates carefully read each revised chapter and suggested editorial changes that enhanced the readability of the text. They are hereby gratefully acknowledged.

LIST OF CONTRIBUTORS

Sankar L. Adhya
Laboratory of Molecular Biology
National Cancer Institute

Robert Cedergren
Department of Biochemistry
University of Montreal

Larry Gold
Department of Molecular, Cellular,
* and Developmental Biology*
University of Colorado

Philip C. Hanawalt
Department of Biological Sciences
Stanford University

Christie A. Holland
Department of Radiation Oncology
University of Massachusetts Medical
* School*

Joseph Ilan
Department of Anatomy
Case Western Reserve University

Judith A. Jaehning
Department of Biology
Indiana University

Thomas Lindahl
Imperial Cancer Research Fund
Clare Hall Laboratories

George M. Malacinski
Department of Biology
Indiana University

Kenneth Marians
Memorial Sloan Kettering

Hans-Peter Müller
Institute of Molecular Biology (II)
University of Zurich

Alexandra Newton
Department of Chemistry
Indiana University

David Parma
Department of Molecular, Cellular,
* and Developmental Biology*
University of Colorado

Barry Polisky
Department of Biology
Indiana University

John Richardson
Department of Chemistry
Indiana University

Walter Schaffner
Institute of Molecular Biology (II)
University of Zurich

Robert L. Sinsheimer
Department of Biological Sciences
University of California, Santa Barbara

Dorothy M. Skinner
Biology Division
Oak Ridge National Laboratory

George M. Malacinski
Professor of Biology
Department of Biology

Birthday: **25 November 1940**

Birth Place: **Norwood, Massachusetts**

Undergraduate Degree: **Boston University**
Majors: Biology and Chemistry 1962

Graduate Degree: **Indiana University,**
Ph.D. 1966, Biochemistry

Postdoctoral Training: **University of**
Washington, 1966–1968, Biochemistry

Present Position: **Indiana University,**
Biology Department

Address: **Bloomington, Indiana**

SKELETAL MUSCLE Is one of the first tissues to differentiate in early vertebrate embryogenesis. Because muscle is so highly specialized and contains such high concentrations of the contractile proteins actin and myosin, it is an excellent model system for investigating the mechanisms that regulate gene expression. Accordingly, my research program focuses on elucidating the structure, transcription, and translation of the myosin-heavy chain genes. The amphibian embryo provides a convenient model system because of its large size, external development, and ready availability, so it is used as the experimental organism. Rather than containing a single myosin gene, however, the genome contains a family of up to a dozen or, in some animals, even more closely related myosin genes, each coding for a unique myosin-heavy chain protein. Some of the genes are expressed only in skeletal muscle, while others are expressed in other muscles as well. Several of the genes appear to be expressed only at certain times in development (e.g., embryo, larva, adult).

By isolating and characterizing the genes of the nucleotide sequence level, and using recombinant DNA techniques to generate novel combinations of coding sequence/upstream sequence, my research program seeks to elucidate the mechanisms that act to control when and where individual myosin genes are expressed.

What advice can you offer to undergraduates who want to become scholars of molecular biology?

Develop your analytical thinking skills. Prospective molecular biologists should take every opportunity during their undergraduate training to (1) engage in problem solving exercises; (2) become proficient in analyzing data (e.g., graphs, charts, etc.); and (3) learn to design experiments. So much of the day to day "goings on" in contemporary molecular biology is data oriented (vs. concept oriented). To be successful, one needs to become highly competent in those aspects of intellectual activity, and come to feel comfortable when engaged in those sorts of endeavors.

1

Systems of and Analytical Approaches to Molecular Biology

The term molecular biology was first used in 1945 by William Astbury, who was referring to the study of the chemical and physical structure of biological macromolecules. By that time, biochemists had discovered many fundamental intracellular chemical reactions, and the importance of specific reactions and of protein structure in defining the numerous properties of cells was appreciated. However, the development of molecular biology had to await the realization that the most productive advances would be made by studying "simple" systems such as bacteria and bacteriophages (bacterial viruses), which yield information about basic biological processes more readily than animal cells. Although bacteria and bacteriophages are themselves quite complicated biologically, they enabled scientists to identify DNA as the molecule that contains most, if not all, of the genetic information of a cell. Following this discovery, the new field of molecular genetics moved ahead rapidly in the late 1950s and early 1960s and provided new concepts at a rate that can be matched only by the development of quantum mechanics in the 1920s. The initial success and the accumulation of an enormous body of information enabled researchers to apply the techniques and powerful logical methods of molecular genetics to the subjects of muscle and nerve function, membrane structure, the mode of action of antibiotics, cellular differentiation and development, immunology, and so forth. Faith in the basic uniformity of life processes was an important factor in this rapid growth. That is, it was believed that the fundamental biological principles that govern the activity of simple organisms, such as bacteria and viruses (organisms that lack an organized nucleus) must apply to more complex cells; only the details should vary. This faith has been amply justified by experimental results.

In this book prokaryotes and eukaryotes will be discussed separately and compared and contrasted. Usually prokaryotes will be discussed first because they are simpler. In keeping with this, we begin by describing the properties of bacteria that are important in molecular biological research.

Bacteria

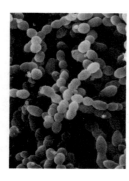

Bacteria are free-living unicellular organisms. They have a single chromosome, which is not enclosed in a nucleus (they are prokaryotes), and compared to eukaryotes they are simple in their physical organization. For all practical purposes, a bacterium can be thought of as a solution of several thousand chemicals and a few organized particles enclosed in a rigid cell wall.

Bacteria have many features that make them suitable objects for the study of fundamental biological processes. For example, they are grown easily and rapidly and, compared to cells in multicellular organisms, they are relatively simple in their needs. The bacterium that has served the field of molecular biology best is *Escherichia coli* (usually referred to as *E. coli*), which divides every 20 minutes at 37°C under optimal conditions. Thus, a single cell becomes 10^9 bacteria in about 20 hours. Bacteria can be grown in a **liquid growth medium** or on a solid surface. A population growing in a liquid medium is called a bacterial **culture.** If the liquid is a complex extract of biological material, it is called a **broth.** If the growth medium is a simple mixture containing no organic compounds other than a carbon source such as a sugar, it is called a **minimal medium.** A typical minimal medium contains each of the ions Na^+, K^+, Mg^{2+}, Ca^{2+}, NH_4^+, Cl^-, HPO_4^{2-}, and SO_4^{2-}, and a source of carbon such as glucose, glycerol, or lactate. If a bacterium can grow in a minimal medium—that is, if it can synthesize *all* necessary organic substances such as amino acids, vitamins, and lipids—the bacterium is said to be a **prototroph.** If any organic substances other than a carbon source must be added for growth to occur, the bacterium is termed an **auxotroph.** For example, if the amino acid leucine is required in the growth medium, the bacterium is a leucine auxotroph; the genetic symbol for such a bacterium is Leu$^-$. A prototroph would be indicated Leu$^+$. Bacteria are frequently grown on solid surfaces. The earliest surface used for growing bacteria was a slice of raw potato. This was later replaced by media solidified by gelatin. Because many bacteria excrete enzymes that digest gelatin, an inert gelling agent was sought. **Agar,** which is a gelling agent obtained from a variety of seaweed and used extensively as a thickening

liquid growth medium
culture
broth

minimal medium

prototroph
auxotroph

petri dish
plate
plating

agent in Japanese cuisine, is resistant to bacterial enzymes and has been universally used. A solid growth medium is called a nutrient agar if the agar was added to a broth rich in nutrients. Otherwise, it is called minimal agar. Solid media are typically placed in a **petri dish.** In lab jargon a petri dish containing a solid medium is called a **plate** and the act of depositing bacteria on the agar surface is called **plating.**

bacterial colony

A bacterium growing on an agar surface divides. Since most bacteria are not very motile on a solid surface, the progeny bacteria remain very near the location of the original bacterium. The number of progeny increases so much that a visible cluster of bacteria appears. This cluster is called a **bacterial colony** (Figure 1-1). Colony formation allows one to determine the number of bacteria in a culture. For instance, if 100 cells are spread over the surface of a plate, 100 colonies will appear the next day.

Plating is a method for determining if a bacterium is an auxotroph. This is done in the following way. Minimal agar and nutrient agar plates are prepared. Several hundred bacteria are plated on each plate and the plates are incubated overnight in an oven. Several hundred colonies are subsequently found on the nutrient agar because it contains so many substances that it can satisfy the requirements of nearly any bacterium. If colonies are also found on the minimal agar, the bacterium is a prototroph; if no colonies are found, it is an auxotroph and some required substance is not present in the minimal agar. Minimal plates are then prepared with various supplements. If the bacterium is a leucine auxotroph, the addition of leucine alone will enable a colony to form. If both leucine and histidine must be added, the bacterium is auxotrophic for both of these substances.

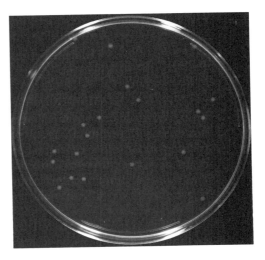

Figure 1-1
A petri dish containing colonies of *E. coli* grown on agar. (Courtesy of Robert Haynes.)

METABOLIC REGULATION IN BACTERIA

Bacteria are precisely regulated and highly efficient organisms. For example, they rarely synthesize substances that are not needed. Thus the enzymatic system for synthesizing the amino acid tryptophan is not formed if tryptophan is present in the growth medium, but when the tryptophan in the medium is used up, the enzymatic system will be rapidly formed. The systems responsible for utilization of various energy sources are also efficiently regulated. A well-studied example is the metabolism of the sugar lactose as an alternate carbon source to glucose. Control of both tryptophan synthesis and lactose degradation are two examples of **metabolic regulation.** This very general phenomenon will be explored extensively throughout the book. Both simple and complex regulatory systems will be seen, all of which act to determine how much of a particular compound is utilized and how much of each intracellular compound is synthesized at different times and in different circumstances. This will demonstrate the length to which the so-called simple cells have gone to utilize limited resources efficiently and to optimize their metabolic pathways for efficient growth.

metabolic regulation

Bacteriophage

bacteriophage
phage

Bacteria are subject to attack by viruses. These small particles are called **bacteriophage,** or simply **phage,** and they are capable of growing only inside bacteria. Phage★ have been the object of choice for a great many types of experiments because they are much simpler than bacteria in their structures (usually having between two and ten components) as well as their life cycles and yet possess the most essential, if minimal, attributes of life.

Most phages contain only a few different types of molecules, usually several hundred protein molecules of one to ten types (depending on the complexity of the phage) and one nucleic acid molecule. The protein molecules are organized in one of three ways. In the most common mode the protein molecules form a protein shell called the **coat** or **phage head,** to which a **tail** is generally attached; the nucleic acid molecule is contained in the head (Figure 1-2). Another form of a phage is a tailless head. The least common form is a filament in which the protein molecules form a tubular structure in which the nucleic acid is embedded. Phages are known that contain double-stranded DNA (the most common variety), single-stranded DNA, single-stranded RNA, and double-stranded RNA (least common).

coat
phage head
tail

Phage are parasites and cannot multiply except in a host bacterium.

★ The plural word phages refers to different species; the word phage is both singular and plural and in the plural sense refers to particles of the same type. Thus, T4 and T7 are both phages, but a test tube might contain either 1 T7 phage or 100 T7 phage.

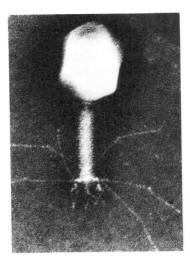

Figure 1-2
An *E. coli* T4 phage. The DNA is contained in the head. Tail fibers come from the pronged plate at the tip of the tail. (Courtesy of Robley Williams.)

Thus, a phage must be able to enter a bacterium, multiply, and then escape. There are many ways by which this can be accomplished. However, a basic life cycle is outlined below and depicted in Figure 1-3.

The life cycle of a phage begins when a phage particle adsorbs to the surface of a susceptible bacterium. The phage nucleic acid then leaves

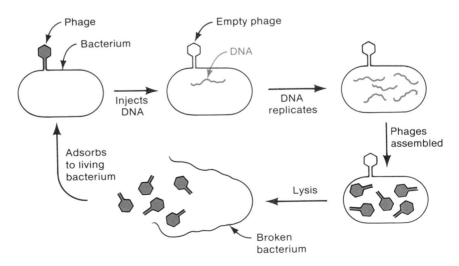

Figure 1-3
A schematic diagram of the life cycle of a typical bacteriophage. The number of phage released usually ranges from 20 to 500.

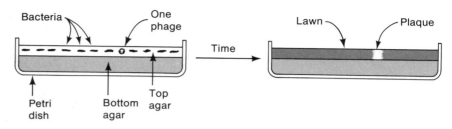

Figure 1-4

Schematic drawing of plaque formation. Bacteria grow and form a translucent lawn. There are no bacteria in the vicinity of the plaque, which remains transparent.

the phage particle through the phage tail (if the phage has a tail) and enters the bacterium through the bacterial cell wall. In a complicated but understandable way the phage essentially converts the bacterium to a phage-synthesizing factory. Within about an hour, the time varying with the phage species, the infected bacterium bursts or **lyses** and several hundred progeny phage are released. The suspension of newly synthe-sized phage is called a **phage lysate.**

lysis

Phage multiply faster than bacteria. A typical bacterium doubles in about half an hour, while a single phage particle gives rise to more than 100 progeny in the same time period. Each of these phage can then infect more bacteria, and those released in this second cycle of infection can infect even more. Thus, in two hours there are four cycles of infection for both a bacterium and a phage yet a single bacterium has become $2^4 = 16$ bacteria and a single phage becomes $100^4 = 10^8$ phage particles.

By using procedures that make the bacterial cell membrane and cell wall permeable, infection can be initiated with free DNA rather than by phage particles. This technique, which is called **transfection,** is ex-ceedingly useful, as it allows an experimenter to alter the DNA molecule either chemically or physically and then study the effect of the change by infecting a cell with the altered DNA. Transfection is an essential procedure in genetic engineering and has been widely used in studying animal and plant viruses.

The life cycle of a phage is highly regulated, but in a slightly different way from the metabolic regulation of a bacterium. Phages are totally dependent on the metabolism of their host bacteria, so the regulatory systems of the hosts usually control the phage's basic metabolic processes such as energy generation and synthesis of the precursors of DNA, RNA, and proteins. The job of an infecting phage is to reproduce itself by synthesizing its own nucleic acid and structural proteins, and finally to cause the bacterial cell wall to break so that progeny phage can escape. This requires that the various steps in phage production be regulated in time. Study of this regulation has been an important part of molecular biological research and has yielded a great deal of information about basic processes in all cells. Some examples will be seen in Chapter 14.

Phage are counted by a technique called the **plaque assay** (Figure

Figure 1-5
Plaques of *E. coli* phage T4. Two types of plaques are present. The smaller plaques are made by wild-type phage; the larger plaques are those of an *rII* mutant. (Courtesy of A. H. Doermann.)

1–4). About 10^8 bacteria plus a phage sample are added to warm liquid agar, which is then poured onto solid agar where it hardens. The bacteria multiply, forming a turbid layer in the agar called a lawn. While the bacteria are multiplying, each phage adsorbs to one cell and initiates an infection, resulting in the release of about 100 phage per cell, which remain localized in the agar. These progeny phage adsorb to nearby bacteria and multiply again; several cycles of infection occur, giving rise to a clear transparent hole in the turbid layer. This clear area is called a **plaque** (Figure 1-5). Since each phage forms one plaque, the individual phage can be counted. Different phage mutants often produce characteristic identifiable plaques, making the technique useful for genetic analysis.

Yeasts

Yeasts are unicellular organisms that have been used for millennia for producing wine and beer (Figure 1-6 (a,b)). A great deal of early biochemical research was carried out with yeasts rather than bacteria, work stimulated mainly by interest in understanding and improving beer.

Yeast cells are propagated in the laboratory and counted in much the same way as bacteria. They grow in liquid suspensions in either chemically defined media or in complex broths. They also grow on a solid surface to form colonies. The multiplication mechanism of all but the fission yeasts differs from the simple splitting of a mature bacterium in that yeast cells do not divide but multiply by budding (Figure 1-6(a)).

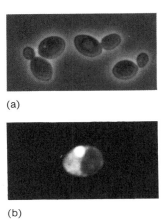

(a)

(b)

Figure 1-6
(a) A light micrograph of the yeast *Saccharomyces cerevisiae.* Many cells are budding. (Courtesy of Breck Byers.) (b) A fluorescence micrograph of a single cell stained with a fluorescent dye that binds to nucleic acids. The bright spot is the nucleus. The dark region is the vacuole, a liquid-filled sac that is free of nucleic acids.

That is, each mother cell produces a daughter cell (a bud) by outgrowth from the cell wall of the mother. The daughter cell grows and matures and can then also produce progeny by budding. The mother cell can bud many times.

In the past, most of the research effort in molecular biology has been with bacteria and viruses, because of their simplicity compared to eukaryotes. In recent years technological advances have made possible efficient and informative study of eukaryotes. The yeast *Saccharomyces cerevisiae* has been an important object of study: it has the genetic organization of eukaryotes and uses many regulatory strategies that are similar to higher organisms, yet the ease of handling and speed of growth are those encountered with typical microorganisms. Of particular interest also is the fact that *Saccharomyces* has both haploid and diploid phases. Haploid cells are of two types, which can mate sexually to form a stable diploid cell line. In particular nutritional conditions diploids undergo meiosis, generating haploid sexual spores. This mating system allows detailed genetic experiments to be performed and makes yeast a useful system for the study of both genetic recombination and the mechanism of meiosis.

Other unicellular eukaryotes, namely the alga *Chlamydomonas* and the protozoan *Tetrahymena*, also possess many of the attributes of yeast and are being used more frequently in eukaryotic molecular biology.

Animal Cells

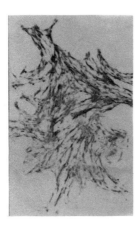

Figure 1-7
A microcolony of Chinese hamster fibroblasts that have been growing for a few days on a glass surface. (Courtesy of Theodore Puck.)

As more information about fundamental mechanisms in bacteria has been gained, increasing efforts have been made to study animal cells. Research with animal cells is exciting because we are beginning to understand such complex processes as hormonal regulation and the development of an egg into an adult organism, and to gain some insight into the differences between normal cells and cancer cells. However, research with animal cells has proceeded much more slowly than work with bacteria. There are two reasons for this. First, animal cells divide every 24 to 48 hours, whereas many bacteria divide every 25 to 50 minutes, so experiments with animal cells often take much longer than experiments with bacteria. Second, bacteria growing in culture are not significantly different from bacteria in nature, but experiments with animal cells require that the cells be removed from the animal and often separated from one another. Cells treated in this way have lost the normal route for receiving nutrients and are definitely in an unnatural state. Many growth media that enable cells to grow in culture have been developed. They have been designed to keep the cells alive as long as possible, but they do not always maintain a normal state for the cell. The most obvious difference is that most cells taken from an organism die within a few weeks. Furthermore, during that period of time the cells grow and divide, which differs from the normal quiescent (or inactive) state of cells in living organisms. A few cells survive and grow indefinitely; these are said to have been *immortalized* and have generated an **established cell line.** In time, these cells develop abnormalities in chromosome number (often extra chromosomes) and behave somewhat like tumor cells (lose their specialized traits). A complete discussion of the problems of working with animal cells is beyond the scope of this book; suffice it to say that a variety of techniques are available that make possible experiments with animal cells. A discussion of this phenomenon can be found in almost any text on cell biology. Figure 1-7 shows cells growing in culture.

Genetics and Molecular Biology

Genetics is an abstract and logical system by which the genetic factors determining certain properties of living cells can be located with respect to one another and by which changes in these properties can be expressed in quantitative terms. Genetics alone cannot prove anything about molecular mechanisms because the conclusions it leads to are independent of the molecular basis of biological phenomena. However, genetic analyses have been the major source of intuition and subsequent strong inference about molecular mechanisms and have forced scientists to consider phenomena that might otherwise be ignored. Genetics-based arguments can be so suggestive of a particular mechanism, and the suggested mechanism has so frequently turned out to be the correct one, that molecular biologists often view alternative and conflicting ideas as not even worthy of consideration. The following sections describe a few of the more common uses of genetics in molecular biology.

USES OF MUTANTS: SOME EXAMPLES

Some of the most significant advances in molecular biology have come about through the analysis of mutants. In the following paragraphs the kinds of approaches that have been taken are described.

1. A *mutant defines a function.* For example, the intake of Fe^{3+} ions by bacteria might be by passive diffusion through the cell membrane, or some system might be responsible for the process. Wild-type *E. coli* can take in the Fe^{3+} ion from a 10^{-5} M solution, but mutants have been found that cannot do so unless the ion concentration is very high. This finding indicates that a genetically determined system for Fe^{3+} intake exists, although the observation does not tell what this system is. Temperature-sensitive mutants (mutants that exhibit a defect only above a particular temperature) are especially useful in defining functions. For example, temperature-sensitive mutants of *E. coli* have been isolated that

fail to synthesize DNA. These mutants fall into at least ten distinct classes, suggesting that there may be at least ten different steps, each involving separate proteins, which are required for DNA synthesis. One of those mutants lacks a particular protein that is normally located in the cell membrane; the interpretation of such an experimental result is not unambiguous but does suggest that there is some connection between DNA synthesis and the cell membrane.

2. *Mutants can introduce biochemical blocks that aid in the elucidation of metabolic pathways.* The metabolism of the sugar galactose, for example, requires the activity of three distinct genes called *galK*, *galT*, and *galE*. If radioactive galactose ([^{14}C]Gal) is added to a culture of *gal*$^+$ cells, many different radioactive compounds can be found as the galactose is metabolized. At very early times after addition of [^{14}C]Gal, three related compounds are detectable: [^{14}C]galactose 1-phosphate (Gal-1-P), [^{14}C]uridine diphosphogalactose (UDP-Gal), and [^{14}C]uridine diphosphoglucose (UDP-Glu). Different mutant genes will block different steps of the metabolic pathway. If the cell is a *galK*$^-$ mutant, the [^{14}C]Gal label is found only in galactose. Thus the *galK* gene is known to be responsible for the first metabolic step. If the mutant *galT*$^-$ is used, Gal-1-P accumulates. Thus the first step in the reaction sequence is found to be the conversion of galactose to Gal-1-P by the *galK* gene product (namely, the enzyme galactokinase). If a *galE*$^-$ mutant is used, some Gal-1-P is found but the principal radiochemical is UDP–Gal. Thus the biochemical pathway must be

The identity of X cannot be determined from these genetic experiments.

3. *Mutants enable one to learn about metabolic regulation.* Many mutants have been isolated that alter the amount of a particular protein that is synthesized or the way the amount synthesized responds to external signals. These mutants define regulatory systems. For example, the enzymes corresponding to the *galK*, *galT*, and *galE* genes are normally not present in bacteria but appear only after galactose is added to the growth medium. However, mutants have been isolated in which these enzymes are always present, whether or not galactose is also present. This indicates that some gene is responsible for turning the system of enzyme production on and off, and this regulatory gene must be responsive to the presence and absence of galactose.

4. *Mutants enable a biochemical entity to be matched with a biological function or an intracellular protein.* For many years an *E. coli* enzyme called DNA polymerase I was studied in great detail. Purified polymerase I is

capable of synthesizing DNA *in vitro,* so it was believed that this enzyme was also responsible for *in vivo* bacterial DNA synthesis. However, an *E. coli* mutant (*polA⁻*) was isolated in which the activity of polymerase I was reduced 50-fold yet the mutant bacterium grew and synthesized DNA normally. This observation suggested strongly that polymerase I could not be the only enzyme that synthesized intracellular DNA. Indeed, biochemical analysis of cell extracts of the *polA⁻* mutant showed the existence of two other enzymes, polymerase II and polymerase III, which could, when purified, also synthesize DNA. In further study, a temperature-sensitive mutation in a gene called *dnaE* was found to be unable to synthesize DNA at 42°C, although synthesis was normal at 30°C. The three enzymes, polymerases I, II, and III, were isolated from cultures of the *dnaE⁻* mutant and each enzyme was assayed. Although polymerases I and II were active at both 30°C and 42°C, polymerase III was active at 30°C but not at 42°C, so that polymerase III was determined to be the product of the *dnaE gene* and the enzyme responsible for intracellular DNA synthesis.

5. *Mutants locate the site of action of external agents.* The antibiotic rifampicin prevents synthesis of RNA. When first discovered, it was not known whether rifampicin might act by preventing synthesis of precursor molecules (by binding to DNA and thereby preventing the DNA from being transcribed into RNA) or by binding to RNA polymerase, the enzyme responsible for synthesizing RNA. Mutants were isolated that were resistant to rifampicin. These mutants were of two types—those in which the bacterial cell wall was altered so that rifampicin could not enter the cell (an uninformative type of mutant) and those in which the RNA polymerase was slightly altered. The finding of the latter mutants proved that the antibiotic acts by binding to RNA polymerase.

6. *Mutants can indicate relations between apparently unrelated systems.* Bacteriophage λ, which normally adsorbs to and grows in *E. coli,* fails to adsorb to a bacterial mutant unable to metabolize the sugar maltose. Such failure is not associated with mutants incapable of metabolizing other sugars or with any other phages, and this knowledge implicated some product or agent of maltose metabolism in the adsorption of λ.

Genetic Analysis of Mutants

A great deal of information about molecular mechanisms can be obtained by combining separately isolated mutations in various combinations. This is done by **genetic recombination.** Another important use of genetic recombination is the construction of an array that indicates the

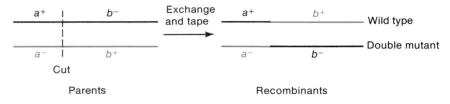

Figure 1-8
A schematic diagram showing genetic exchange.

positions of genes on a chromosome with respect to one another. When this is done by genetic techniques, the array is called a **genetic map.** **Physical maps** have also been constructed by using various physical techniques, and some elegant experiments have shown that the gene positions in the two maps are identical. Another genetic technique is **complementation.** By this procedure it is possible to determine the number of genes responsible for a particular phenotype and to distinguish genes from regulatory sites. In the following sections both genetic recombination and complementation will be reviewed.

complementation

GENETIC RECOMBINATION AND GENETIC MAPPING

Genetic recombination includes a variety of phenomena in which genetic loci are rearranged. Here we will be concerned only with the process of combining two genes or two mutations, initially on two chromosomes, onto a single chromosome. The molecular basis is quite complex; for our purposes it is sufficient to refer to the "scissors–and-tape" mechanism. In this, one assumes that two chromosomes align with one another, that a cut is made in both chromosomes at random but matching points, and that the four fragments are then joined together to form two new combinations of genes. This is a naive and incorrect model; it enables one to account for only the simpler features of gene exchange, but these features are in fact the only ones that are of concern at this time. The process is depicted in Figure 1-8. Two parental chromosomes having the genotypes a^+b^- and a^-b^+ pair, are cut, and are then joined to form two recombinant chromosomes whose genotypes are a^+b^+ (wild type) and a^-b^- (double mutant).

In genetic mapping, distance along a chromosome between two recombining genetic loci (or mutations) determines the recombination frequency. As long as the two loci are not too near one another, the recombination frequency is proportional to distance, because chromosomal cuts are made at random. Thus, in the following crosses between chromosomes, the genotypes of which are $a^+b^-c^-$ and $a^-b^+c^+$, and the genes of which are in alphabetical order and equally spaced.

$$\frac{a^+ \quad b^- \quad c^-}{\times}$$
$$a^- \quad b^+ \quad c^+$$

there will be twice as many a^+c^+ recombinants as a^+b^+ recombinants, because loci a and c are twice as far apart as loci a and b.

Because the recombination frequency is proportional to distance, recombination frequency can be used to determine the arrangement of genes on the chromosome. This can be seen in a simple example (Figure 1-9). Consider three genes a, b, and c, whose arrangement is unknown. Using the notation $p \times q = m$ percent to denote a recombination frequency of m percent between genes p and q, we assume it has been shown that $a \times b = 1$ percent and that $b \times c = 2$ percent. The two arrangements shown in the figure are consistent with these values and can be distinguished by determining the recombination frequency between a and c. Let us assume that this is 1 percent. If that is the case, only arrangement II is possible. The order $c\ a\ b$ for these genes and the relative separation constitute a genetic map.

Any number of genes can be mapped in this way. For instance, consider a fourth gene, d, in the preceding example. If $d \times b = 0.5$ percent, d must be located 0.5 unit either to the left or to the right of b. If $a \times d = 1.5$ percent, d is clearly to the right of b and the gene order is $c\ a\ b\ d$. If $a \times d = 0.5$ percent, the gene order would be $c\ a\ d\ b$.

COMPLEMENTATION

Complementation analysis is a technique that can be used to characterize a mutation. When one is dealing with multiple organisms that have identical mutant phenotypes, it would be helpful to know if these organisms have mutations of the same gene, or if it is possible that a single mutation on one of several genes might result in the same mutant phenotype. Simply put, complementation analysis allows one to tell whether two organisms (or strains of organisms) that exhibit identical mutant phenotypes have mutations of the same gene. It also can be used to elucidate the number of genes involved in a metabolic pathway.

Complementation testing is based on the fact that genes code for proteins—usually enzymes. If a gene should become mutated, that gene

Figure 1-9
Two arrangements of three genes for which $a \times b = 1$ percent and $b \times c = 2$ percent. See text for discussion.

might not make a completely functional product, resulting in a mutant phenotype. The key criterion for performing complementation analysis is that two copies of the gene(s) in question must be within the same cell. Performing this procedure in bacteria (which are haploid) necessitates the creation of a "merodiploid" or "partial diploid" cell (one with the "normal" genome plus an extra copy of the gene(s) in question). This extra genetic material is introduced into the cell by methods discussed in later chapters. This "merodiploid" is not necessary in yeast, because two haploid cells, with certain restrictions, can "combine" and form a diploid cell. Suppose that we have a hypothetical strain of yeast that synthesizes an enzyme we will refer to as enzyme x. Enzyme x catalyzes the conversion of lactate to pyruvate, and is composed of two different protein molecules. If either of the two protein molecules (call them a and b) are incorrectly made, the yeast cell will not be able to convert lactate to pyruvate. Therefore, if either gene 1 (which codes for a) or gene 2 (which codes for b) is mutated, x will not be functional—unable to convert lactate to pyruvate. As it happens, in our strains of wild-type yeast, we find two colonies that are unable to convert lactate to pyruvate (which tells us that they have acquired at least one mutation such that a functional enzyme x is not made). We also have a strain of yeast with the same mutant phenotype, but in this case it is known that gene 1 is mutated, so that an incorrect protein a is synthesized. We can now use complementation analysis to determine which genes (1 or 2) are mutated in our two unknown colonies. Haploid cells from one of the colonies with an unknown mutation are combined with haploids from the strain with a mutation on gene 1 and are induced to form diploids. At this point, two situations can occur. Suppose the unknown mutant has a mutation on gene 2, so that proteins b is not functional (but a is, because nothing is wrong with gene 1). When the two cells combine to form one daughter cell, the daughter cell will have one copy of every gene from each parent—meaning it has two copies of gene 1 (one that is normal, and one from the known mutant strain that doesn't produce a functional protein a), and two copies of gene 2 (a normal copy from the known strain, and a mutated one from the unknown). Will the daughter cell be able to convert lactate to pyruvate (will it produce both functional a and b)? The answer is yes. To have any functional protein a, the cell must have an unmutated gene 1, and it does, so normal a is produced. The fact that the cell also happens to have one mutated copy of gene 1 in this case has no effect. To have functional b, the cell must have an unmutated gene 2, which it does. It makes no difference that the functional genes came from different parent cells—what is important is that the diploid daughter cell has a copy of each. Suppose the daughter cells produced by the mating of our known strain with the second, unknown strain still displayed the mutant phenotype (their mutations did not complement). When two mutations do not complement, one of three explanations exists: the mutations are mutations of the same gene, either mutation is in a regulatory site for the other, or either of the mutations is in a gene that produces an inhibitor of the other.

Rapid Progress in Molecular Biology

Direct experimentation has proven to be a powerful driving force in molecular biology. A variety of conceptual approaches have emerged that guide the thought processes of today's molecular biologist. Although certainly not exclusive to molecular biology, they are employed by most molecular biologists on an almost daily basis as data are collected, interpretations are formulated, conclusions are drawn, and additional experiments are designed. The following are examples of what constitutes the **logic of molecular biology.**

THE EFFICIENCY ARGUMENT

During the hundreds of millions of years that living organisms have been evolving, the rigors of competition for survival in an environment where resources are limited has led to a natural selection for efficiency. Accordingly, when attempting to understand how a mechanism or process works, molecular biologists intuitively favor the simplest, most efficient scheme over others that appear more complex, contrived, or cumbersome. This approach has been especially useful for understanding the molecular mechanisms involved in regulating bacterial metabolism, RNA, DNA, and protein synthesis, as well as phage life cycles. But for analogous processes in the cells of higher plants and animals, where intense competitive pressure between individual cells for survival is lacking, simple, efficient mechanisms do not always exist.

QUANTITATIVE ASSESSMENT

The molecular biologist is constantly searching for quantitative relationships. In the process of unraveling a molecular mechanism, an assessment of the timing, chemical balance, and flow of components in space is routinely made in order to establish whether a proposed mechanism or hypothesis is feasible. Often, proposed pathways or schemes can be either approved or disregarded based on quantitative assessments.

MODEL BUILDING

Models usually represent explanations or schemes that explain (often in the form of a diagram or cartoon) a complex hypothesis. In this context, they represent conceptual entities rather than the 3-D hardware associated with molecular model kits for constructing larger-than-life representations of the structure of one or another macromolecule (e.g., a protein). For example, models have been developed to explain how an antigen stimulates the immune system to produce antibodies that react specifically with the challenge antigen.

Models are generally regarded as being good only if they can be subjected to direct experimental test. That situation generally applies to the usefulness of almost any hypothesis. Molecular biologists are, therefore, constantly revising and refining their models as the interpretation of new experimental results call for modification.

PARALLELISM

Molecular biologists have great faith in the universality of basic biological processes. As a starting point in an attempt to understand a molecular mechanism or process, explanations that are satisfactory for simple systems (e.g., bacteriophages) are often scaled up to fit similar phenomena in more complex systems (e.g., mammalian cells). This approach does not always work, as has been demonstrated for messenger RNA processing, which is inherently different in eukaryotic cells. Nevertheless, newly discovered phenomena in molecular biology are often initially explained in terms of similar processes that had been analyzed earlier.

STRONG INFERENCE

Like all scientific enterprises, molecular biology is a human endeavor. Accordingly, intuition, which we might define here as a sort of comprehension based more on experience and common sense than on a conscious and elaborate line of logical argument, is often employed to develop a set of explanations or possible models for a particular molecular event. Strong inference is based on the exclusion of all but one final alternative explanation for a phenomenon. Initially, one states all the reasonable explanations for a particular phenomenon. Then by direct experimentation various possible explanations are ruled out, one by one. The final alternative, which cannot be ruled out, and which satisfies our common sense, is considered to be correct. This strong inference approach accounts for the relatively common use of the phrase "it is likely that" when experimental data are being interpreted.

Summary

Molecular biologists use a number of biological systems as tools. Bacteria are prokaryotes, which grow rapidly, either in liquid medium as a culture or on solid medium as colonies. Organisms capable of growing on minimal media, containing only inorganic salts and a carbon source, are called prototrophs; auxotrophs require additional compounds. Bacteria are often used to study metabolic processes. Bacteriophage are viruses which consist of only a few proteins and genetic material encoding a few genes. Phage grow inside bacteria and escape by lysing the cell. They are often used to study regulatory events. Yeasts are eukaryotes, which can be grown much like bacteria. They have a sexual lifestyle and allow the study of many eukaryotic-specific events. Established animal cell lines grow indefinitely in culture, have long division times, and are altered from their normal state in a living organism, but still maintain characteristics of differentiated cells. Molecular biologists also use genetics as a tool. Mutations can help define cellular functions, elucidate metabolic pathways, define regulatory elements, match enzymatic activity to a specific protein, locate the site of action of external agents, and indicate relationship between different systems.

When making conclusions, molecular biologists usually try to find the simplest, most efficient explanation of the data, using knowledge about the physical limitations of the cell and from simpler systems. Models, conceptual diagrams of a process, are made and tested until only one possible explanation remains.

Drill Questions

1. a. Is a cell in which the DNA is enclosed in a nuclear membrane a prokaryote or a eukaryote?
   b. Classify the following cells as either prokaryotes or eukaryotes: bacteria, fungi, algae, yeasts, amoebae, wheat cells, human liver cells.

2. Match the terms with the definitions:
   Terms:
   1. Minimal medium. 2. Prototroph. 3. Plating. 4. Auxotroph.
   Definitions:
   A. A cell that can grow in a minimal medium.
   B. Depositing bacteria or phage on an agar surface for growth.
   C. A cell that requires for growth organic substances other than an organic carbon source.
   D. A growth medium containing salts and only one organic compound, which is used as a carbon source.

3. Define haploid and diploid.

4. What is a partial diploid and what are its uses?

5. What is meant by transfection?

6. List several differences between animal cells within a living organism and those grown in culture. What is an additional difference in an established cell line?

7. Mutants that fail to synthesize a substance x have been found in four complementation groups, none of which are cis-acting (residing in the same chromosome as another mutation, for instance in a regulatory site). How many proteins are required to synthesize x?

8. What is a model? What makes a particular model a good one?

Problems

1. A bacterium divides every 35 minutes. If a culture containing 10^5 cells per ml is grown for 175 minutes, what will be the cell concentration?

2. Four genes $kyuA$, $kyuB$, $kyuC$, and $kyuQ$ are known to be required to synthesize substance q, and each biochemical reaction can be detected. The reaction sequence is $P \rightarrow B \rightarrow C$

→ A → Q in which the product of a gene *kyuX* is needed to synthesize substance *x*. Addition of ^{14}C-P yields ^{14}C-Q.

  a. A mutant is found for which addition of ^{14}C-P yields ^{14}C-A but not ^{14}C-Q. In what gene is this mutation?

  b. Another mutant is found for which there is no conversion of ^{14}C-P to any other substance. Furthermore, addition of ^{14}C-A fails to yield ^{14}C-Q. What kind of mutant is this one?

3. In *E. coli* phage β some mutations in the phage gene *g* are compensated for by an additional mutation in gene *h*, and some mutations in gene *h* are compensated for by mutations in gene *g*. What do these facts indicate?

4. The following recombination frequencies have been observed between the four genes *a*, *b*, *c*, and *d*: *a* × *c*, 2%; *b* × *c*, 13%; *b* × *d*, 4%; *a* × *b*, 15%; *c* × *d*, 17%; *a* × *d*, 19%.

  a. Construct a genetic map showing the gene order and their genetic map distances (use the recombination frequencies for map distances).

  b. In the cross *aBd* × *AbD*, what is the frequency of production of *ABD* progeny?

5. You are interested in the biosynthetic pathway of compound *x* and isolate ten different (independently isolated) mutants of *E. coli* that require compound *x* for growth. The mutations are mapped and their approximate positions are given below:

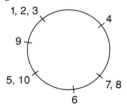

  a. What is the minimum number of genes involved in the synthesis of compound *x*?

  b. Why must your answer be the minimum estimate?

6. The complementation data shown below are observed. The numbers refer to particular mutations. The symbols + and − indicate that the two mutations do and do not complement, respectively. How many genes are represented? Assign the mutations to the genes.

	Mutants						
	1	2	3	4	5	6	7
1	−	+	+	+	+	+	−
2		−	+	+	+	+	+
3			−	+	+	−	+
4				−	+	+	+
5					−	+	+
6						−	+
7							−

Conceptual Questions

1. What are some of the considerations (both pro and con) that an experimenter might take into account when choosing a biological system in which to study a specific phenomenon?

2. You have at your disposal a set of 100 mutations in a complex process, which almost certainly involves a number of proteins and is probably highly regulated. Assuming you have the capability to perform a complete genetic analysis, what types of information about this process could be obtained? What techniques would you use? What would be the expected results for each type of information?

3. Which aspects of the logic of molecular biology have you encountered outside of science? Why are those aspects so widely used to formulate an understanding of an event or process?

2

Macromolecules

A typical cell contains 10^4–10^5 different kinds of molecules. Roughly half of these are small molecules—namely, inorganic ions and organic compounds whose molecular weights usually do not exceed several hundred. The others are polymers so massive (molecular weights from 10^4 to 10^{12}) that they are called **macromolecules.** These molecules are of three major classes: proteins, nucleic acids, and polysaccharides. They are polymers of amino acids, nucleotides, and sugars, respectively.

Our approach in these introductory chapters is to review the general properties of macromolecules. Macromolecules represent good examples of the axiom "The whole is greater than the sum of its parts." The properties of macromolecules, we will soon realize, are very different from the properties of their component amino acids, nucleotides, and sugars. Whereas those individual components alone contribute little to the functioning of the living cell, their polymers have extraordinary functions. Some of them (e.g., proteins) catalyze chemical reactions, while others (e.g., nucleic acids) store information and transfer it from one generation to the next. Indeed, a living cell's inventory of macromolecules gives it its identity. Cells differ from one another in size, shape, and function. Those features are largely generated by macromolecules.

Knowledge of the properties of macromolecules is essential for understanding living processes. This and the next two chapters will emphasize nucleic acids and proteins. Information about other types of macromolecules will be given as needed in other chapters. The Appendix includes additional information about basic chemistry, which serves as a background for those chapters.

Chemical Structures of the Major Classes of Macromolecules

In this unit we examine the chemical structure of proteins, nucleic acids, and polysaccharides—in particular, the monomers and the chemical linkages by which they are joined. Physical properties of these macromolecules and the interactions that exist between various parts of a single macromolecule are described in later units.

PROTEINS

A protein is a polymer consisting of several amino acids (a **polypeptide**). Each amino acid can be thought of as a single carbon atom (the α carbon) to which there is attached one carboxyl group, one amino group, and a side chain denoted R (Figure 2-1). The side chains are generally carbon chains or rings to which various functional groups may be attached. The simplest side chains are those of glycine (a hydrogen atom) and alanine (a methyl group). The complete chemical structures of each amino acid are given in Figure 2-2. The ionization state each would have within a living cell is shown.

To form a protein the amino group of one amino acid reacts with the carboxyl group of another; the resulting chemical bond is called a **peptide bond** (Figure 2-3). Amino acids are joined together in succession to form a linear **polypeptide chain** (Figure 2-4). When the number of peptide bonds exceeds about 15 (the number is arbitrary), the polypeptide is called a **protein.**

The two ends of every protein molecule are distinct. One end has a free —NH₂ group and is called the **amino terminus;** the other end

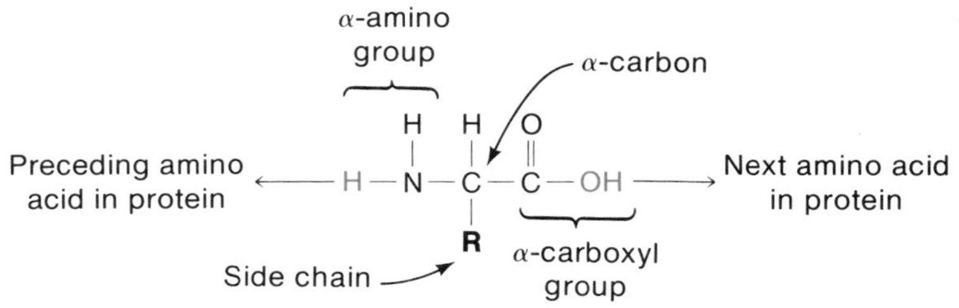

Figure 2-1
Basic structure of an α-amino acid. The NH₂ and COOH groups are used to connect amino acids to one another. The red OH of one amino acid and the red H of the next amino acid are removed when two amino acids are linked together (see Figure 2-3).

has a free —COOH group and is the **carboxyl terminus.** The ends are also called the N (or NH_2) and C termini, respectively.

The amino-acid side chains usually do not engage in covalent bond formation; an exception is the —SH group of cysteine, which often forms a disulfide (S—S) bond by reaction with a second cysteine in the vicinity within the same or different polypeptide chain.

NUCLEIC ACIDS

A nucleic acid is a polynucleotide—that is, a polymer consisting of nucleotides. Each nucleotide has the three following components (Figure 2-5):

1. A cyclic five-carbon sugar. This is ribose, in the case of ribonucleic acid (RNA), and deoxyribose, in deoxyribonucleic acid (DNA). The difference in the structure of ribose and 2'-deoxyribose is shown in Figure 2-5.

2. A purine or pyrimidine base attached to the 1'-carbon atom of the sugar by an *N*-glycosylic bond. The bases, which are shown in Figure 2-6, are the purines, adenine (A) and guanine (G), and the pyrimidines, cytosine (C), thymine (T), and uracil (U). DNA and RNA contain A, G, and C; however, T is found only in DNA and U is found only in RNA.

3. A phosphate attached to the 5' carbon of the sugar by a phosphoester linkage. This phosphate is responsible for the strong negative charge of both nucleotides and nucleic acids.

A base linked to a sugar is called a **nucleoside;** thus *a nucleotide is a nucleoside phosphate.*

The nucleotides in nucleic acids are covalently linked by a second phosphoester bond that joins the 5'-phosphate of one nucleotide and the 3'-OH group of the adjacent nucleotide (Figure 2-7). Thus the phosphate is esterified to both the 3'- and 5'-carbon atoms; this unit is often called a **phosphodiester group.**

POLYSACCHARIDES

Polysaccharides are polymers of sugars (most often glucose) or sugar derivatives. These are very complex molecules because sometimes covalent bonds occur between many pairs of carbon atoms. This has the effect that one sugar unit can be joined to more than two other sugars, which results in the formation of highly branched macromolecules. These branched structures are sometimes so enormous that they are almost macroscopic. For example, the cell walls of many bacteria and plant cells are single gigantic polysaccharide molecules.

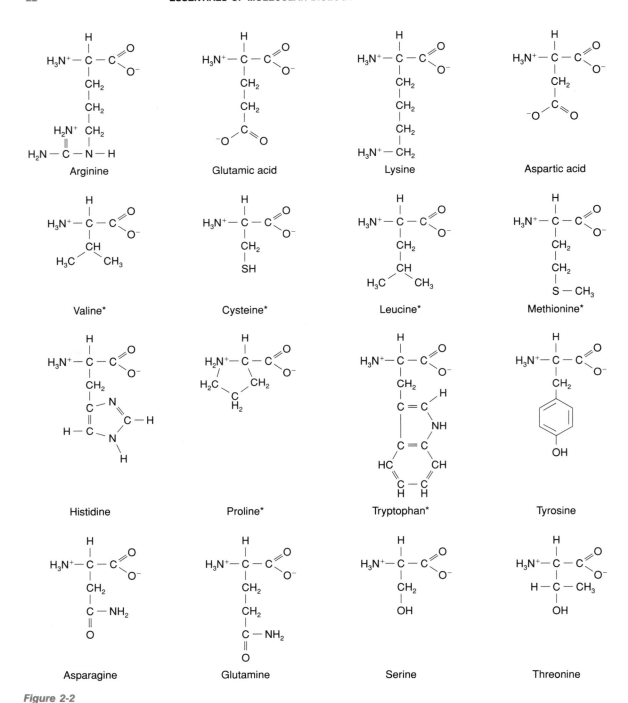

Figure 2-2
Chemical structures of the amino acids at physiological pH. Asterisks denote amino acids which contain nonpolar (hydrophobic) side chains.

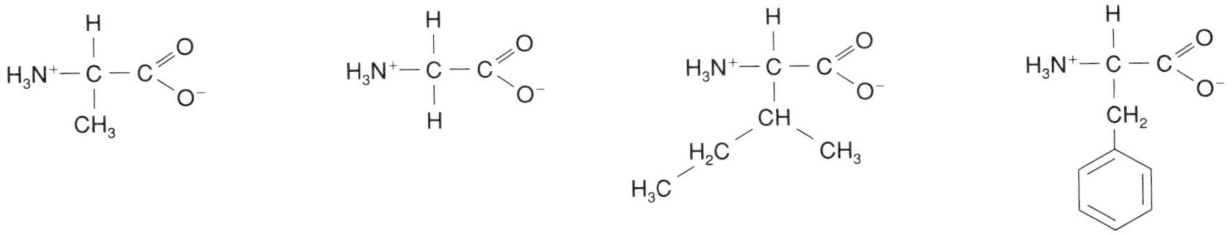

Alanine*   Glycine   Isoleucine*   Phenylalanine*

Figure 2-2
(continued)

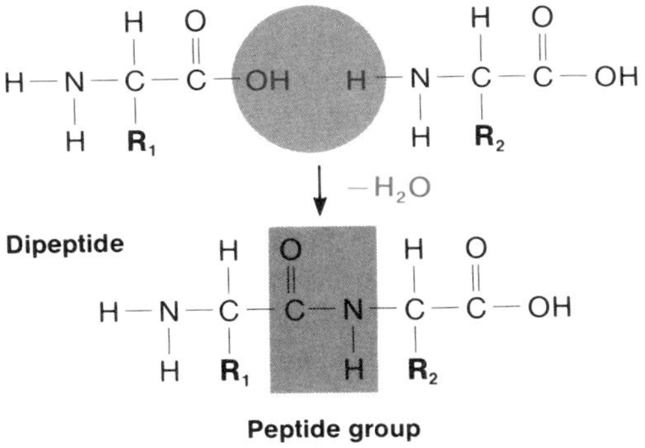

Dipeptide

Peptide group

Figure 2-3
Formation of a dipeptide from two amino acids by elimination of water (shaded circle) to
form a peptide group (shaded rectangle).

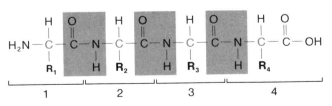

Figure 2-4
A tetrapeptide showing the alternation of α-carbon atoms (unshaded) and peptide groups
(shaded). The four amino acids are numbered below.

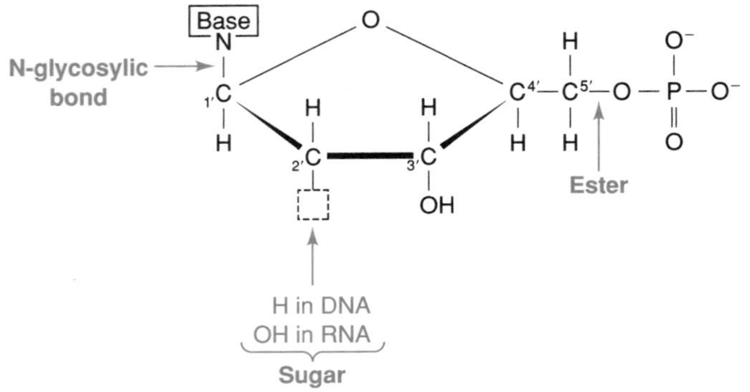

Figure 2-5
Structure of a mononucleotide. (The carbon atoms in the sugar ring are numbered).

Noncovalent Interactions that Determine the Three-dimensional Structures of Proteins and Nucleic Acids

The biological properties of macromolecules are mainly determined by activating noncovalent interactions that result in each molecule acquiring a unique three–dimensional structure. These noncovalent interactions are much weaker than typical covalent bonds, and for any given type of

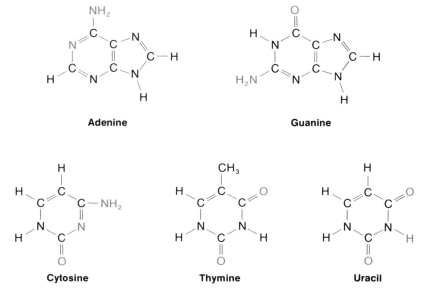

Adenine   Guanine

Cytosine   Thymine   Uracil

Figure 2-6
The bases found in nucleic acids. The weakly charged groups are shown in red.

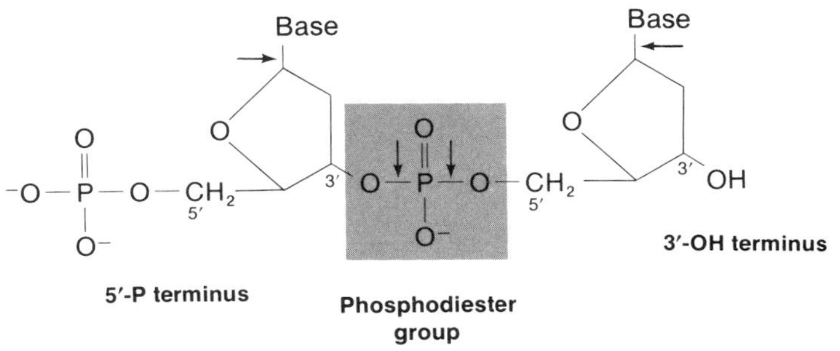

Figure 2-7
The structure of a dinucleotide. The vertical arrows show the bonds in the phosphodiester group about which there is free rotation. The horizontal arrows indicate the *N*-glycosylic bond about which the base can freely rotate. A polynucleotide would consist of many nucleotides linked together by phosphodiester bonds.

macromolecule several of these weaker interactions may play a role in determining three-dimensional structure. In this section the noncovalent interactions that are important and the chemical groups responsible for them are described.

THE RANDOM COIL

Linear polypeptide and polynucleotide chains contain several bonds about which there is free rotation. In the absence of any intrastrand interactions★ each monomer would be free to rotate with respect to its adjacent monomers; this is limited only by the fact that atoms cannot occupy the same space. The three-dimensional configuration of such a chain is called a **random coil;** it is a somewhat compact and globular structure that *changes shape continually* owing to constant bombardment by solvent molecules.

Few, if any, nucleic acid or protein molecules exist in nature as random coils because there are many interactions between elements of the chains. These interactions are hydrogen bonds, hydrophobic interactions, ionic bonds, and van der Waals interactions.

HYDROGEN-BONDING

The most common hydrogen bonds found in biological systems are shown in Figure 2-8. In RNA, intrastrand hydrogen bonds cause the polynucleotide strand to be folded, and therefore more compact than expected for long, single-stranded molecules. In DNA, interstrand hydrogen bonds are responsible for the double-stranded helical structure.

★ An intrastrand interaction is an interaction between two regions of the same strand. An interstrand interaction is between different strands. It is important to remember the distinction between these similar words.

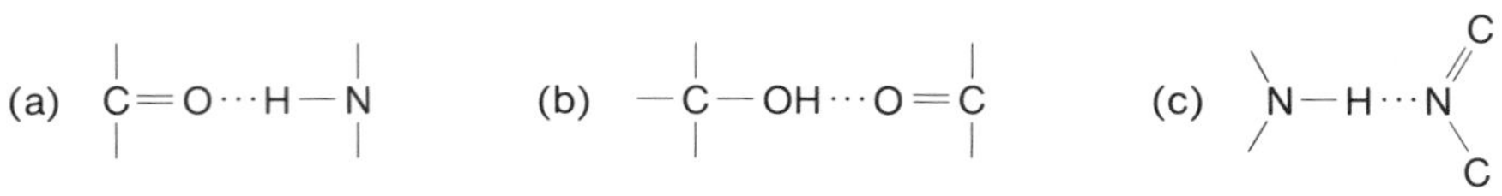

(a) $C=O\cdots H-N$ (b) $-C-OH\cdots O=C$ (c) $N-H\cdots N\begin{smallmatrix}C\\C\end{smallmatrix}$

Figure 2-8
Structures of three types of hydrogen bonds (indicated by three dots). (a) A type found in proteins and nucleic acids. (b) A weak bond found in proteins. (c) A type found in DNA and RNA.

In proteins, intrastrand hydrogen-bonding occurs between a hydrogen atom on a nitrogen adjacent to one peptide bond and an oxygen atom adjacent to a different peptide bond. This interaction gives rise to several particular polypeptide chain configurations, which will be discussed shortly.

THE HYDROPHOBIC INTERACTION

A hydrophobic interaction is an interaction between two molecules (or portions of molecules) that are somewhat insoluble in water. These two molecules (*which may be different*) are poorly soluble in water, which tends to repulse them. In response to that repulsion, they tend to associate.

Many components in nucleic acids and proteins have the hydrophobic property just described. For example, the bases of nucleic acids are planar organic rings carrying localized weak charges (Figure 2-6). The localized charges are sufficient to maintain solubility, but the large, poorly soluble organic ring portions of the bases tend to cluster, minimizing the contact with water molecules. The most efficient kind of clustering is one in which the faces of the rings are in contact, an array known as **base-stacking** (Figure 2-9). Note that since the bases are

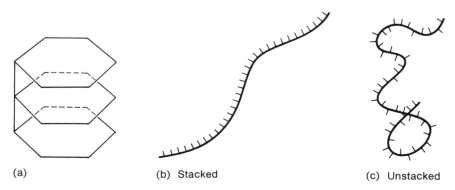

(a)   (b) Stacked   (c) Unstacked

Figure 2-9
The meaning of stacking. (a) Three stacked bases. (b) Structures of stacked and unstacked polynucleotides. The stacked polynucleotide is more extended because the stacking tends to decrease the flexibility of the molecule.

adjacent in the chain, stacking gives some rigidity to single polynucleo-tide strands and tends to extend them rather than allow a random coil to form. In the following chapter we will see how stacking is of major importance in determining nucleic acid structure. Many amino–acid side chains are very poorly soluble, and this causes (1) the benzene ring of phenylalanine to stack and (2) the hydrocarbon chains of alanine, leucine, isoleucine, and valine to form tight unstacked clusters. Since these amino acids are not necessarily adjacent, *hydrophobic interactions tend to bring distant hydrophobic parts of a polypeptide chain together*. Figure 2-2 indicates the hydrophobic amino acids.

IONIC BONDS

An ionic bond is the result of attraction between unlike charges. Several amino–acid side chains have ionized, negatively charged carboxyl groups (aspartic and glutamic acids) and positively charged amino groups (lysine, histidine, and arginine) that can form such bonds. These charges tend to bring together distant parts of the chain. Ionic interactions can also be repulsive—namely, between two like charges; thus, it would be unlikely for a polypeptide chain to fold in such a way that two aspartic acids are very near one another. Ionic bonds are the strongest of the noncovalent interactions. However, they are destroyed by extremes of pH, which can change the charge of the groups, and by high concen-trations of salt, the ions of which shield the charged groups from one another.

VAN DER WAALS ATTRACTION

Van der Waals forces exist between all molecules and are a result of both permanent dipoles and the circulation of electrons. The dependence of the attractive force between two atoms is proportional to $1/r^6$, in which r is the distance between their nuclei. Thus, the attraction is a very weak force and is significant only if two atoms are very near one another (1 to 2 Å apart).

   Van der Waals forces are very weak and are easily overcome by thermal motion; in general, the force between two atoms will not keep them in proximity. However, if interactions of *several* pairs of atoms are combined, the cumulative attractive force can be great enough to withstand being disrupted by thermal motion. Thus, two molecules can attract one another if several of their component atoms can mutually interact. However, because of the $1/r^6$-dependence, the intermolecular fit must be nearly perfect. This means that *two molecules can bind to one another if their shapes are complementary*. This is true also of two separate regions of a polymer—that is, the regions can hold together if their shapes match. Sometimes the van der Waals attraction between two regions is not large enough to effect this; however, it can significantly

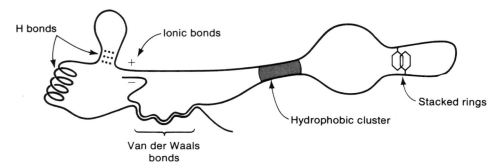

Figure 2-10
A hypothetical polypeptide chain showing attractive (black) and repulsive (red) interactions.

strengthen other weak interactions such as the hydrophobic interaction, if the fit is good.

SUMMARY OF THE EFFECTS OF NONCOVALENT INTERACTIONS

The effect of noncovalent interactions is to constrain a linear chain to fold in such a way that different regions of the chain, which may be quite distant from each other in a chain (if the chain were linear), are brought together. An example of a molecule showing the four kinds of noncovalent folding that have been described is shown in Figure 2-10.

In Chapters 3, 4 and 5 we will discuss the specific structures brought about by these interactions and the physical properties of nucleic acids and proteins.

Macromolecule Isolation and Characterization

Several methods for studying macromolecules are practically standard in molecular biology laboratories. Proteins are usually isolated from extracts of living cells with molecular sieves and chromatography on supports which are either positively or negatively charged. The former method takes advantage of the heterogeneity in molecular weight displayed by different proteins. The latter exploits the fact that most proteins differ from one another in the charge they carry, which is due of course to the unique amino acid composition of individual proteins.

The most convenient and widely employed method for monitoring the protein purification scheme, and also for initial characterization studies is electrophoresis. This method consists of setting the protein sample

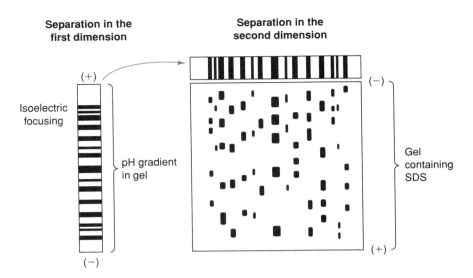

**Separation in the
first dimension**

**Separation in the
second dimension**

(+)

(−)

Isoelectric
focusing

pH gradient
in gel

Gel
containing
SDS

(−)

(+)

Figure 2-11
Two-dimensional gel electrophoresis of proteins. A protein solution (e.g., crude extract of cells) is first subjected to electrophoresis in a tube with a semi-solid gel inside it. The gel contains a pH gradient. This first step separates proteins according to their net charges (isoelectric focusing). Next, that gel is set on top of a rectangular slab containing an acrylamide gel which includes the anionic detergent sodium dodecyl sulfate (SDS). SDS molecules interact with amino acids (one SDS molecule/amino acid) to give all proteins a similar charge-to-mass ratio. In this second dimension, proteins are therefore separated during electrophoresis solely according to their molecular size. Using this method, several hundred of the proteins in a crude extract of cells can be identified as individual spots on the rectangular second dimension gel.

on a solid support (usually a gel) and establishing an electric field. Depending on the net charge of specific proteins, positively charged proteins will migrate toward the cathode (negative pole), and negatively charged proteins toward the anode (positive pole). The gel consists of a semi-solid sheet with pores through which the migrating proteins pass on their way toward either pole. Acrylamide, an inert synthetic polymer, is most commonly employed as the gel. By formulating the gel so as to have relatively small pores, smaller proteins will migrate much faster than larger proteins. In fact the most sophisticated gel electrophoresis methods for proteins employ a sequence of charge and size separation methods, as shown in Figure 2-11.

Other protein isolation and characterization methods include ultra-centrifugation, amino acid sequencing, X-ray crystallography, and occasionally, for larger proteins, electron microscopy.

For nucleic acid isolation prior to characterization, gene cloning methods are routinely employed. The simplest protocol is illustrated in Figure 2-12. Pieces of DNA are inserted into a plasmid which is then

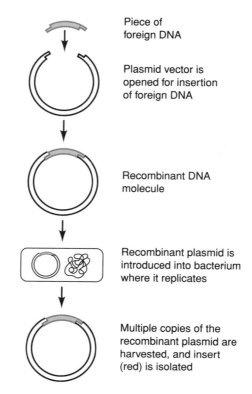

Piece of foreign DNA

Plasmid vector is opened for insertion of foreign DNA

Recombinant DNA molecule

Recombinant plasmid is introduced into bacterium where it replicates

Multiple copies of the recombinant plasmid are harvested, and insert (red) is isolated

Figure 2-12
Procedure for cloning foreign DNA in a plasmid vector.

replicated inside a bacterial host. The duplicated plasmid is then harvested, and the replicated foreign DNA is isolated from the plasmid.

Due to the unusually large size of most nucleic acids, a molecular sieve with larger pores than normally used for electrophoresis of proteins is employed (Figure 2-13). Agarose, a plant polysaccharide with various desirable qualities, is commonly employed to make the semi-solid gel support. After electrophoresis, the location of the nucleic acids is easily visualized by soaking the gel in an appropriate dye solution (see Figure 2-14).

Additional nucleic acid isolation and characterization methods include ultracentrifugation, nucleotide sequencing, and electron microscopy. Technological developments associated with these isolation and characterization methods have been the driving force behind the recent rapid accumulation of information on macromolecule structure. From this information on structure, ideas about cellular function of specific macromolecules can often be developed.

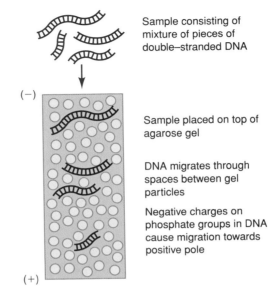

Sample consisting of mixture of pieces of double–stranded DNA

Sample placed on top of agarose gel

DNA migrates through spaces between gel particles

Negative charges on phosphate groups in DNA cause migration towards positive pole

Figure 2-13
Gel electrophoresis of nucleic acids. Double-stranded DNA is shown. RNA (normally single stranded) can also be characterized on agarose gels.

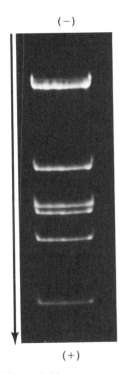

Figure 2-14
A gel electrophoregram of six fragments of *E. coli*: phage DNA. The direction of movement is from top to bottom. The DNA is made visible by the fluorescence of bound ethidium bromide. (Courtesy of Arthur Landy and Wilma Ross.)

Summary

About half of all molecules found in living cells can be termed macromolecules because of their massive molecular weights. The linear structure of the three major classes of macromolecules—proteins, nucleic acids, and polysaccharides—is determined by polymerization of specific monomers. Proteins are built from amino acids joined by peptide bonds that result from the reaction of the carboxyl group of one amino acid with the amino group of another amino acid. Nucleic acids are formed by phosphoester bonds linking the 5'-phosphate of one nucleotide to the 3'-OH group of another. Sugars and their derivatives are the building blocks which form highly balanced polysaccharides. The three-dimensional molecular properties of these macromolecules, however, are determined by noncovalent interactions including hydrogen bonds, hydrophobic interaction, ionic bonds, and van der Waals attractions. The joining of cysteine residues through the formation of disulfide bonds also affects molecular properties of proteins. All of these intrastrand interactions act to constrain a chain so that regions which are often distant in the linear chain can be brought together.

Electrophoresis, a standard method used for studying macromolecules in molecular biology laboratories, allows charged macromolecules to be separated in an electric field while travelling on a solid support such as an acrylamide or agarose gel. In the case of protein separation, sequential electrophoresis can be performed. The first electrophoretic run separates proteins on the basis of their differing net charges, while the second run causes a separation by size. Nucleic acids are also routinely separated using gel electrophoresis followed by visualization using appropriate dyes.

Drill Questions

1. What are the monomers of which proteins, nucleic acids, and polysaccharides are composed?

2. What groups in an amino acid become linked in forming a polypeptide chain? What is the linkage called?

3. What chemical groups are at the ends of a protein molecule or a polypeptide chain? Do polypeptide chains or proteins contain these groups at positions other than their termini?

4. Which nucleic acid base is unique to DNA? to RNA?

5. What chemical groups on nucleotides become covalently linked to form a nucleic acid? What is the linkage created called?

6. What is the difference between a nucleoside and a nucleotide?

7. In simple terms, what is the nature of a hydrophobic interaction?

8. What is the strongest and the weakest noncovalent force involved in stabilizing macromolecules?

9. All DNA molecules move in the same direction when electrophoresed. Why? Is this generally true of proteins also? Explain.

10. In SDS–electrophoresis of proteins, what two features of the technique cause protein molecules to move in a single direction?

Problems

1. A mixture of different proteins is subjected to electrophoresis in three polyacrylamide gels, each having a different pH value. In each gel five bands are seen. a) Can one reasonably conclude that there are only five proteins in the mixture? Explain. b) Would the conclusion be different if a mixture of linear DNA fragments were being sized?

2. Name the kinds of bonds in proteins in which each of the following amino acids might participate: cysteine, lysine, isoleucine, glutamic acid?

3. a) If a particular macromolecule that is very compact in 0.01 M NaCl expands considerably in 0.5 M NaCl, what forces are probably important in determining the overall size and shape of the molecule? b) If a macro-molecule that is a near-random coil in 0.2 M NaCl becomes very extended and rigid in 0.01 M NaCl what forces are probably involved in the extension?

4. A protein consisting of one polypeptide chain contains four cysteines. If all were engaged in disulfide bonds and if all possible pairs of cysteines could be joined, how many different theoretical structures would be possible? What would determine which was the naturally occurring structure?

5. Amino acids can be separated by electrophoresis on paper. A pH is chosen to make some amino acids positively charged and others negatively charged. Thus, some amino acids will move toward the anode and others toward the cathode, their charge principally determining the direction and rate of movement. Often amino acids having the same charge (such as alanine and valine) also separate; in other words, they move at different rates, though in the same direction. What determines the difference in rate, and would alanine or valine move faster?

Conceptual Questions

1. With an understanding of the various types of noncovalent interactions that determine protein three-dimensional structure, predict which of these interactions would be most prevalent in proteins located in different body tissues. For example, what noncovalent interactins would most likely be found in proteins spanning lipid cell membranes? In proteins that transport iron throughout the body?

2. When studying a protein one can gain information on a) the protein's three-dimensional structure and the interactions that determine it, as well as, b) the order of amino acids comprising the protein. How could each set of information be used to determine an unknown protein's function?

3. Upon collecting a new microorganism, what information might you gain from studying its macromolecules that would help you compare it with previously known microorganisms? How could you discover this information?

Robert Cedergren
Professor of Biochemistry
Département de biochimie

Birthday: **27 November 1939**

Birth Place: **Minneapolis, Minnesota**

Undergraduate Degree: **University of Minnesota**
Major: Chemistry 1961

Graduate Degree: **Cornell University, Ph.D. 1966, Organic Chemistry**

Postdoctoral Training: **Cornell University, 1966, Biochemistry and Molecular Biology; University of Toronto, 1967, Chemistry**

Present Position: **Université de Montreal, Biochemistry Department**

Address: **Montréal, Québec Canada**

TWO KEY ISSUES in molecular biology guide our research efforts: how long polymers, such as proteins and nucleic acids, perform their biological roles and how they evolved to perform these roles. One function that preoccupies us at the present time is the relation between the structure and the function of the hammerhead catalytic domain of RNA. The discovery of catalytic RNA, the ribozymes, shattered the traditional belief that only enzymes could have a catalytic activity. The ribozymes thus offer new and unique opportunities to study the activity of biological polymers. In addition ribozymes have revived interest in the origin of life on this planet. The RNA world hypothesis, based on the catalytic and information-carrying potential of RNA, proposes the existence of cellular life without proteins and nucleic acids. In order to fully understand the subtleties of RNA structure, we are actively pursuing the use of computer-aided RNA modeling. Finally the primary sequence comparison of RNAs from a variety of organisms allows predictions about the structural motifs required for different biological activities. Sequence comparisons also provide a probe of evolutionary relationships between organisms. Our studies make use of a wide variety of chemical, informatic, and genetic techniques.

In what direction is the field of molecular biology moving?

Molecular biology is still in its "information gathering" stage. The future must therefore involve the bringing together of these bits of information into coherent theories on structure–function relationships, for it is only when detailed molecular mechanisms are known that the promise of the biological sciences can be fully realized. This gathering phase will give way to the comprehension phase due in large part to emerging computer technologies in modeling macromolecules, pattern recognition, and experimental design.

3

Nucleic Acids

In this chapter you will learn:

1. About the physical and chemical nature of DNA and RNA molecules, the differences between them, and the conformations in which they may exist.

2. About various biochemical techniques used in laboratories to examine and characterize DNA molecules.

Deoxyribonucleic acid (DNA) is the single most important molecule in living cells and contains all of the information that a cell needs to live and to propagate itself. In this chapter its remarkable structure and many of its physical and chemical properties are described together with those of ribonucleic acid (RNA), a class of molecules intimately linked to the use or expression of genetic information.

We will focus in this chapter on the structural properties of nucleic acids. Realizing that genes consist of simple polymers of nucleotide sequences, we will first learn how the individual atoms in DNA are covalently linked to form the double helix. Our study of the structural properties of DNA will then reveal how nucleic acids store information and convert that information into the functional units (i.e., proteins) that give various cells their individual characteristics.

Along the way, we will learn several modern techniques that are routinely used in molecular biology laboratories to study nucleic acids. Soon we will appreciate why such spectacular advances have been made in understanding gene structure and function. Not only is it possible to establish the nucleotide sequence of virtually any gene, it is possible to biochemically synthesize genes using special chemical methods.

The development of those methods is in large part responsible for such sensational breakthroughs as the gene cloning and transgenic animals described later in this textbook.

Physical and Chemical Structure of DNA

In the previous chapter the structure of nucleotides and how they are joined to form a polynucleotide, or a nucleic acid, were described. In this section we show how nucleotides in a polynucleotide chain interact with one another to produce complex three-dimensional structures.

In early physical studies of DNA many experiments indicated that the molecule is an extended chain having a highly ordered structure. The most important technique was x-ray diffraction analysis, by which

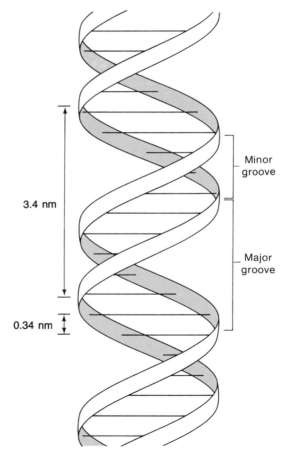

Figure 3-1
Diagrammatic model of the DNA double helix in the common B form. Parallel lines represent the base pairs, of which there are 10 per complete turn. The base pairs are perpendicular to the axis of the helix.

information was obtained about the arrangement and dimensions of various parts of the molecule. The most significant observations were that the molecule is helical and that the bases of the nucleotides are stacked with their planes separated by a spacing of 0.34 nanometers (nm).

Chemical analysis of the molar content of the bases (generally called the **base composition**) adenine, thymine, guanine, and cytosine in DNA molecules isolated from many organisms provided the important fact that [A] = [T] and [G] = [C], in which [ ] denotes molar concentration, from which followed the corollary [A + G] = [T + C] or [purines] = [pyrimidines].

double-stranded helix

James Watson and Francis Crick combined chemical and physical data for DNA with a feature of the x-ray diffraction diagram that suggested that two helical strands are present in DNA and showed that the two strands are coiled about one another to form a **double-stranded helix** (Figure 3-1). In this model the sugar-phosphate backbones follow a helical path at the outer edge of the molecule and the bases are in a helical array in the central core. The bases of one strand are hydrogen-bonded to those of the other strand to form the purine-to-pyrimidine base pairs A · T and G · C (Figure 3-2).

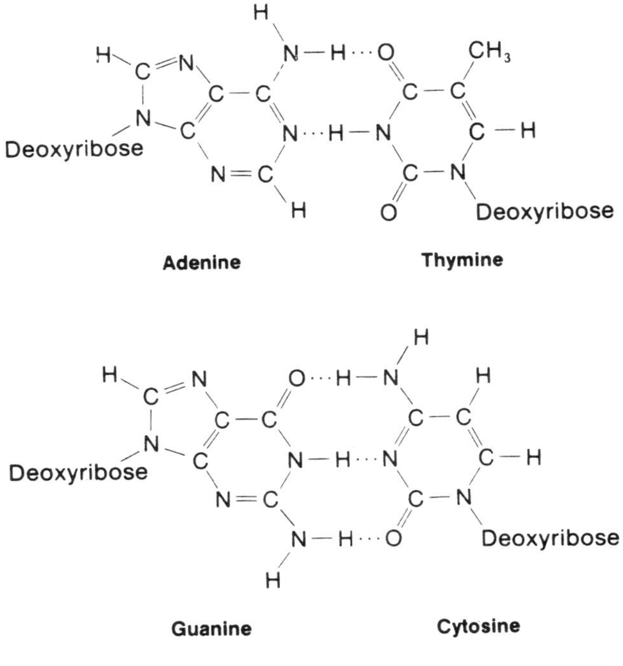

Figure 3-2
The two common base pairs of DNA. Note that hydrogen bonds (dotted lines) are between the weakly charged polar groups noted in Figure 2-6.

The two bases in each base pair lie in the same plane and the plane of each pair is perpendicular to the helix axis. The base pairs are rotated 36° with respect to each adjacent pair, so that there are ten pairs per helical turn. The diameter of the double helix is 2.0 nm and the molecular weight per unit length of the helix is approximately 2×10^6 Daltons per micrometer.

The helix creates two external helical grooves, a deep wide one (the **major groove**) and a shallow narrow one (the **minor groove**); both of these grooves are large enough to allow protein molecules to come in contact with the bases.

The double helix shown in Figure 3-1 is a right-handed helix. A right-handed helix is the arc described by your right hand when you touch your nose with your thumb; a left-handed helix is that described by your left hand. Naturally occurring DNA molecules are generally right-handed helices.

Base-pairing is one of the most important features of the DNA structure because it means that the base sequences of the two strands are complementary (Figure 3-3); thus an A in one strand is paired with a T in the other. Likewise a C in one strand is paired with a G in the other. This has deep implications for the mechanism of DNA replication because, in this way, the replica of each strand is given the base sequence of its complementary strand.

The two polynucleotide strands of the DNA double helix are antiparallel—that is, the 3'-OH terminus of one strand is adjacent to the 5'-P (5'-phosphate) terminus of the other (Figure 3-4). The significance of this is twofold. First, in a linear double helix there is one 3'-OH and one 5'-P terminus at each end of the helix. The second and generally more significant point is that the orientations of the two strands are different. That is, if two nucleotides are paired, the sugar portion of one nucleotide lies upward along the chain, whereas the sugar of the other nucleotide lies downward. We will see in Chapter 7 that this structural feature poses an interesting constraint on the mechanism of DNA replication.

Because the strands are antiparallel, a convention is needed for stating the sequence of bases of a single chain. The convention is to write a sequence with the 5'-P terminus at the left; for example, ATC denotes the trinucleotide 5'-P—ATC—3'-OH.

It is always true that [A] = [T] and [G] = [C], but there is no rule governing the ratio of total concentrations of guanine and cytosine to those of adenine and thymine in a DNA molecule; in fact, there is enormous variation in this ratio. The usual way that the base composition of the DNA molecule from a particular organism is expressed is not by a ratio, but by the percent of all bases that are G · C pairs; that is ([G] + [C]) × 100/[all bases]. This is termed the G + C content, or percent G + C. The base composition of hundreds of organisms has been determined. Generally speaking, the percent of G + C content is

Figure 3-3
A diagram of a DNA molecule showing complementary base-pairing of the individual strands. The light lines represent hydrogen bonds. The numbers orient the strands.

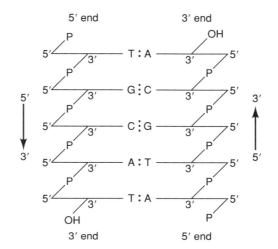

Figure 3-4
A stylized drawing of a segment of a DNA duplex showing the antiparallel orientation of the complementary chains. The arrows indicate the 5' to 3' direction of each strand. The phosphates (P) join the 3' carbon of one deoxyribose (horizontal line) to the 5' carbon of the adjacent deoxyribose. The dots between the pairs indicate the number of hydrogen bonds.

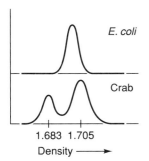

Figure 3-5
The distribution of DNA isolated from the bacterium *E. coli* and the crab *Cancer borealis* after density-gradient centrifugation. Analysis of the curves shows that (1) in common with other bacteria, *E. coli* DNA has a narrow density range; (2) crab DNA consists of two discrete fractions of DNA: one with a higher density than *E. coli* DNA (and therefore with a higher G–C content) and one with a lower density. This minor band is called satellite DNA. (From N. Sueoka. *J. Mol. Biol.*, 3:31–40. Copyright 1961: Academic Press Inc. [London] Ltd.)

near 50% for most higher organisms, and has a very small range from one species to the next (49–51% for the primates). For the lower organisms, the percent G + C varies widely from one genus to another. For example, in bacteria the extremes are 27% for the genus *Clostridium* and 76% for the genus *Sarcina*. *E. coli* DNA is around 50% G + C. Proceeding toward more complex organisms, the variation becomes smaller. For instance, for most plants and animals, the extremes are 48% and 52%. Because the density of DNA is linearly related to base composition, a convenient way to measure the mean base composition is by density gradient centrifugation. When DNA of prokaryotes is broken randomly into about 1000 fragments, the bands that result after centrifugation are usually very narrow. This indicates that the mean base composition does not vary much from fragment to fragment (Figure 3-5, top).

However, when the DNA of the crab is examined, two discrete bands are observed (Figure 3-5, bottom). The main band accounts for 70% of the total DNA and is 52% G + C. The smaller band, having 3% G + C, is called **satellite DNA.**

The phenomenon of satellite DNA has been observed often during the analysis of the DNA from eukaryotes; the G + C content, however, can be higher or lower than that of the main fraction. Analysis has shown that satellite DNA is usually highly repetitive, consisting sometimes of 10^6 copies of a single sequence.

Alternate DNA Structures

DNA structure is very rich in variety. Although the double-stranded B helix is the predominant form, other helical and some completely unexpected forms can be found. Among these structural variants is the **A helix** (the predominant form of double stranded RNA) which has 11 bases per turn rather than 10, and has the plane of the bases tilted 30° with respect to the helical axis. This helix, favored under conditions of dehydration, is wider and shorter than the B helix, and the distinction between the major and minor grooves is reduced.

The **Z helix** (so-named because the backbone has a zig-zag shape) is a more radical departure from the B helix theme. Although double-stranded, the Z helix is left-handed (Figure 3–6) and has 12 base pairs per turn. Since the length of a turn is 4.5 nm rather than 3.4 nm (B helix), Z DNA appears longer and slimmer-looking than the B helix. In the very long DNA molecules found in cells, different conformations can coexist in the same molecule. Regions of the Z conformation can be found interspersed among regions of essentially B conformation; this fact raises the question as to what determines the conformation of DNA. In the case of Z DNA, a primary structure or sequence of G alternating with C favors the Z DNA conformation. The presence of regions of Z DNA near genes on the same molecule can influence their expression.

Further variations of DNA structure involve the "strandedness" of the DNA. For example, single-stranded DNA occurs in some viruses. In these DNA molecules the rule of double-stranded complementarity, $[G] + [A] = [C] + [T]$, does not hold. The structure of this DNA is much more irregular—the single strand folds back on itself. Short, double-stranded helical regions are formed between complementary regions of the molecule. These duplexes are separated by loop structures, and residual single-stranded regions. Single-stranded DNA is more dense than double-stranded DNA as a result of its more contorted and compacted structure. Alternating helix–loop structures are also a feature of single-stranded RNA molecules.

Triple- and even four-stranded DNA helices have also been observed. Triple-stranded helices are found in regions of DNA that have repeating stretches of purines alternating with stretches of pyrimidines complementary to the purine stretches. Under these circumstances, one strand of a repeating unit folds back into the major groove of the preceding repeating unit. This conformation leaves one strand of the repeating unit unpaired. The four-stranded helix is characterized by G-rich regions of DNA. This structure is often found at the end of chromosomes in the telomeres, and may be important in the process of meiosis.

DNA structure can thus be best described as a number of variations on the B helix theme. Quite clearly the structure adopted by a particular

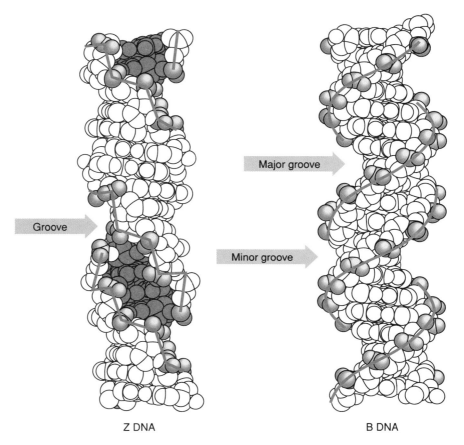

Z DNA   B DNA

Figure 3-6
Space filling models of Z DNA and B DNA. The dark spheres are the phosphate groups, and the heavy lines link adjacent phosphates. The zig-zag shape of the Z DNA backbone is clearly visible. To verify if the helices are right- or left-handed; rotate the book 180° and look at the helices. (From Rich et al. *Am. Rev. Biochem.* 53:795. Copyright 1984: Annual Reviews Inc. [Palo Alto])

region of a DNA molecule depends on the base sequence in that region. This is simply another example where the primary sequence influences the three-dimensional structure of a biological polymer.

Circular and Superhelical DNA

The rather crude techniques originally used to isolate and study DNA suggested that DNA had the form of a linear double helix. The overall length of DNA makes it extremely sensitive to shear forces during preparation. Typical DNA preparations are often fragmented (both strands broken) or nicked (single strand broken). However as improvements in technique were made, it became clear that many DNAs are actually circular, with the ends of the helix covalently joined. This observation raised novel topological questions concerning the effect of the circular state on the helical structure.

Because of the 5′, 3′ polarity of the strands, the 5′ terminus of one strand can only join its own 3′ end to close the circle. As a result circular, double-stranded DNA is in fact two circles of single-stranded DNA which are twisted around each other resembling a piece of twine whose two ends are joined. However, consider the result of giving the double helix exactly one 360° twist before joining the ends. If the twist is in the same direction as the helix it would have the effect of tightening the double helix. Alternatively a twist in the opposite direction of the helix would tend to open or loosen the helix. Circular DNA which has incorporated one or more twists to increase the number of times one strand crosses the other is said to be positively **supercoiled** or **super-helical.** A negative superhelix results from decreasing the number of times the strands cross. Compensation for the higher energy state of supercoiling DNA is effected by modifying the dimensions or the pre-ferred conformations of the double helix.

supercoiling

One major structural effect of positive or negative supercoiling is that the helical axis curves slightly to maintain the 10–base periodicity of the B helix, which would otherwise be modified. The bent helical axis can in fact be observed by electron microscopy (Figure 3–7). Other

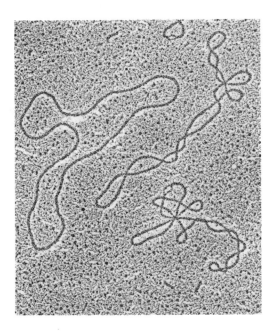

Figure 3-7
Circular and supercoiled DNA of phage PM2 as imaged by an electron microscope. (Courtesy of K. G. Murti.)

effects are more dramatic: highly negatively-superhelical DNA can be accommodated by conformational changes in the molecule. The ultimate accommodation is the Z DNA structure which can compensate for two negative turns per left-handed turn. The formation of short bubbles of unpaired, single-stranded DNA can be further stabilized by **cruciform** structures (Figure 3-8). Cruciforms are conformations which can be adopted by highly symmetrical regions of DNA. Because of the 5′ to 3′ polarity of the strands and the complementarity of bases, cruciforms occur in regions of **palindromic** (inverted repeat) DNA. Supercoiling thus favors large, local deviations from the standard conformation. However, since DNA structures are dynamic, many regions undergo a constant conformational flux between a normal B helix and other forms.

topoisomerase

That superhelical DNA exists offers a hint that the structural variations discussed above have some biological function. One of the first surprising developments was the discovery of enzymes called **topoisomerases** that modify the superhelicity of DNA. Of course, the only way to modify the superhelicity of closed circular DNA is to break one or both strands and twist the ends with respect to each other before linking them together again; topoisomerases are able to accomplish this. In addition, considerable information is accumulating that relates the presence of DNA palindromes to protein binding sites. Since most DNA in its natural state is negatively supercoiled, it is likely that some palindromes are in the cruciform conformation, and therefore accessible to binding by other molecules. Likewise, single-stranded bubbles generated by negative supercoiling can serve as regions for protein binding.

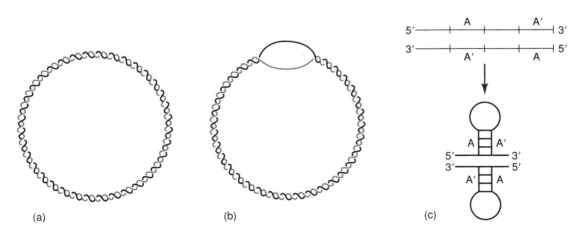

(a)   (b)   (c)

Figure 3-8
a) A non-supercoiled covalent circle of DNA. b) A negatively supercoiled circle of DNA. The single-stranded bubble allows the remainder of the molecule to adopt the standard B helix. c) A cruciform structure of DNA existing in a palindromic region (an inverted repeat). Region A is complementary to A′, so, due to the symmetry of an inverted repeat, A can base-pair with A′ of the second strand or with the A′ sequence of its own strand.

DENATURATION OF DNA

The helical structure of nucleic acids is determined by the thermodynamically-favored stacking between adjacent bases in the same strand, and the double-stranded structure of the helix is maintained by hydrogen-bonding between the bases of the base pairs. This conclusion has come from studies of the thermal stability of DNA molecules.

The free energies of the noncovalent interactions described in Chapter 2 are not much greater than the energy of thermal motion at room temperature, so that at elevated temperatures the three-dimensional structures of both proteins and nucleic acids are disrupted. A macromolecule in a disrupted state, in which the molecules are in a nearly random conformation, is said to be **denatured;** the ordered state, which is presumably that originally present in nature, is called **native.** A transition from the native to the denatured state is called **denaturation.** When double-stranded DNA or native DNA is heated, the bonding forces between and within the strands are disrupted, and the two strands can separate; thus, *denatured* DNA is single-stranded.

denaturation

A great deal of information about structure and stabilizing interactions has been obtained by studying nucleic acid denaturation. This is done by measuring some property of the molecule that changes as denaturation proceeds—for example, the absorption of ultraviolet light. For DNA, denaturation can be detected by observing the increase in the ability of a DNA solution to absorb ultraviolet light at a wavelength of 260 nm. A standard measure of such absorption is a logarithmic ratio of transmitted light and incident light called the **absorbance A;** at 260 nm, this is written A_{260}. Nucleic acid bases absorb 260-nm light strongly, and the amount of light absorbed by a given number of bases depends on their proximity to one another. When the bases are highly ordered (very near one another), as in double-stranded DNA, A_{260} is lower than that for the less ordered state in single-stranded DNA. In particular, if a solution of double-stranded DNA has a value of $A_{260} = 1.00$, a solution of single-stranded DNA at the same concentration has a value of $A_{260} = 1.37$. This relation is often described by stating that a solution of double-stranded DNA becomes **hyperchromic** when heated. The corresponding value for free bases is 1.60, so that absorbance can also be used to measure degradation of nucleic acids.

If a DNA solution is slowly heated and the A_{260} is measured at various temperatures, a melting curve such as that shown in Figure 3-9 is obtained. The following features of this curve should be noted:

1. The A_{260} remains constant up to temperatures well above those encountered by living cells in nature.
2. The rise in A_{260} occurs over a relatively narrow range of 6–8°C.
3. The maximum A_{260} is about 37% higher than the starting value.

The state of a DNA molecule in different regions of the melting curve is illustrated in Figure 3-9. Before the rise begins, the molecule is fully

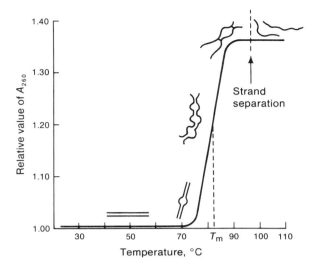

Figure 3-9
A melting curve of DNA showing T_m and possible molecular configurations for various stages of melting.

melting temperature (T_m)

double-stranded. In the rise region, base pairs in various segments of the molecule are broken; the number of broken base pairs increases with temperature. In the initial part of the upper plateau a few base pairs remain to hold the two strands together until a critical temperature is reached at which the last base pair breaks and the strands separate completely. A convenient parameter to characterize a melting transition is the temperature at which the rise in A_{260} is half complete. This temperature is called the **melting temperature;** it is designated T_m.

A great deal of information is obtained by observing how T_m varies with base composition and experimental conditions. For example, it has been observed that T_m increases with percent G + C, which has been interpreted in terms of the greater number of hydrogen bonds in a G · C pair (three) than in an A · T pair (two). That is, a higher temperature (more energy) is required to disrupt a G · C pair than an A · T pair, because the double-stranded structure is stabilized, at least in part, by hydrogen bonds. In addition, reagents that increase the solubility of the bases and therefore decrease hydrophobic interaction, such as base stacking, also lower T_m. These results imply that a polynucleotide would tend to have a three-dimensional structure that maximizes the contact of the high soluble phosphate group with water and minimizes contact between the hydrophobic bases and water. This explains why, in DNA, the sugar-phosphate chain is on the outside in contact with water, and the bases are on the inside in a stacked array. Clearly, stacked bases are more easily hydrogen-bonded, and correspondingly, hydrogen-bonded bases, which are oriented by the bonding, stack more easily. Thus, the two interactions act cooperatively to form a very stable structure. If one of the interactions is eliminated, the other is weakened; this explains why T_m drops so markedly after the addition of a reagent that destroys either type of interaction.

A very effective denaturant is NaOH, for the charge of several groups engaged in hydrogen-bonding is changed at high pH; and a base bearing such a group loses its ability to form these bonds. At a pH greater than 11.3 all hydrogen bonds are eliminated, and DNA is completely denatured. Since DNA is quite resistant to alkaline hydrolysis, this procedure is the method of choice for denaturing DNA, as high temperatures promote phosphodiester bond cleavage.

A repulsive force is also present in DNA, namely, that between the negatively charged phosphate groups in complementary strands. These charges are neutralized by bound cations such as Na^+ and Mg^{2+}. As a result, the repulsion is so great in distilled water that strand separation occurs.

Renaturation

renaturation

A solution of denatured DNA can be treated in such a way that native DNA re-forms. The process is called **renaturation** or **reannealing** and the re-formed DNA is called **renatured DNA.** Renaturation has proved to be a valuable tool in molecular biology since it can be used to demonstrate genetic relatedness between different organisms, to detect particular species of RNA, to determine whether certain sequences occur more than once in the DNA of a particular organism, and to locate specific base sequences in a DNA molecule.

Two requirements must be met for renaturation to occur:

1. The salt concentration must be high enough that electrostatic repulsion between the phosphates in the two strands is eliminated—usually 0.15 to 0.50 M NaCl is used.

2. The temperature must be high enough to disrupt the random, intrastrand hydrogen bonds present in single-stranded DNA. However, the temperature cannot be too high, or stable interstrand base-pairing will not occur and be maintained. The optimal temperature for renaturation is 20–25° below the value of T_m.

Since strands can separate during denaturation, renatured DNA does not contain its own original single strand; renaturation is thus a random mixing process. For example, if DNA labeled with the nonradioactive isotope ^{15}N is mixed with DNA containing the normal isotope ^{14}N and then denatured and renatured, three kinds of double-stranded DNA molecules will result—25% will have ^{14}N in both strands, 25% will have ^{15}N in both strands, and 50% will have one ^{14}N strand and one ^{15}N strand.

Let us now consider how renaturation actually takes place. After denaturation, the DNA solution is composed of a mixture of essentially single-stranded fragments. In order to renature, a fragment must find its complementary partner in the solution. If the concentrations of the fragment and its partner are low, they will take much more time to find each other than if the two fragments are in high concentration. This process can be followed by observing the absorption of the solution at 260 nm (A_{260}). Since during renaturation the bases order themselves, the A_{260} decreases with time. The **rate of decrease** is the rate of renaturation, which is related to the concentration of individual fragments in the solution. The **value** of A_{260} in the solution gives the overall concentration of DNA. Renaturation kinetics offers a powerful tool for the analysis of the presence of repeated sequences in a DNA sample.

A long DNA molecule is shown in Figure 3-10. The sequence ABC/A'B'C' represents 3% of the total number of base pairs in the sample; however, since it is repeated 5 times this sequence makes up 15% of the total weight of the DNA sample. When the molecule is broken into smaller pieces, there are five times more of the fragments originating from the repetitive sequence than from the remainder of the DNA. Because of its higher concentration in the solution, the repetitive DNA renatures faster than single copy DNA. Such a DNA would have a renaturation curve such as that shown in Figure 3-10(b). Thus, this curve, which is obtained with DNA from a single organism, contains

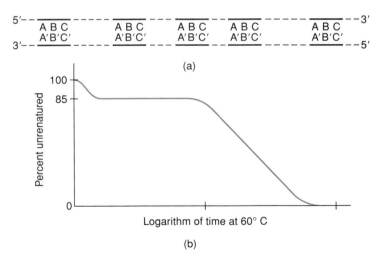

Figure 3-10
(a) A hypothetical DNA molecule having a base sequence that is 3% of the total length of the DNA and is repeated five times. The dashed lines represent the nonrepetitive sequences; they account for 85% of the total length. (b) A renaturation curve for the DNA shown in (a). Time is logarithmic in order to keep the curve on the page. The y axis—percent unrenatured DNA—is equivalent to the unit, relative value of A_{260}, used in previous figures with $A_{260} = 1.37$ equal to 100% (percent unrenatured-percent single stranded).

two components. The more rapidly renaturing component accounts for 15% of A_{260}, from which one can conclude that 15% of the DNA (by weight) contains a repeating sequence. From the kinetics one can determine the number of bases in the repeating sequence (called its **complexity**), and thus the number of copies of this sequence and the number of bases in a single nonrepeating sequence.

Repeated sequences are found in most eukaryotes but not generally in unicellular organisms. These sequences do not always show up as satellite bands (see Figure 3-5) because their base compositions may not be very different from the average base composition of the bulk of the DNA. The fraction of the DNA that is repetitive and the size of the repeated sequences can be determined by renaturation. Figure 3-11 shows a comparison between the renaturation kinetics of a bacterial and mouse DNA. Close examination of these curves from many species shows that there are four classes of sequences—**unique** (single copy), **slightly repetitive** (1 to 10 copies), **middle repetitive** (10 to several hundred copies), and **highly repetitive** (several hundred to several million copies). The boundaries of these classes are arbitrary, so one must be careful when making generalizations about the classes. The unique sequences account for most of the genes of the cell, but often for only a few percent of the DNA. The slightly repetitive class consists of genes that exist in a few copies; in some cases the sequences are not perfectly repetitive: the gene products may have slight differences in amino acid composition—for example, the various forms of hemoglobin. The middle-repetitive sequences are of two types—clustered and dispersed. The

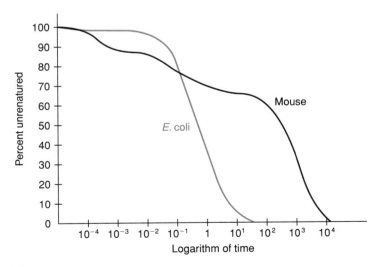

Figure 3-11

Renaturation curves for *E. coli* and mouse DNA. The sequences are unique in *E. coli*, so the curve has a single step. In mouse DNA 10% consists of about one million copies of a 300-base-pair sequence, 30% consists of 10^3–10^4 copies of several sequences, and 60% consists of unique sequences.

clustered sequences usually represent genes that exist in multiple copies in order to increase the amount or the rate of synthesis of the gene product. Some of the dispersed sequences may have regulatory function. However, the major fraction consists of 20–60 copies of sequences, containing 2000–7000 nucleotide pairs, that are able to move from one location to another in a chromosome and between chromosomes. These are called transposable elements and are discussed in detail in Chapter 12. The highly repetitive sequences include the short sequences in satellite DNA and sequences of normal length in certain genes that exist in very large numbers, such as those making ribosomal RNA, which is needed for all protein synthesis. There are in fact very few different highly repetitive sequences, but the number of copies is so great that these sequences may account for 20% or more of the mass of the DNA.

Hybridization

hybridization

The technique wherein renatured DNA is formed from separate single-stranded samples is called **hybridization.** This renaturation technique can be done on thin filters (**membrane filters**) made of nitrocellulose, which bind single-stranded DNA very tightly but fail to bind either double-stranded DNA or RNA. They provide an important technique for measuring hybridization. Renaturation is carried out on a filter as shown in Figure 3-12. A sample of denatured DNA is filtered. The single strands bind tightly to the filter along the sugar-phosphate backbone, but the bases remain free. The filter is then placed in a vial with a solution containing a small amount of radioactive denatured DNA and a reagent that prevents additional binding of single-stranded DNA to

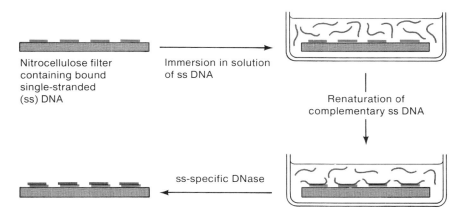

Nitrocellulose filter
containing bound
single-stranded
(ss) DNA

Immersion in solution
of ss DNA

Renaturation of
complementary ss DNA

ss-specific DNase

Figure 3-12
Method of hybridizing DNA to nitrocellulose filters containing bound, single-stranded DNA. In the final step the filter is washed to remove excess DNA.

the filter. After a period of renaturing, the filter is washed. Radioactivity is found on the filter only if renaturation has occurred.

Renaturation has been combined with the technique of electron microscopy in a procedure that allows the localization of common, distinct, and missing sequences in DNA. This procedure is called **heteroduplexing.** Consider the DNA molecules #1 and #2 shown in Figure 3-13(a). If these molecules differ in sequence only in one region (A and A'), then when a mixture of the two molecules is denatured and renatured, hybrid molecules having unpaired single strands are produced, as shown in Figure 3-13(b). Figure 3-14 shows an actual electron micrograph of a heteroduplex. Measurement of the lengths of the single- and double-stranded regions yields the endpoints of the regions of nonhomology.

Another important use of filter hybridization is the detection of sequence complementarity between a single strand of DNA and an RNA molecule—this is called **DNA–RNA hybridization,** and it is the method of choice to detect an RNA molecule that has been copied from a particular DNA molecule. In this procedure, a filter to which single strands of DNA have been bound, as above, is placed in a solution containing radioactive RNA. After renaturation the filter is washed and hybridization is detected by the presence of radioactive RNA on the filter.

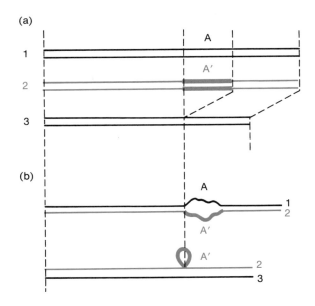

Figure 3-13
(a) Three DNA molecules to be heteroduplexed. Sequences A and A' of molecules 1 and 2 differ. Neither sequence is present in molecule 3. The dashed lines indicate reference points. (b) Heteroduplexes resulting from renaturating molecules 1 and 2 or 2 and 3. Hybridization of 1 and 2 shows the regions A and A' as a single-stranded bubble. Hybridization of 2 and 3 shows that sequence 3 has a deleted region with respect to 2.

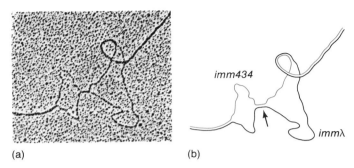

Figure 3-14
An electron micrograph of a portion of a heteroduplex between two *E. coli* phage λ molecules λ*imm434* and λ*immλ*, which have two non-complementary regions on either side of a short complementary region. (a) The micrograph. (b) An interpretive drawing. The arrow points to the common sequence. (Courtesy of W. Szybalski and B. Westmoreland.)

The Structure of RNA

ribosomal RNA (rRNA)
transfer RNA (tRNA)
messenger RNA (mRNA)

A typical cell contains about ten times as much RNA as DNA. The high RNA content is due mainly to the large variety of roles played by RNA in the cell. There are three primary types of RNA—**ribosomal RNA (rRNA),** by far the most abundant, **transfer RNA (tRNA),** and **messenger RNA (mRNA).** Each one of these general classes is actually composed of several unique species: 3 or 4 rRNAs, up to 50 tRNAs, and over 1000 mRNAs.

RNA has many of the structural characteristics of DNA. That is, both molecules are linear polynucleotide chains. However, RNA has the following distinctions: ribose replaces deoxyribose, U replaces T; and with the exception of some viruses, RNA is single-stranded. Most RNAs therefore resemble the conformation of single-stranded DNA, i.e. they have stem/loop or hairpin structures because the chains fold back on themselves, creating loops and small base-paired stretches between complementary regions (Figure 3-15). RNA is defined in terms of its primary structure (the order of the nucleotides in the chain), its secondary structure (the pattern of base-pairing), and its tertiary structure (the conformation of the molecule in three dimensions).

Due to the essential relationship of RNA structure to its functions, considerable research has been devoted to the prediction of the base-pairing pattern. Since the techniques involved in prediction are empirical, secondary structures, although extremely useful, should only be considered as a model. Two methods of prediction which are available as computer programs utilize either the number of base-pairs in the molecule or the overall free energy of the molecule using assigned energy values for different base-pairs. The fact that G–C base pairs are held together by three hydrogen bonds favors their presence in the model

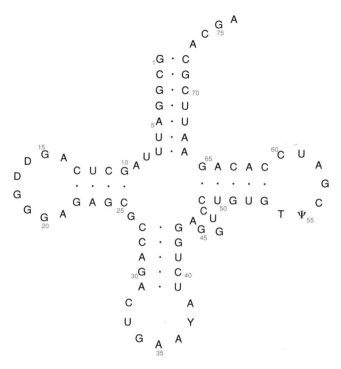

Figure 3-15
The proposed secondary structure of a tRNA. D, T, Y and Ψ are modified nucleotides often found in tRNA.

over the presence of A–U pairs. Another approach is based on the availability of sequences of the same molecule from several organisms. Base-pairs are proposed between positions in sequences which have undergone compensatory base changes—potential base-paired positions in one sequence must be base-paired in a second, even though the identity of the nucleotides at these positions are different (Figure 3-16).

Hydrolysis of Nucleic Acids

With few exceptions (e.g. tRNA), the structures of nucleic acids are of such size that they are difficult to characterize. The strategy that has emerged to circumvent this problem involves breaking down the polymers into smaller fragments, since both DNA and RNA can be hydrolyzed either chemically or enzymatically.

At very low pH (1 or less), phosphodiester hydrolysis of both DNA and RNA occurs. This is accompanied by breakage of the *N*-glucosylic bond between the base and the sugar (see Figure 2-5) so that free bases are produced. At high pH the behavior of DNA is strikingly different from that of RNA. DNA remains very resistant to hydrolysis at pH 13

nuclease
exonuclease
endonuclease

Figure 3-16
Compensatory base changes that support secondary structure. In sequences from organisms 1 and 2, it was found that potential base-pairs U–A and G–C occupy similar positions. The fact that the sequences are different at these positions but the bases are complementary supports the hypothesis that these are indeed paired positions.

(about 0.2 phosphodiester bonds broken per million bonds per minute at 37°C), whereas the presence of the 2′-OH groups renders RNA very sensitive to alkaline hydrolysis; at 37°C free ribonucleotides are produced within minutes at pH 11.

A variety of enzymes called **nucleases** hydrolyze nucleic acids. They usually show chemical specificity and are classified as either deoxyribonucleases (DNase) or ribonucleases (RNase). Many DNases act only on single-stranded or only on double-stranded DNA, although some act on either kind. Furthermore, nucleases which act only at the end of a nucleic acid, removing a single nucleotide at a time, are called **exonucleases,** and they may also be specific for the 3′ or 5′ end of the strand. Most nucleases act within the strand and are **endonucleases;** some of these are even more specific, in that they cleave only between particular bases.

The availability of base-specific endonucleases opened the door to the determination of the primary structure of many RNAs. RNase Ti, which cleaves an RNA chain 3′ to a G, and pancreatic RNase, which cleaves 3′ to an U or C, were particularly useful. Since the tertiary structures of many of these ribonucleases are now known, these enzymes provide valuable insight into how proteins interact with nucleic acids (Figure 3-17).

In contrast, work on DNA structures lagged far behind RNA structures because of a lack of base-specific DNases. This situation has dramatically changed, however. As we shall see in the next section, it is now much easier to determine the sequence of DNA than RNA. The breakthrough for DNA occurred when restriction endonucleases were discovered. Restriction nucleases cleave double-stranded DNA at specific sequences that are several nucleotides long. They are also largely responsible for the development of the genetic engineering techniques discussed in Chapter 13.

One of the most remarkable and unexpected discoveries of recent years is that some RNAs have nucleolytic activity. Previously, the dogma of biochemistry was that only proteins could have catalytic activity. The RNA enzymes, called **ribozymes,** are able to cleave specific phosphodiester bonds in a manner analogous to protein enzymes. An example is provided later (see Chapter 9, Figure 9-9).

Sequencing Nucleic Acids

Throughout this book, nucleotide sequences of DNA molecules will be described as a means of understanding the kinds of physical and chemical signals that are used by cells. This section describes how these sequences are determined.

DNA sequence determination is based on the high resolution separation of polynucleotide fragments on polyacrylamide gels (Chapter 2).

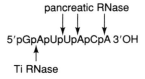

pancreatic RNase

5'pGpApUpUpApCpA 3'OH

Ti RNase

Figure 3-17
Hydrolysis of RNA. The position of digestion by Ti RNase and pancreatic RNase are shown. The sequence includes a "p" for the phosphate between nucleosides to illustrate which bond is cleaved.

Two methods are used to generate fragments. In the **Maxam-Gilbart procedure,** fragments are generated using chemical reagents which are specific for each nucleotide; in the **Sanger procedure,** which is outlined here, polynucleotide fragments are produced by an ingenious use of the enzyme DNA polymerase. Since the activity of DNA polymerase is detailed in Chapter 7, only a superficial description of the process is given here.

DNA polymerase uses triphosphate derivatives of deoxyribonucleotides to construct polynucleotides. This enzyme also requires the presence of a short primer polynucleotide and a longer single-stranded template fragment from the DNA to be sequenced (Figure 3-18a). Normally, double-stranded DNA is made continuously in this solution until all the nucleotide triphosphates are consumed. Consider the same reaction where a dideoxyribonucleotide (a nucleotide derivative lacking hydroxyl groups at both the 2' and 3' carbons) is added; these nucleotide analogs are incorporated in the growing nucleotide chain, but the resulting chain can not be extended, since there is no 3' terminal hydroxide upon which to attach the following nucleotide (Figure 3-18b). When dideoxy G is used at 1/100 of the concentration of G, an average of one analog is incorporated into the newly-synthesized strand for every 100 Gs.

If three new reactions are prepared where each contains one of the remaining dideoxy nucleotides (i.e., A, T or C) then four reaction mixtures would exist—one for each nucleotide—and each would contain a complex mixture of polynucleotide fragments differing in length, depending upon the point at which a dideoxynucleotide was incorporated. After the reaction, each mixture would be analyzed by gel electrophoresis. The fragments would separate based on their lengths—the shortest migrating the farthest. As can be seen in Figure 3-19, each lane has a

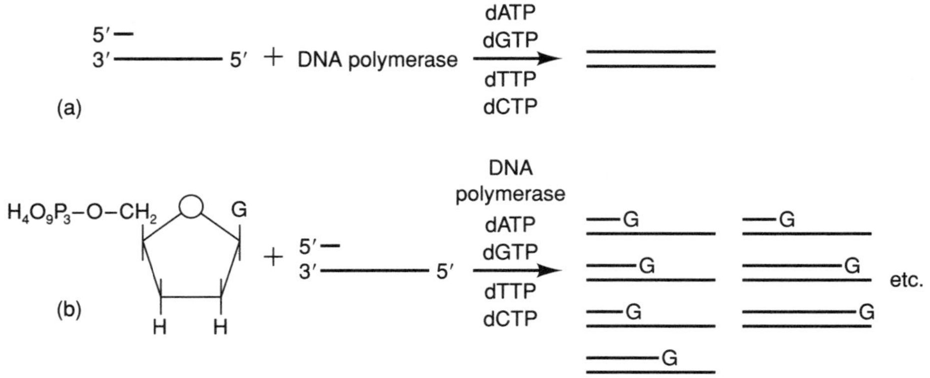

$$
\begin{array}{l}
5'- \\
3'\text{———— } 5' + \text{DNA polymerase} \\
\end{array}
\quad
\xrightarrow[\substack{\text{dTTP} \\ \text{dCTP}}]{\substack{\text{dATP} \\ \text{dGTP}}}
\quad
=====
$$

(a)

$H_4O_9P_3-O-CH_2$ ⬠ G

(b)   H   H

$+$
$\begin{array}{l} 5'- \\ 3'\text{———— } 5' \end{array}$

$\xrightarrow[\substack{\text{dTTP} \\ \text{dCTP}}]{\substack{\text{DNA} \\ \text{polymerase} \\ \text{dATP} \\ \text{dGTP}}}$

—G   —G
—G   ———G
—G   ———G   etc.
——G ———G

Figure 3-18
(a) DNA polymerase reaction. ATP—adenosine triphosphate GTP—guanosine triphosphate, TTP—thymidine triphosphate. CTP—cytidine triphosphate. (b) The same reaction with added dideoxy G. Chain termination takes place at each position where dideoxy G is incorporated.

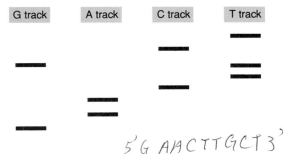

Figure 3-19
The gel pattern of DNA sequencing reactions. The determined sequence is
5'GAACTTGCT3'. This means that the template sequence (which is what we are
interested in knowing) is 5'AGCAAGTTC3'.

different pattern, that represents the positions of the nucleotide whose dideoxy derivative was used. In this example, the G lane has the shortest fragment, and therefore the first position in the newly-made sequence is G. The second fastest moving band is in the A track, so A is the second nucleotide. Continuation of this reading allows the definition of the entire sequence of the DNA. A word of caution: the determined sequence is the complement of the template sequence: a G in the template directs the incorporation of a C in the growing chain, and the determined sequence has the opposite 3', 5' orientation from the template sequence, so that when the gel pattern tells us that there is a G at the 5' terminus of the newly synthesized strand, we know that there is a C at the 3' terminus of DNA strand we are sequencing.

In a typical experiment, many copies of an isolated single strand of DNA are divided into four samples. Primers, the four types of nucleotide triphosphates, and one type of dideoxy nucleotide are added to each sample. In order to analyze the very small amounts that are generally available, either radioactive nucleotides or primers are used; analysis is then done with x-ray films, an example of which is shown in Figure 3-20. This technique is obviously limited by the capacity of a gel to separate fragments differing in length by one nucleotide; however, with care, a sequence of over 600 nucleotides can be determined on one gel. This method has provided complete sequences of DNA from viruses, mitochondria and chloroplasts. The complete sequencing of bacterial DNA (4.5×10^6 nucleotides) is underway, and the biological project of the century has begun—the sequencing of human DNA!

Synthesis of DNA

One of the techniques responsible for the explosive progress made in molecular biology in recent years is the chemical synthesis of DNA. Years of research on the optimal way to synthesize DNA have produced

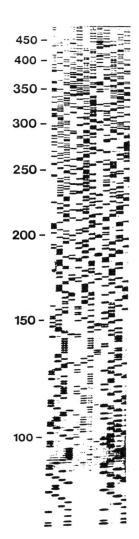

450 –
400 –
350 –
300 –
250 –
200 –
150 –
100 –

Figure 3-20
Sequences of DNA are stored and freely available from DNA Databanks in Europe, the United States and Japan via international computer networks.

a method which is now available in most laboratories. The technique is based on two chemical principles: the activation of the phosphate group so that it will combine with another nucleotide at a reasonable speed, and the protection of reactive groups on the nucleotides to prevent unwanted side reactions (Figure 3-21). Activation is accomplished using the condensing agent tetrazole on highly reactive nucleoside phosphoramidites.

Protection is the process wherein a chemical group is temporarily added to critical positions of the nucleotide to ensure that during the formation of the internucleotide bond only the correct 3′-5′ phosphodiester bonds are made. Amino groups on the bases, for example, react under conditions of phosphoramidite activation so they must be made chemically inert. In addition 5′-OH groups must be modified to control which atoms are used for the phosphodiester bond. Later in the synthesis, protecting groups have to be removed to produce the desired unmodified DNA. In addition, the example in Figure 3-21 makes use of an insoluble resin to avoid purification after each step in the synthesis.

The synthesizing strategy starts when the first nucleotide is added to a resin via its 3′-OH. To ensure that only this OH reacts, the 5′-OH is protected with a trityl group. This group is acid–sensitive so that it can be removed easily to make the 5′-OH available for subsequent steps. The resin is a porous glass preparation; the trityl group is generally the mono- or dimethoxy–derivative, and resins for all four nucleosides are commercially available.

After removal of the trityl group, the next nucleotide (in the form of its protected phosphoramidite derivative) is added in the presence of tetrazole. The internucleotide bond formation is virtually instantaneous; however the product is a *phosphite* rather than a *phosphate* ester. The phosphate bond is then oxidized to phosphate using iodine, the trityl group is removed, and the product is ready for the next cycle.

Now the advantage of the resin can be exploited. Since the product is insoluble in the reaction medium at all times during synthesis, purification is achieved by simple filtration of the resin. Another process packs the resin in a column and passes the various reagents, one after the other, through the column. In this way, the process is continuous. At the termination of the synthesis, the DNA is cleaved from the resin and all the remaining protective groups are removed.

This process has now been completely automated. Reagents are stored in bottles which are connected to the column with tubing and a series of valves. A computer controls which valves are opened and shut— the chemist simply keys in the desired sequence. The cycle of nucleotide addition has been reduced to around five minutes, and the synthesizing machines (gene machines) are capable of producing oligonucleotides over 100 units long. The tedious work involved in the optimization of each step must be fully appreciated, since unwanted side products accumulate very rapidly when yields at each step fall below 99%. Recently, a scheme for the chemical synthesis of RNA has been implemented. RNA syn-

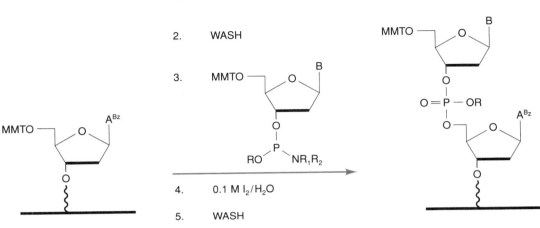

1.   5% TCA

2.   WASH

3.   MMTO

4.   0.1 M I$_2$/H$_2$O

5.   WASH

Figure 3-21
Synthesis of DNA. The first protected nucleoside is linked to the insoluble resin via the 3'-OH. Acid hydrolysis of the monoethoxytrityl group (MMT) is done in 5% trichloroacetic acid. The following nucleotide is added as its phosphoramidite derivative (R = methyl; R$_1$ and R$_2$ are isopropyl groups). Iodine is used to oxidize the phosphite to phosphate and the cycle is repeated. Notice that the synthesis is done in the 3' to 5' direction.

thesis is complicated by the fact that RNA is much more chemically labile than DNA, and the additional 2'-OH must be protected during the synthesis.

Summary

DNA consists of deoxyribonucleotides linked together by phosphodiester bonds connecting the 5'-carbon of one nucleotide with the 3'-carbon of the next nucleotide. DNA commonly exists as an antiparallel double strand. In double-stranded DNA, every base of one strand is hydrogen-bonded to a base of the other strand, and since purines pair only with pyrimidines, then the concentration of purines should equal that of pyrimidines—which can be expressed as [A + G] = [C + T]. Double-stranded DNA usually assumes the B form of a double helix, with ten base pairs per turn, 0.34 nm between each adjacent base pair, and bases oriented perpendicular to the helical axis. In addition to the B form, DNA also exists in an A form (a right-handed helix with 11 base pairs per turn, and bases tilted 30° with respect to the helical axis), and a Z form (a left-handed helix with 12 base pairs per turn). DNA can also form a double-stranded, covalently-closed circle. These circular molecules are often coiled into a superhelix—the formation of which is catalyzed by enzymes called topoisomerases. The hydrophobic properties and hydrogen-bonding abilities of the bases serve to hold the double-stranded molecule together. However, this double-stranded conformation is sensitive to the temperature and ion concentration of its environment. At a sufficiently high temperatures, the hydrogen bonds between bases are broken and the molecule is denatured. DNA made single-stranded will spontaneously renature if the temperature is high enough to disrupt any intrastrand base-pairing and if the salt concentration is sufficient to counteract the electrostatic repulsion between the phosphates. Renaturation kinetics can be used to determine the number of bases in a sequence that is repeated within a molecule and the number of copies of that sequence that are present.

Three primary classes of RNA exist—rRNA, mRNA, and tRNA. RNA is usually single-stranded, although some molecules—tRNA, for example—have extensive secondary structure. Enzymes called nucleases, catalyze the hydrolysis of phosphodiester bonds linking adjacent nucleotides. DNA molecules can be sequenced using the Sanger method, in which synthesis of a copy of the strand to be sequenced is halted at specific positions by the incorporation of dideoxy derivatives of specific nucleotides. The fragments are separated by gel electrophoresis, and the sequence is determined from the banding pattern of the gel. Also, nucleic acid molecules can be synthesized to include virtually any nucleotide sequence.

Drill Questions

1. The base composition of a double-stranded DNA molecule has been determined, and the [T] to [G] ratio was found to be 0.2. What fraction of the bases is A?

2. A new virus is discovered, the genome of which is found to be 28% A, 23% T, 24% C, and 25% G. What does this tell you about the genome of the virus?

3. In a double-stranded DNA molecule, what is the relationship between [A] + [G] and [C] + [T], and between [A] + [T] and [G] + [C]?

4. In what circumstances does a double-stranded DNA molecule form an A helix? a B helix? a Z helix?

5. Could a supercoiled DNA molecule be formed by:
   a) joining the ends of a linear DNA molecule and twisting the resulting circle?
   b) twisting the ends of a linear DNA molecule and then linking the ends together?
   c) linking together the ends of a linear DNA molecule?

6. What non-covalent interactions are involved in maintaining the double-helical conformation of DNA?

7. Why are salt concentration and temperature important to the renaturation of DNA?

8. What would happen if double-stranded DNA were placed in distilled water?

9. A DNA molecule is composed of 4800 base pairs. How many helical turns are present? How long is the linear, double-helical form of this molecule?

10. Could an exonuclease remove nucleotides from a circular DNA molecule?

Problems

1. The DNA molecules below are denatured and then allowed to reanneal. Which of the two is least likely to re-form the original structure? What would prevent that molecule from doing so?
   (a) ATATGTATATATAGAT
       TATACATATATATCTA
   (b) GCCTATACGTGCACCA
       CGGATATGCACGTGGT

2. Which of the two molecules shown above would have the highest T_m? Why? Which would require a higher temperature for renaturation?

3. The following gel pattern was obtained in an attempt to sequence a DNA molecule using the Sanger method. What is the sequence of the DNA molecule in question?

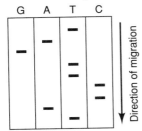

4. A single-stranded DNA molecule is hybridized with a DNA molecule that is identical except that the central 10% of the molecule has been deleted. What is the structure of the heteroduplex that would result? What if that missing 10% had been replaced with a segment of nonhomologous DNA of equal length?

5. DNA from one species of bacteria is labelled with ^{15}N and is hybridized with an equal concentration of unlabeled (^{14}N) DNA from another species of bacteria. The result is that 15% of the TOTAL amount of renatured DNA has a hybrid density. What fraction of base sequences are common to the two species?

6. The melting curve below was obtained from a solution of DNA. What does this curve tell you about the composition of the solution?

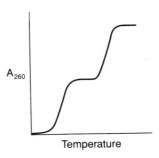

Conceptual Questions

1. What features of the DNA double helix facilitate its replication? What features might make its replication complicated?

2. In what ways does RNA represent the ideal molecule for converting the information encoded in a nucleotide sequence into the structural and catalytic macromolecules of a cell?

3. With modern DNA sequencing techniques it is now feasible to sequence all the DNA in a human cell. Social, moral, and ethical considerations can however be raised. Should such a major sequencing endeavor nevertheless be encouraged?

Alexandra Cynthia Newton
Professor of Chemistry
Department of Chemistry

Birthday: **9 January 1958**

Birth Place: **Cape Town, South Africa**

Undergraduate Degree: **Simon Fraser University, Canada**
Major: Biochemistry 1980

Graduate Degree: **Stanford University, Ph.D. 1986**

Postdoctoral Training: **University of California, Berkeley, 1986–1988**

Present Position: **Indiana University, Chemistry Department**

Address: **Bloomington, Indiana**

RESEARCH IN MY laboratory is directed towards understanding how cells sense, respond, and adapt to extracellular information. We are examining how signals are received at the cell surface, transferred across the plasma membrane into the cell, and processed within the cell.

We are examining the chemistry of signal transduction using fluorescence, biochemical, and molecular biological techniques. One protein of interest to us is the Ca^{2+}/phospholipid-dependent protein kinase C. This unusual membrane protein plays a critical role in the transduction of a wide variety of extracellular signals. This protein's ability to covalently modify proteins and high concentration in the brain suggests it may be involved in memory. Furthermore, uncontrolled activation of the enzyme results in tumor promotion. We are examining how protein kinase C structure and function are regulated. For example, experiments are aimed at elucidating the nature of the reversible interaction of the kinase with the plasma membrane and determining how specific phospholipids regulate protein conformation and protein function. Also of interest are the effects of ligand binding and protein phosphorylation on the interaction of receptors with the plasma membrane. The insulin receptor is ideally suited to these studies: it is activated by ligand (insulin) binding and its function is regulated by phosphorylation. A third area of research involves the role of protein kinase C in phototransduction. When protein kinase C is activated in the retina, the phosphorylation state of rhodopsin, the receptor for light, changes. We are examining the effect of phosphorylation by protein kinase C on rhodopsin structure and function, with the goal of understanding how protein kinase C transduces the light signal.

What were the major influences or driving forces that led you to a career in molecular biology?

The major driving force that led me to a career in biochemistry has been a life-long curiosity about nature. Many summers snorkeling in the Greek Islands convinced me that I wanted to become a marine biologist. However, my first Chemistry courses introduced me to what was an even more delightful pre-occupation: the molecular aspects of life. So my present research on the chemistry of fundamental processes, such as vision and cancer, is really a continuation of a life-long fascination in the biology and chemistry of life.

4

The Physical Structure of Protein Molecules

All proteins are polymers of amino acids, yet each species of protein molecule has a unique three-dimensional structure determined principally by the amino acid sequence of that protein, in contrast with DNA, which has a universal structure—the double helix. This makes the study of proteins very complex but, on the other hand, the diversity of protein structures enables these molecules to carry out the thousands of different processes required by a cell and makes proteins fascinating objects to study.

Proteins have been called the "footsoldiers" of the living cell. Their functions are the most diverse of all types of macromolecules. As we shall learn in this chapter, size, shape, and macromolecular composition vary tremendously among proteins. Those diverse features generate a wide range of functional properties. Some proteins carry out structural functions (e.g., stabilize a living cell's shape), while others speed up the rate of chemical reactions (e.g., enzymes).

In order to understand the manner in which proteins manage to carry out such diverse functions, we will learn how individual amino acids contribute a structural feature to a protein. Beginning with the amino acid sequence, we will review twisting, folding, and aggregation; each of which contributes to a protein's function. Then we will examine several examples of proteins that display widely different properties and hence carry out diverse functions.

The study of the detailed structure of proteins is beyond the scope of this book, being heavily dependent on a variety of optical techniques, especially the mathematically complex technique of x-ray diffraction. For this reason the discussion of proteins in this chapter will be a survey.

Basic Features of Protein Molecules

In Chapter 2 the basic chemical features of protein molecules were described. That is, a protein is a polymer of amino acids in which carbon atoms and peptide groups alternate to form a linear polypeptide chain and specific groups—the amino-acid side chains—project from the α-carbon atom (Figure 2-1). This linear sequence of amino acids is called the **primary structure** of the protein. The term "linear" requires careful consideration, for, as will be seen in the following sections, a polypeptide chain is highly folded and can assume a variety of three-dimensional shapes; each shape in turn consists of several standard elementary three-dimensional configurations and other configurations which may be unique to that molecule.

In Chapter 3, it was seen that nucleic acid molecules are very large, having molecular weights often as high as 10^{10}. Protein molecules are much smaller; in fact, the molecular weight of a typical protein molecule is comparable to that of the smallest nucleic acid molecules, the transfer RNA molecules. The molecular weights of hundreds of different proteins have been measured. Typical polypeptide chains have molecular weights ranging from 15,000 to 70,000. The average molecular weight of an amino acid is 110, which means that typical polypeptide chains contain some 135 to 635 amino acids.

The length of a typical polypeptide chain, if it were fully extended, would be 1000 to 5000 Å. A few of the longer fibrous proteins, such as myosin (from muscle) and tropocollagen (from tendon) are in this range—with lengths of 1600 Å and 2800 Å, respectively. However, most proteins are highly folded and their longest dimension usually ranges from 40 to 80 Å. This folding constitutes the **secondary structure** of a protein and is described in the next section.

The Folding of a Polypeptide Chain

A fully extended chain, if it were to exist, would have the configuration shown in Figure 4-1. This stretched out form has no biological activity. Rather, protein function arises because polypeptides fold in complex ways to yield precise three-dimensional structures. The folded structure is stabilized by many non-covalent interactions (e.g., H-bonding, ionic interactions, van der Waals attractions). Physical properties of the peptide bond and the amino acids in the polypeptide chain govern the manner of this folding. Three rules based on these physical properties can be described:

1. The peptide bond has a partial double-bond character (Figure 4-2) and hence is constrained to be planar and rigid. Free rotation occurs

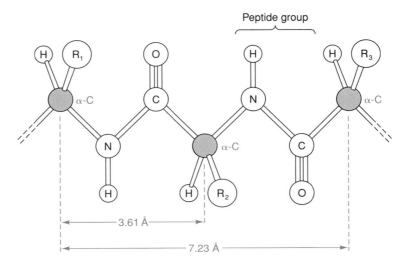

Figure 4-1
The configuration of a hypothetical fully extended polypeptide chain. The length of each amino acid residue is 3.61 Å; the repeat distance is 7.23 Å. The α-carbon atoms are shown in red. Side chains are denoted by R.

only between the α carbon and the peptide unit. Thus, the polypeptide chain is flexible but is not as flexible as would be the case if there were free rotation about all of the bonds.

2. The side chains of the amino acids cannot overlap. Thus, the folding can never be truly random, because certain orientations are forbidden.

3. Two charged groups having the same sign will not be very near one another. Thus, like charges tend to cause extension of the chain.

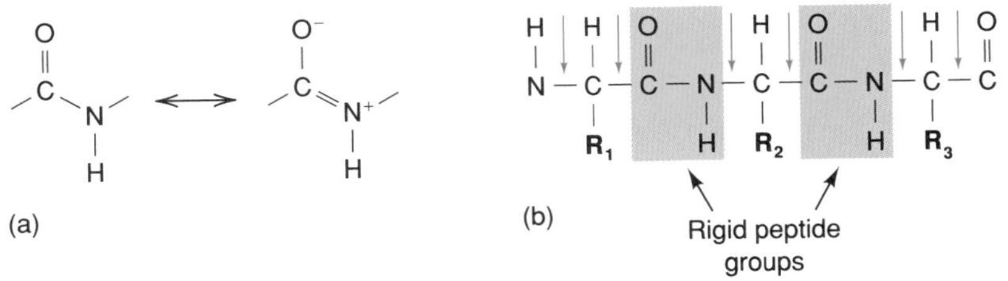

(a)

(b)

Rigid peptide groups

Figure 4-2
The planarity of the peptide group. (a) Two forms of the peptide bond in equilibrium. In the form at the right the double bond creates a rigid unit. Owing to the resonance, the peptide group has a partial double-bond character and is rigid. (b) The peptide groups in a protein molecule. They are planar and rigid, but there is freedom of rotation about the bonds (red arrows) that join peptide groups to α-carbon atoms (red).

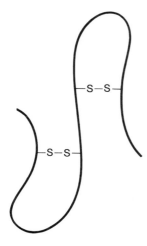

Figure 4-3
A polypeptide chain in which four cysteines are engaged in two disulfide bonds.

However, in addition to these rules, folding behaves according to the following general tendencies:

1. Amino acids with polar side chains tend to be on the surface of the protein in contact with water.

2. Amino acids with nonpolar side chains tend to be internal. Very hydrophobic side chains tend to cluster. Figure 2-2 in Chapter 2 indexes the nonpolar amino acids.

3. Hydrogen bonds tend to form between the carbonyl oxygen of one peptide bond and the hydrogen attached to a nitrogen atom in another peptide bond.

4. The sulfhydryl group of the amino acid cysteine tends to react with an —SH group of a second cysteine to form a covalent S—S (**disulfide**) bond. Such bonds pose powerful constraints on the structure of a protein (Figure 4-3).

These tendencies should not be considered invariable because there are many exceptions. Nevertheless they indicate that the structure of a protein may be changed markedly by a single amino acid substitution — for example, a polar amino acid for a nonpolar one; similarly, the change in structure might be minimal if one nonpolar amino acid replaces another nonpolar one. This notion will be encountered again in Chapter 11 when mutations are considered.

The three-dimensional shape of a polypeptide chain is a result of a balance between all of the rules and tendencies just described and can be very complex. However, in examining many polypeptide chains, it has become apparent that certain geometrically regular arrays of the chain are found repeatedly in different polypeptide chains and in different regions of the same chain. These are the arrays resulting from hydrogen-bonding between different peptide groups. These arrays are described in the next section.

The α Helix and β Structures

In the absence of any interactions between different parts of a polypeptide chain, a random coil would be the expected configuration. However, hydrogen bonds easily form between the H of the N—H group and the O of the carbonyl group. H-bonding results in a variety of ordered structures of the polypeptide chain called secondary structures, the most common being the α helix and the β structures elucidated by Pauling and Corey in the early 1950s.

In an **α helix** the polypeptide chain follows a helical path that is stabilized by hydrogen-bonding between peptide groups. Each peptide group is hydrogen-bonded to two other peptide groups, one three units ahead and one three units behind in the chain direction (Figure 4-4). All

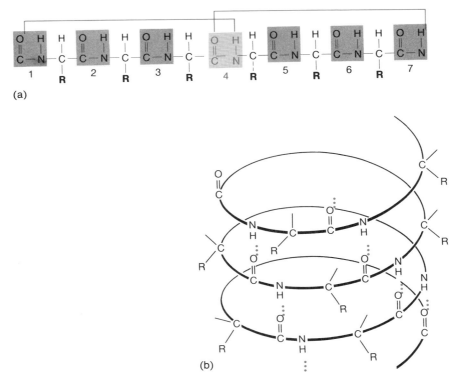

(a)

(b)

Figure 4-4
Properties of an α helix. (a) The two hydrogen bonds in which peptide group 4 (red) is engaged. The peptide groups are numbered below the chain. (b) An α helix drawn in three dimensions, showing how the hydrogen bonds stabilize the structure. The red dots represent the hydrogen bonds. The hydrogen atoms that are not in hydrogen bonds are omitted for the sake of clarity.

α helices have a pitch of 5.4 Å, which is the repeat distance, a diameter of 2.3 Å, and contain 3.6 amino acids per turn. Thus, it is a much tighter helix than the DNA helix. The side chains stick out from the rod-shaped core of the α helix.

In the absence of all interactions other than the hydrogen-bonding just described, the α helix is the preferred form of a polypeptide chain because, in this structure, all monomers are in identical orientation and each forms the same hydrogen bonds as any other monomer. Thus, polyglycine, which lacks side chains and hence cannot participate in any interactions other than those just described, is an α helix.

If all monomers are not identical or if there are secondary interactions that are not equivalent, the α helix is not necessarily the most stable structure. While some amino acid side chains can stabilize the α helix, others are responsible for preventing α-helicity. A striking example of the disruptive effect of a side chain on an α helix is evident with the synthetic polypeptide polyglutamic acid, a polypeptide containing only glutamic acid. At a pH below about 5, the carboxyl group of the side chain is not ionized and the molecule is almost purely an α helix. However, above pH 6, when the side chains are ionized, electrostatic repulsion totally destroys the helical structure.

If the amino acid composition of a real protein is such that the helical structure is extended a great distance along the polypeptide backbone, the protein will be somewhat rigid and fibrous (not all rigid fibrous proteins are α helical, though). This structure is common in many structural proteins, such as the α-keratin in hair.

Another hydrogen-bonded configuration is the **β structure.** In this form, the molecule is almost completely extended (repeat distance = 7 Å) and hydrogen bonds form between peptide groups of polypeptide segments lying adjacent and parallel with one another (Figure 4-5(a)). The side chains lie alternately above and below the main chain.

Two segments of a polypeptide chain (or two chains) can form two types of β structure, which depend on the relative orientations of the

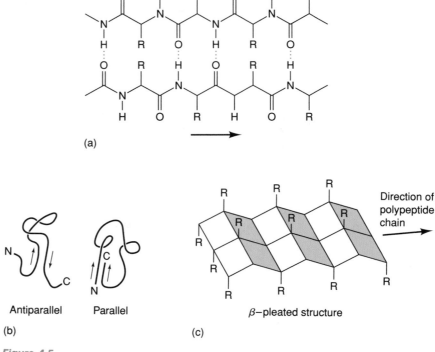

(a)

Antiparallel   Parallel

(b)

(c)

β-pleated structure

Direction of polypeptide chain

Figure 4-5
β structures. (a) Two regions of nearly extended chains are hydrogen-bonded (red dots) in an antiparallel array (arrows). The side chains (R) are alternately up and down. (b) Antiparallel and parallel β structures in a single molecule. (c) A large number of adjacent chains forming a β pleat.

segments. If both segments are aligned in the N-terminal-to-C-terminal direction or in the C-terminal-to-N-terminal direction, the β structure is **parallel.** If one segment is N-terminal-to-C-terminal and the other is C-terminal-to-N-terminal, the β structure is **antiparallel.** Figure 4–5(b) shows how both parallel and antiparallel β structures can occur within a single polypeptide chain.

When many polypeptides interact in the way just described, a pleated structure results called the **β-pleated sheet** (Figure 4–5(c)). These sheets can be stacked and held together in rather large arrays by van der Waals attractions and are often found in fibrous structures such as silk.

Protein Structure

Few proteins are pure α helix or β structure; usually regions having each structure are found within a protein. Since these configurations are rigid, a protein in which most of the chain has one of these forms is usually long and thin and is called a **fibrous protein.** In contrast are the quasi-spherical proteins called the **globular proteins** in which α helices and β structures are short and interspersed with randomly coiled regions and compact structures.

The fibrous proteins are typically responsible for the structure of cells, tissues, and organisms. Some examples of structural proteins are collagen (the protein of tendon, cartilage, and bone), elastin (a skin protein), and spectrin (a protein that contributes to cell shape). Some of the fibrous proteins are not soluble in water—examples are the proteins of hair and silk.

The catalytic and regulatory functions of cells are performed by proteins that have a well defined but deformable structure. These are the globular proteins, of which the catalytic proteins, or enzymes, are the most widely studied. Globular proteins are compact molecules having a generally spherical or ellipsoidal shape. Large segments of the polypeptide backbone of a typical globular protein are α-helical. However, the molecule is extensively bent and folded. Usually, the stiffer α-helical segments alternate with very flexible randomly coiled regions, which permit bending of the chain without excessive mechanical strain. Numerous segments of the chain, which might be quite distant along the backbone, form short parallel and antiparallel β structures (Figure 4–5(b)); these also are responsible in part for the folding of the backbone (Figure 4–5(c)). The extensive folding of the backbone is usually called

tertiary structure

the **tertiary structure** or **tertiary folding.**

A very important distinction can be made between secondary and tertiary structure; namely,

Secondary structure results from hydrogen-bonding between peptide groups that are close together in the sequence of the polypeptide chain. Tertiary structures are formed from the folding of the secondary structures (α helices, β sheets) to yield three-dimensional structures.

The most prevalent interactions responsible for tertiary structure are the following:

1. Ionic bonds between oppositely charged groups in acidic and basic amino acids. For example, glutamic acid forming an ionic interaction with lysine.

2. Hydrogen bonds between H-bond donors and acceptors in amino acid side chains. For example, the hydroxyl group in tyrosine and a carboxyl group of aspartic or glutamic acids.

3. Hydrophobic clustering between the hydrocarbon side chains of nonpolar residues such as phenylalanine, leucine, isoleucine, and valine.

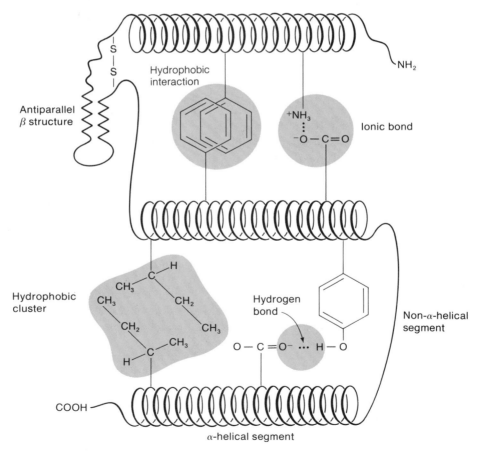

Figure 4-6
A hypothetical globular protein having several types of side-chain interactions.

4. Metal-ion coordination complexes between amino, hydroxyl, and carboxyl groups, ring nitrogens, and pairs of SH groups.

Hydrophobic clustering (item 3 above) is the most important stabilizing feature.

Figure 4-6 shows a schematic diagram of a hypothetical protein (in two dimensions) in which several of these interactions determine the structure. This figure should be examined carefully, for it indicates the role of different features of a protein molecule in determining the overall conformation of the molecule. One can see the following:

1. Disulfide bonds bring distant amino acids together.

2. Hydrophobic interactions bring distant amino acids together.

3. Hydrogen bonds sometimes bring distant amino acids together, but usually a single hydrogen bond makes a more subtle change in position.

4. Electrostatic interactions bring amino acids together or keep them apart, depending on the signs of the charges.

5. A β structure brings distant segments of the polypeptide backbone together and creates rigidity.

6. An α helix makes adjacent regions of a polypeptide backbone stiff and linear.

7. Van der Waals attractions produce specific interactions between clusters of amino acids that may or may not be nearby in the polypeptide chain.

Figure 4-7 shows the structures of a number of globular proteins; some are predominantly α-helical (a) some are predominantly β sheet (b), and some are mixed α helix and β sheet (c).

Proteins with Subunits

A polypeptide chain usually folds so that nonpolar side chains are internal—that is, isolated from water. However, it is rarely possible for a polypeptide chain to fold in such a way that *all* nonpolar groups are internal. Thus, it is often the case that nonpolar amino acids on the surface form clusters in an effort to minimize contact with water. A protein molecule having a large hydrophobic patch can further reduce contact with water by pairing with a hydrophobic patch on another protein molecule. Similarly, if a molecule has several distantly located hydrophobic patches, a structure consisting of several protein molecules in contact effectively minimizes contact with water. The protein is then said to consist of identical **subunits;** this is in fact a very common phenomenon, with two, three, four, and six subunits occurring most

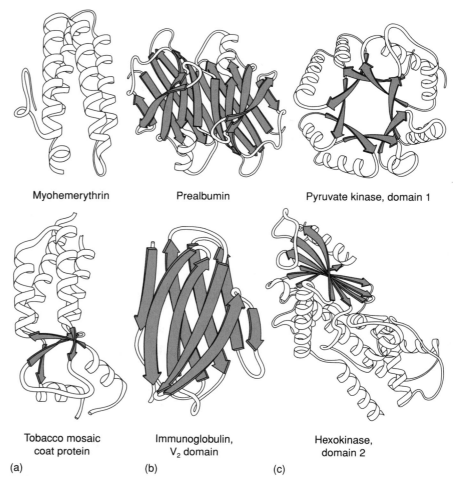

Myohemerythrin   Prealbumin   Pyruvate kinase, domain 1

Tobacco mosaic   Immunoglobulin,   Hexokinase,
coat protein   V$_2$ domain   domain 2

(a)   (b)   (c)

Figure 4-7
Idealized drawings of the tertiary structures of several globular proteins. (a) Myohemerythrin and tobacco mosaic coat protein are mostly α helical. (b) Prealbumin and the
immunoglobulin V$_2$ domain are predominantly β sheet. (c) Pyruvate kinase, domain 1, and
hexokinase, domain 2, have α helical and β sheet structures. (Adapted and redrawn from
Mathews and van Holde, *Biochemistry*. Menlo Park, CA, Benjamin Cummings, 1990.)

frequently. A multisubunit protein may also contain unlike subunits,
and this is quite common. For example, hemoglobin, the oxygen carrier
of blood, consists of four subunits, two each of two different types;
likewise, RNA polymerase, which catalyzes synthesis of RNA, has five
subunits of which four are different; and DNA polymerase III, which
synthesizes DNA in *E. coli*, consists of ten different subunits.

Multisubunit proteins have certain advantages, particularly with respect to economy of synthesis and regulation of enzymatic activity; a
detailed presentation of the value of subunits can be found in *MB*, pages
196–197. An example of a multisubunit protein, whose physical and

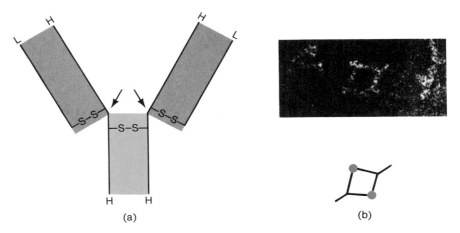

Figure 4-8
(a) The subunit structure of immunoglobulin G. There are two L subunits and two H subunits. Disulfide bonds join each L strand to an H strand and join the two H strands to one another. Treatment with papain cleaves the H strands at the arrows, yielding two F_{ab} units (shaded black) and one F_c unit (shaded red). (b) An electron micrograph showing two immunoglobulin G molecules joined at the antigen binding site. The joining is a result of binding two antigen molecules not visible in the micrograph but shown as red dots in the interpretive drawing. (Courtesy of R. C. Valentine.)

biological properties are well known, is immunoglobulin G. It will be described in some detail in the following paragraphs; its synthesis is discussed in Chapter 16.

antibody

antigen

The immunoglobulins (**antibodies**) are the proteins of the immune system. Their function is to interact with specific foreign molecules (**antigens**) and thereby render them inactive. This interaction is called the **antigen–antibody reaction.** The best-understood immunoglobulin is immunoglobulin G (**IgG**); other classes of immunoglobulins are IgA, IgM, IgD, and IgE. In this section, we shall examine only the IgG class, itself comprising several subclasses defined by slight structural differences.

Cleavage of the disulfide bonds of IgG yields two polypeptide chains whose molecular weights are about 25,000 and 50,000. The lighter polypeptide is called an **L chain;** the heavier one is an **H chain.** IgG is a tetramer containing two L chains and two H chains. A schematic structure of IgG is shown in Figure 4-8. Experimentally, IgG can be cleaved with the proteolytic enzyme papain, which causes each of the H chains to break, as shown in the figure, thus producing three separate subunits. The two units that consist of an L chain and a fragment of the H chain equal in mass to the L chain are called F_{ab} fragments (the subscript stands for antigen-binding). The third unit, consisting of two equal segments of the H chain, is called the F_c (the subscript stands for "constant" or "common") fragment.

Two sites on an IgG molecule can bind antigen. Each site is at the end of an F_{ab} segment, as shown in Figure 4-9. The F_c segment is not

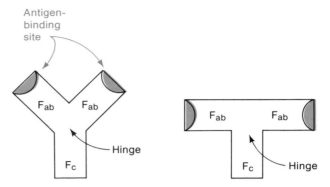

Figure 4-9
Immunoglobulin G is a Y-shaped molecule. It contains a hinge that gives it segmental flexibility.

involved in antigen–antibody binding but is used in later processes needed to destroy the antigen.

The amino acid sequences of many subclasses of IgG molecules, each capable of combining with only a single kind of antigen molecule, have been determined. All immunoglobulins are structurally similar inasmuch as L and H chains both have what is called a **variable (V)** and a **constant (C)** region.

The V and C regions have separate functions in IgG. The V regions confer antigen-binding specificity, whereas the C regions are responsible for the overall structure of the molecule and for its recognition by other components of the immune system.

The C regions of the L chains (C_L) of all types of IgG molecules have an identical amino acid sequence. Likewise, the C regions of all H chains (C_H) are identical for all IgG, though different from the C_L sequences. The V regions differ from one type of IgG to the next, however. The arrangement of the C_L, V_L, C_H, and V_H regions is shown in Figure 4–10.

Some of the similarities between amino acid sequences in different parts of an IgG molecule are quite striking. For example, the C region of the H chain can be divided into three segments: C_H1, C_H2, and C_H3, whose amino acid sequences are quite similar though not identical to one another. Furthermore, the sequence of C_L resembles the C_H sequences, though generally they are different. The V_L and V_H sequences also are nearly the same. This means that the IgG molecule, which already has twofold symmetry, consists of four domains, each of which has twofold symmetry, as shown in Figure 4–10, in which the intensity of the shading indicates two homologous regions. The symmetry of the molecules is made especially evident in Figure 4–11, which shows the three-dimensional structure of IgG. A feature of IgG, not clearly shown in the figure, is that the antigen-binding site is formed by joining parts

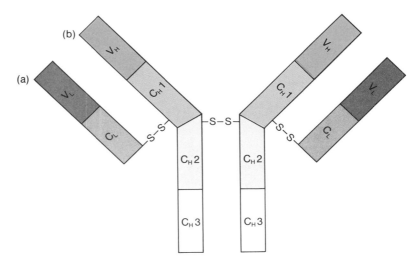

Figure 4-10
Model of an antibody molecule showing arrangement of the variable and constant regions of the IgG (a) L chain and (b) H chain.

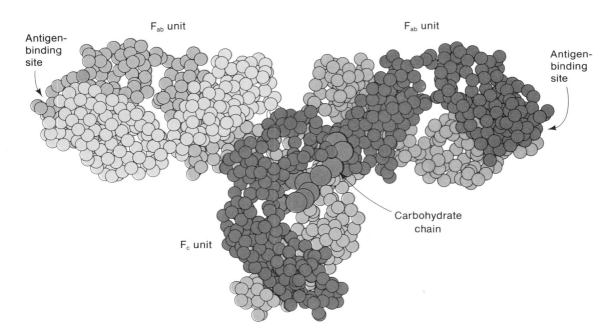

Figure 4-11
Schematic drawing of the three-dimensional structure of an IgG molecule. One of the H chains is shown in dark pink, the other in dark gray. One of the L chains is shown in light pink, the other in light gray. (After E. W. Silverton, M. A. Navia, and D. R. Davies. *Proc. Nat. Acad. Sci.*, 1977 74: 5142.)

of different subunits, namely, the V_H and V_L regions. Formation of a binding site by a subunit interaction is a property of many multisubunit proteins.

The structures of some multisubunit proteins are quite complex. For example, collagen, the protein of tendon, cartilage, bone, and skin, is formed by the mutual binding of three polypeptide chains as a triple helix and the interaction of numerous triple helices to form a tough fiber.

Enzymes

Enzymes are special proteins able to catalyze chemical reactions. Their catalytic power exceeds all manmade catalysts. A typical enzyme accelerates a reaction 10^8- to 10^{10}-fold, though there are enzymes that increase the reaction rate by a factor of 10^{15}. Enzymes are also highly specific in that each catalyzes only a single reaction or set of closely related reactions. Furthermore, only a small number of reactants, often only one, can participate in a single catalyzed reaction. Since nearly every biological reaction is catalyzed by an enzyme, these clearly require a very large number of distinct enzyme molecules.

The detailed mechanism of catalysis by particular enzymes is beyond the scope of this book. However, all enzymes have certain general features, the knowledge of which is important for understanding molecular biological phenomena. These features are described in this unit.

THE ENZYME-SUBSTRATE COMPLEX

In any reaction that is enzymatically catalyzed, one reactant always forms a tight complex with the enzyme. This reactant is called the **substrate** of the enzyme; in descriptive formulas, it is denoted S. The complex between the enzyme E and the substrate is called the enzyme-substrate or **ES complex.**

active site

The substrate binds to a region of the enzyme called the **active site,** and it is here that the chemistry occurs. The active site is often a cleft in the enzyme, with some side chains of amino acids contributing to the binding of the substrate and others to the catalysis of the reaction. The extraordinary selectivity in enzyme catalysis is almost entirely a result of specificity of enzyme-substrate binding. After the ES complex forms, the substrate is usually altered in some way that facilitates further reaction. When the substrate is in the altered state, the ES-complex is said to be active and is usually denoted **(ES)★.** The **(ES)★** complex then engages in one or a series of transformations, which result in conversion of the substrate to the product and dissociation of the product from the enzyme. The extent to which ES forms is determined by the strength of binding between E and S; this is called the **affinity** of E and S.

THEORIES OF FORMATION OF THE ENZYME-

SUBSTRATE COMPLEX

Catalysis by an enzyme occurs in several steps—binding of the substrate, conversion to the product, and release of the product. The initial step, formation of the ES complex, is in principle easy to understand and is often considered in thinking about molecules. The subsequent rearrangements are chemical phenomena and will not be discussed in this book.

There are two major theories of enzyme binding, the **lock-and-key** and **induced-fit** models (Figure 4-12). In the lock-and-key model, the shape of the active site of the enzyme is complementary to the shape of the substrate. In the induced-fit model, the enzyme changes shape upon binding the substrate, and the active site has a shape that is complementary to that of the substrate only *after* the substrate is bound. For every enzyme-substrate interaction examined to date, one of these two models applies. However, enzyme catalysis is generally best described by the induced-fit model. It is often the case, though, that the substrate itself undergoes a small change in shape; in fact, the strain to which the substrate is subjected is often the principle mechanism of catalysis—that is, the substrate is held in an enormously reactive configuration.

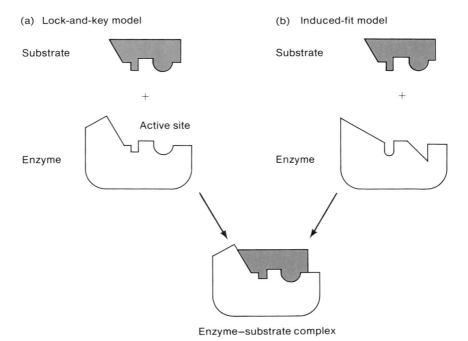

Figure 4-12
Two models for enzyme-substrate binding. (a) The lock-and-key model. The active site of the enzyme by itself is complementary in shape to that of the substrate. (b) The induced-fit model. The enzyme changes shape upon binding a substrate. The active site has a shape complementary to that of the substrate only after the substrate is bound.

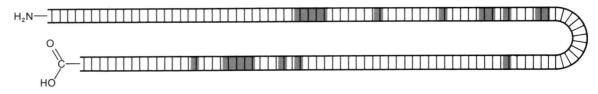

Figure 4-13
Schematic diagram of the amino acid sequence of lysozyme showing that the amino acids (red) in the active site are separated along the chain. Folding of the chain brings these amino acids together.

MOLECULAR DETAILS OF AN ENZYME-SUBSTRATE COMPLEX

The first detailed analysis of enzyme-substrate binding was carried out using hen egg-white lysozyme. This enzyme cleaves certain bonds between sugar residues in some of the polysaccharide components of bacterial cell walls and is responsible for maintaining sterility within eggs. The amino acid sequence of lysozyme is shown in Figure 4-13. The 19 amino acids that are part of the active site are printed in red; it should be noticed that they form widely separated clusters along the chain. Only when the chain is folded do they come into proximity and form the active site. The folding of the chain is shown in Figure 4-14. The deep cleft indicated by the arrow is the active site. This is seen more clearly in the space-filling model shown in Figure 4-15. The substrate

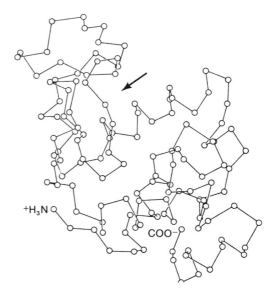

Figure 4-14
Three-dimensional structure of lysozyme. Only the α-carbon atoms are shown. The active site is in the cleft indicated by the arrow. (Courtesy of Dr. David Phillips.)

Figure 4-15
A space-filling molecular model of the enzyme lysozyme. The arrow points to the cleft that accepts the polysaccharide substrate. (C atoms are black; H, white; N, gray; O, gray with slots.) (Courtesy of John A. Rupley.)

is a hexasaccharide segment that fits into the cleft and is distorted upon binding. The enzyme itself changes shape when the substrate is bound.

Another enzyme, yeast hexokinase A, has been studied in order to examine what structural changes occur on substrate binding; these changes are now well documented. This is shown in the pair of space-filling models shown in Figure 4-16.

Summary

The primary structure of a polypeptide is the linear sequence of amino acids connected by peptide bonds. Secondary structure involves H-bonding between nearby peptide bonds. In the α helix, each peptide group H-bonds with groups three ahead and behind itself. Each turn contains 3.6 amino acids. The side chains stick out from the helical core. The β sheet is more extended, and the H-bonding occurs between peptide units in two or more adjacent polypeptide chains that can be parallel or antiparallel to one another. The side chains extend alternately above and below the sheet. Tertiary structure is the overall folding of a polypeptide in which secondary structure domains are positioned via long-range interactions which may be covalent, e.g., disulfide bonds, ionic (repulsion or attraction of charged side chains), hydrophobic (clustering of nonpolar side chains), or H-bonds. Fibrous proteins have long, uninterrupted sections of α helix or β sheet; globular proteins are more spherical, with short stretches of secondary structure. Generally, polar side chains are on the surface near water, and nonpolar side chains are internal, away from water. Multisubunit proteins contain several poly-

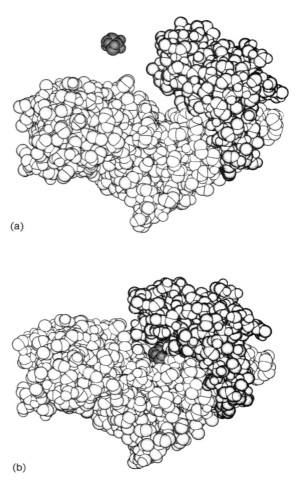

Figure 4-16
Structure of yeast hexokinase A. All atoms except hydrogen are shown separately.
(a) Free hexokinase and its substrate, glucose. (b) Hexokinase complexed with glucose.
Note that glucose binding causes two domains to move together and close the binding
site cleft. (With permission, from J. Bennett and T. A. Steitz. *J. Mol. Biol.*, 140: 211.
Copyright 1980: Academic Press Inc. [London] Ltd.)

peptides, e.g., antibodies that contain two heavy and two light chains.
The antigen-binding site is formed by the interaction of two chains, a
common property for binding and active sites. Enzymes are protein
catalysts that speed up the rate of cellular reactions. The active site is
where substrate is bound, forming an enzyme-substrate complex where
chemical reactions take place. These steps are facilitated by amino acid
side chains, often from separated regions of the polypeptide. Substrate
binding may occur by a lock-and-key mechanism, in which the active
site is of complementary shape to the substrate; or more often by an

induced-fit mechanism, in which both the active site and substrate change shape slightly, thus facilitating catalysis.

Drill Questions

1. (a) Which amino acids are polar and which are nonpolar?
   (b) Is isoleucine more nonpolar than alanine? Why?

2. About which bond in the polypeptide backbone is there no free rotation?

3. Name the kinds of bonds in proteins in which the side chains of each of the following amino acids might participate: cysteine; arginine; valine; aspartic acid.

4. What would you guess to be the environment of a glutamate that is internal? What is the environment of an internal lysine?

5. Which of the following sets of three amino acids are probably clustered within a protein? (a) Asn, Gly, Lys; (b) Met, Asp, His; (c) Phe, Val, Ile; (d) Tyr, Ser, Lys; (e) Ala, Arg, Pro.

6. For a particular enzyme, fifteen amino acid residues are known to be at the active site of the molecule. What would you expect their relative positions to be in the primary sequence of the protein?

7. What is meant by the primary, secondary, and tertiary structure of a protein?

8. What would be the most likely general tertiary structure for a protein with very long α helical or β structure domains? With many short regions of α helix and/or β structure?

9. What is meant by induced-fit? How does this facilitate catalysis?

Problems

1. State whether the following polypeptides would aggregate to form a multisubunit protein consisting of identical subunits and give the reason for your conclusion.
   (a) The folded molecule contains distinct surface regions with complementary shapes, and no polar amino acids are nearby.
   (b) There is a large hydrophobic cluster in a crevice just below the surface.
   (c) The protein has a large hydrophobic patch flanked by two lysines.
   (d) The surface has a region where positively and negatively charged amino acids alternate in a linear array.

2. Is an enzyme likely to have a very rigid configuration? Explain.

3. In general, a diploid cell containing the $^+$ and $^-$ alleles encoding an enzyme E has an E^+ phenotype. You have isolated a particular mutant that yields an E^- phenotype even when the $^+$ allele is present. Suggest a possible explanation for the observed phenomenon.

4. Many proteins consist of several identical subunits. Some of these proteins have a single binding site, whereas others have several identical binding sites. What might you predict about the locations of the binding sites in these two classes of multisubunit proteins?

5. Assuming you had a way to look at the exact structure of the active site of an enzyme or binding protein, how might you distinguish between the lock-and-key and induced-fit models of substrate binding? Explain the expected results.

Conceptual Questions

1. How might the various amino acid side chains (functional groups) that are associated with the active site of an enzyme chemically aid in the binding or catalysis of the substrate?

2. Enzymatic reactions are very fast, accelerated up to 10^{15}-fold. Suggest ways that you might slow this down in the laboratory in order to be able to study the mechanism of a particular enzyme activity.

3. It is generally estimated that in a typical cell 1,000–2,000 different chemical reactions are catalyzed by enzymes. Why do you suppose it is necessary that each of those reactions have its own special enzyme?

5

Macromolecular Interactions and the Structure of Complex Aggregates

Interactions between macromolecules underly most biological phenomena. Structural components both within individual cells and in between the cells associated with various tissues and organs invariably consist of assemblies of macromolecules. For example, nucleic acids are often associated with proteins as in chromosomes (which are DNA-protein complexes), viral nucleic acids are encased in protein shells, and bone and cartilage are complex assemblies of proteins and other macromolecules. Proteins can also interact with lipids to produce membranes such as those which separate the contents of a cell from the environment, and which separate different intracellular components from one another. Finally, polysaccharides form extraordinarily complex structures such as the cell walls of bacteria and of plants.

The study of such complex structures and how they are formed is called structural biology. A few structures are almost completely understood and many more are actively being studied. In this chapter only a few structures will be described. These have been selected by two criteria—they are of general importance, or they illustrate general principles, and their structures are reasonably well understood. We will begin by reviewing the structure of one macromolecular complex—the *E. coli* chromosome.

In this chapter you will learn:
1. About the organization of prokaryotic and eukaryotic genomes.
2. Several characteristics of proteins that interact with DNA.
3. About several aspects of biological membranes.

A Complex DNA Structure: The *E. Coli* Chromosome

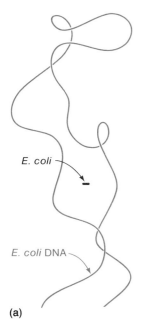

E. coli

E. coli DNA

(a)

(b)

All genes of *E. coli*, and presumably of most bacteria, are contained on a single supercoiled circular DNA molecule. The total length of the circle is about 1300 μm. A cylindrical bacterium has a diameter and a length of about 1 and 3 μm respectively (Figure 5-1); clearly the bacterial DNA must be highly folded when it is in a cell.

When *E. coli* DNA is isolated by a technique that avoids both breakage of the molecule and denaturation of proteins, a highly compact structure known as a bacterial chromosome, or **nucleoid,** is found. This structure contains protein and a single supercoiled DNA molecule. Some RNA is also present in the isolated nucleoid, but it is probably not an essential component, possibly becoming associated with the structure during the isolation procedure. An electron micrograph of the *E. coli* chromosome is shown in Figure 5-2. The particular feature of the structure to note is that the DNA is in the form of loops, which are supercoiled and emerge from a dense protein–containing structure, sometimes called the **scaffold.** The physical dimensions of the isolated chromosome are

Figure 5-1
(a) Schematic diagram showing the relative sizes of *E. coli* and its DNA molecule, drawn to the same scale except for the width of the DNA molecule, which is enlarged approximately 10^6 times. (b) An *E. coli* cell, whose DNA has bound an added fluorescent dye. (Courtesy of Todd Steck and Karl Drlica.)

Figure 5-2
Electron micrograph of the chromosome of *E. coli* attached to two fragments of a proteinaceous substance, which may be the cell membrane. This single molecule of double-helical DNA is intact and supercoiled. (From H. Delius and A. Worcel. *J. Mol. Biol.*, 82: 108. Copyright 1974: Academic Press, Inc. [London] Ltd.)

affected by a variety of factors, and some controversy exists about the state of the nucleoid within the cell. If the nucleoid is treated with an enzyme that degrades proteins or various detergents that break down protein–protein interactions, the chromosome expands markedly, though it remains much more compact than a free DNA molecule. As proteins are removed, the scaffold is disrupted and the chromosome goes through a series of transitions between forms having different degrees of compactness. The conclusion from these and other observations is that the chromosome is held in compact form by the binding of different regions of the DNA molecule to the scaffold. Whether the scaffold has a well-defined organization and is a true "structure" (as opposed to a disorganized aggregate) is not known.

Treatment of a chromosome with very tiny amounts of a DNase, producing single-strand breaks, followed by sedimentation in a centrifuge of the treated DNA, gives some insight into the physical structure of the chromosome. In Chapter 3 it was explained that if a typical supercoiled DNA molecule receives one single-strand break, the strain of underwinding is immediately removed by free rotation about the opposing sugar–phosphate bond, and all supercoiling is lost. Since a nicked circle is much less compact than a supercoil of the same molecular weight, the nicked circle sediments much more slowly. Thus, one single-strand break causes an abrupt decrease (by about 30%) in the s value. However, if one single-strand break is introduced by a nuclease into the *E. coli* chromosome, the s value decreases by only a few percent. Furthermore, a second break causes a second decrease in the s value, and subsequent breaks cause additional stepwise changes. After about forty-five breaks, the form of the chromosome remains constant. This clearly indicates that free rotation of the entire DNA molecule does not occur when a single-strand break is introduced. The structure of the *E. coli* chromosome that has been deduced from these data is shown in Figure 5-3. The DNA is assumed to be fixed to the scaffold at 45 ± 10 positions, each of which prevents free rotation. Thus, there would be about 45 ± 10 supercoiled loops of DNA. One single-strand break would then cause one supercoiled loop to become open, and each subsequent break would, on the average, open another loop. This notion has been confirmed in electron micrographs of chromosomes in which there are one or two nicks; structures such as that in Figure 5-2, but with one or two open loops, are observed.

The enzyme DNA gyrase, which plays an important role in DNA replication (Chapter 7), is responsible for the supercoiling. If coumermycin, an inhibitor of *E. coli* DNA gyrase, is added to a culture of *E. coli* cells, the chromosome loses its supercoiling in about one generation time. Indirect measurements of the number of binding sites for gyrase on the chromosome indicate that there are roughly forty-five sites. The spatial distribution of these sites is not known, but it is tantalizing to think that there may be one binding site in each supercoiled loop.

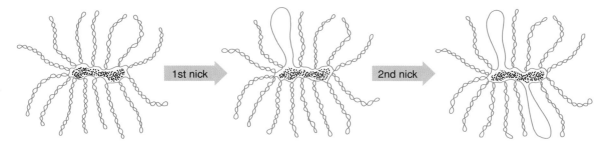

Figure 5-3
A schematic drawing of the highly folded supercoiled *E. coli* chromosome, showing only 15 of the approximately 45 loops attached to the scaffold and the opening of a loop by a single-strand break (nick).

In the next section it will be seen that DNA is arranged in a much more complex way in eukaryotic cells.

Chromosomes and Chromatin

Figure 5-4
Human chromosomes from a cell at metaphase. Each chromosome is partially separated along its long axis prior to separation of the two daughter chromosomes. The constriction is the site of attachment to the mitotic spindle. (Courtesy of Theodore Puck.)

The DNA of all eukaryotes is organized into morphologically distinct units called **chromosomes** (Figure 5-4). Each chromosome contains only a single enormous DNA molecule. For example, the DNA molecule in a single chromosome of the fruit fly *Drosophila* has a molecular weight greater than 10^{10} and a length of 1.2 cm. (Since the width of all DNA molecules is 2.0 nm or 2×10^{-7} cm, the ratio of length to width of *Drosophila* DNA has the extraordinarily high value of 6×10^{6}.) These molecules are much too long to be seen in their entirety by electron microscopy because, at the minimum magnification needed to see a DNA molecule, the field of view is only about 0.01 cm across.

A chromosome is much more compact than a DNA molecule and in fact a DNA molecule cannot spontaneously fold to form such a compact structure because the molecule would be strained enormously. Instead, DNA is made compact by a hierarchy of different types of folding, each of which is mediated by one or more protein molecules.

The DNA molecule in a eukaryotic chromosome is bound to basic proteins called **histones.** The complex comprising DNA and histones is called **chromatin.** There are five major classes of histones—H1, H2A, H2B, H3, and H4—whose properties are listed in Table 5-1. Histones have an unusual amino acid composition in that they are extremely rich in the positively charged amino acids lysine and arginine. The lysine-to-arginine ratio differs in each type of histone. The positive charge of the histones is one of the major features of the molecules, enabling them to bind to the negatively charged phosphates of the DNA. This elec-

Table 5-1

Types of Histones

Type	Lys/Arg Ratio	Location
H1	20.0	Linker
H2A	1.25	Core
H2B	2.5	Core
H3	0.72	Core
H4	0.79	Core

trostatic attraction is apparently the most important stabilizing force in chromatin since, if chromatin is placed in solutions of high salt concentration (e.g., 0.5 M NaCl), which breaks down electrostatic interactions, chromatin dissociates to yield free histones and free DNA. Chromatin can also be reconstituted by mixing purified histones and DNA in a concentrated salt solution and gradually lowering the salt concentration by dialysis.

Reconstitution experiments have been carried out in which histones from different organisms are mixed. Usually, almost any combination of histones works because, except for H1, the histones from different organisms are very much alike. In fact, the amino acid sequences of both H3 and H4 are nearly identical (sometimes one or two of the amino acids differ) from one organism to the next. Histone H4 from the cow differs by only two amino acids from H4 from peas—arginine for lysine and isoleucine for valine—which shows that the structure of histones has not changed in the 10^9 years since plants and animals diverged. Clearly histones are very special proteins indeed.

As a cell passes through its growth cycle, chromatin structure changes. In a resting cell the chromatin is dispersed and fills the entire nucleus. Later, after DNA replication has occurred, the chromatin condenses about 100-fold and chromosomes form. Chromosomes have been isolated and gradually dissociated, and chromosomes at various degrees of dissociation have been observed by electron microscopy. The most elementary structural unit is a fiber 10 nm wide, which appears like a string of beads (Figure 5-5). The beadlike structure, which is chromatin, is also seen when chromatin is isolated from resting cells. The structural hierarchy of a chromosome, which has been deduced from a variety of studies, is shown in Figure 5-6. The higher orders of structure are still somewhat speculative, but a great deal of information is known about the beads themselves, each of which is an ordered aggregate of DNA and histones.

Prior to electron microscopic studies, the effect of various DNA endonucleases on chromatin also suggested that chromatin contains repeating units. Treatment of chromatin with pancreatic DNase I (which cannot attack DNA that is in contact with protein) yields a collection

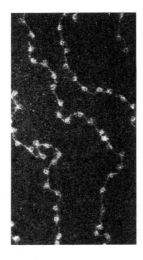

Figure 5-5

Electron micrograph of chromatin. The beadlike particles have diameters of nearly 100 Å. (Courtesy of Ada Olins.)

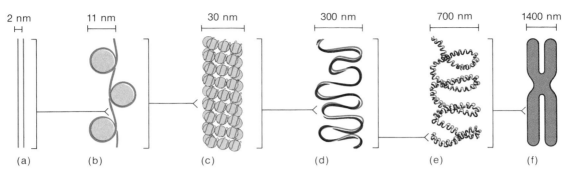

2 nm   11 nm   30 nm   300 nm   700 nm   1400 nm

(a)   (b)   (c)   (d)   (e)   (f)

Figure 5-6
Various stages in the condensation of (a) DNA and (b–e) chromatin in forming (f) a
metaphase chromosome. The dimensions indicate known sizes of intermediates, but
the detailed structures are hypothetical. Note that each diagram represents a magnified
portion of a section of the next diagram in the series.

of small particles containing DNA and histones (Figure 5-7). If, after
enzymatic digestion, the histones are removed, DNA fragments having
roughly 200 base-pairs, or a multiple of 200, are found. After a long
period of digestion the multiple-size units are not found, and all of the
DNA has the 200-base-pair-unit size. Fragments have also been isolated
before removal of the histones and been examined by electron microscopy; it has been found that a fragment containing $200n$ base-pairs consists of n connected beads, indicating that there is a fundamental bead
unit containing about 200 base-pairs.

The beadlike particles are called **nucleosomes.** Each nucleosome is
found to consist of one molecule of histone H1, two molecules each of
histones H2A, H2B, H3, and H4, and a DNA fragment. Treatment of
the nucleosomes (which have been obtained by digestion of chromatin
with pancreatic DNase) with the enzyme micrococcal nuclease gradually
removes additional amounts of DNA. All histones remain associated
with the DNA until the number of base-pairs is less than 160, at which
point H1 is lost. More bases can be removed, but the number cannot
be reduced to less than 140 base-pairs. The structure that remains is
called the **core particle.** It contains an octameric protein disc consisting
of two copies each of H2A, H2B, H3, and H4, around which the 140-
base-pair segment is wrapped like a ribbon (Figure 5-8). Thus a nucleosome consists of a core particle, additional DNA that links adjacent
core particles (**linker DNA**), and one molecule of H1. The H1 binds
to the histone octamer and the linker DNA, causing the linkers extending
from both sides of the core particle to cross and draw nearer to the
octamer, though some of the linker DNA does not come into contact
with any histones. The overall structure of the chromatin fibril is probably a zigzag, as shown in panels (d) and (e) of Figure 5-6. Assembly

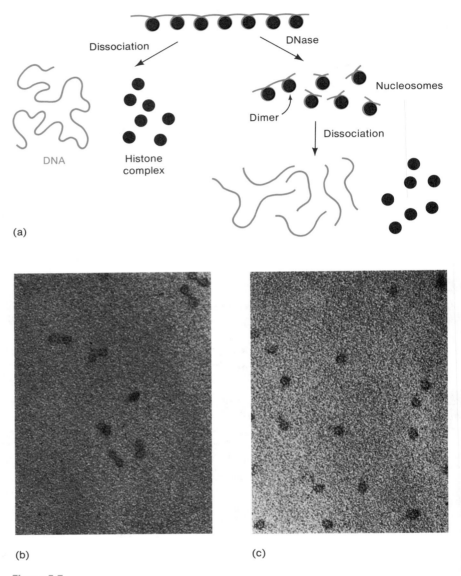

(a)

(b)   (c)

Figure 5-7
(a) The DNase-digestion method for production of individual nucleosomes and 200-base-pair fragments (b, c). (b) Electron micrographs of monomers and dimers. (c) Nucleosomes produced as described in part (a). (From J. T. Finch, M. Noll, and R. D. Kornberg. *Proc. Nat. Acad. Sci.*, 1975. 72: 3321.)

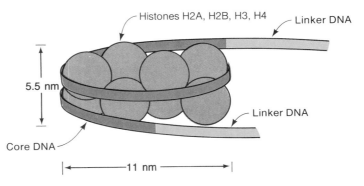

Figure 5-8
Schematic diagram of nucleosome core particle. The DNA molecule is wound 1¾ turns around a histone octamer (2 molecules each of histones H2A, H2B, H3, and H4). Histone H1 (not shown) is bound to the linker DNA. Note that the two linker units point in the same direction.

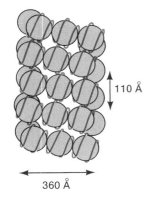

Figure 5-9
A proposed solenoidal model of chromatin. Six nucleosomes (shaded) form one turn of the helix. The DNA double helix (shown in red) is wound around each nucleosome. (After J. T. Finch and A. Klug. *Proc. Nat. Acad. Sci.*, 1976. 73: 1900.)

of DNA and histones is the first stage of shortening of the DNA strand in a chromosome—namely, a sevenfold reduction in length of the DNA and the formation of a beaded flexible fiber 11 nm wide, roughly five times the width of free DNA.

The binding of DNA and histones has been examined and it appears that about 80 percent of the amino acids in the histones are in α-helical regions. Preliminary data suggest that many of the extended α-helical regions lie in the major groove of the DNA helix, and that the complex is stabilized by an electrostatic attraction between the positively charged lysines and arginines of the histones and the negatively charged phosphates of the DNA.

The DNA content of nucleosomes varies from one organism to the next, ranging from 150 to 240 base-pairs per unit. The core particles of all organisms have the same DNA content (140 base-pairs), so that the observed variation results from different sizes of the linker DNA between the nucleosomes (namely, 10–140 nucleotide pairs). Little is known about the structure of linker DNA or whether it has a special function, and the cause of variation of its length is unknown.

In forming a compact chromosome, the DNA molecule is folded and refolded in several ways, as shown in Figure 5-6. The second level of folding is the shortening of the 11-nm fiber to form a solenoidal supercoil with six nucleosomes per turn, called the **30-nm fiber** (Figures 5-6(c) and 5-9); this form has been isolated and well characterized. The remaining levels of organization—folding of the 30-nm fiber and further folding of the folded structure—shown in Figure 5-6(d–f), are less well understood. Electron micrographs of isolated metaphase chromosomes from which histones have been removed indicate that the partially unfolded DNA has the form of an enormous number of loops that seem

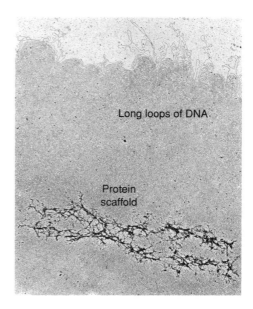

Figure 5-10
Electron micrograph of a segment of a human metaphase chromosome from which the histones have been chemically removed. The dense network near the bottom of the figure is the protein scaffold to which the DNA loops are attached and on which the chromosome is assembled. (Courtesy of Ulrich Laemmli.)

to extend from a central core or scaffold, composed of nonhistone chromosomal proteins, as was seen for the *E. coli* chromosome (Figure 5-10).

The compaction of DNA and protein into chromatin and ultimately into the chromosome greatly facilitates the distribution of the genetic material during nuclear division. We will see in later chapters that certain variations in chromatin structure affect gene expression.

Interaction of DNA and Proteins That Recognize a Specific Base Sequence

In the formation of chromatin the histones do not bind to specific base sequences. Rather, they recognize only the general DNA structure. There are, however, numerous proteins that bind only to particular sequences of bases, and we will encounter many such proteins in later chapters of this book (for example, see Chapter 14). These proteins are of three types: (1) polymerases, which initiate synthesis of DNA and RNA from particular base sequences; (2) regulatory proteins, which turn on and off the activity of particular genes; and (3) certain nucleases, which cut phosphodiester bonds between a single pair of adjacent nucleotides contained in unique sequences of four to six nucleotides. Although the structures of only a few of the binding regions in DNA-protein complexes have been elucidated, a general pattern seems to be emerging that applies both to sequence-specific binding and to nonspecific binding. This pattern is the following:

1. The base sequence of the DNA has some sort of twofold symmetry, allowing binding of the protein to two sites.

2. The protein–DNA contact region is on only one side of the DNA molecule.

3. The protein usually has one or more α-helical regions and these regions bind to the DNA within two segments of the major groove of the DNA.

4. The protein usually is in the form of a dimer having twofold symmetry; the monomers are arranged so that the α helices are in a symmetric array that matches the symmetry of the base sequence.

5. Amino acids of the α-helical regions are in contact with bases in particular positions; in sequence-specific binding, the correct bases must be present at those positions. A hydrogen-bonding complementarity between those specific bases and amino acid side chains of the protein provides correct matching.

6. Positively charged amino acids often form ionic bonds with the phosphate groups, which are negatively charged. In sequence-specific binding these do not confer specificity, but stabilize the interaction.

A detailed analysis of the structure of a DNA-protein complex is beyond the scope of this book. However, some of the features just listed will be illustrated by an examination of the complex between the Cro protein of *E. coli* phage λ and the operator sequence in λ DNA. Cro is a small protein, consisting of 66 amino acids; it forms a dimer and binds to a DNA sequence containing 17 nucleotide pairs (Figure 5-11). Each monomer of the Cro protein contains three α-helical regions, one of which is in direct contact with DNA bases in the Cro–DNA complex. Figure 5-12 shows the binding region of the DNA molecule, with the bases and phosphate groups that contact the Cro protein shown. These sites have been identified by treating the complex with reagents that attack the sites in free DNA but that cannot alter a region of the DNA that is in contact with Cro. Analysis of the chemical structure of the DNA after exposure of the complex to these reagents showed that certain bases and phosphates were unaltered and, hence, are in the contact

TATCACCGCAAGGGATA
ATAGTGGCGTTCCCTAT

Figure 5-11
The base sequence of one of the λ Cro protein binding sites. This sequence is called *oR3*. The adenines and guanines shown in Figure 5-12 are printed in red.

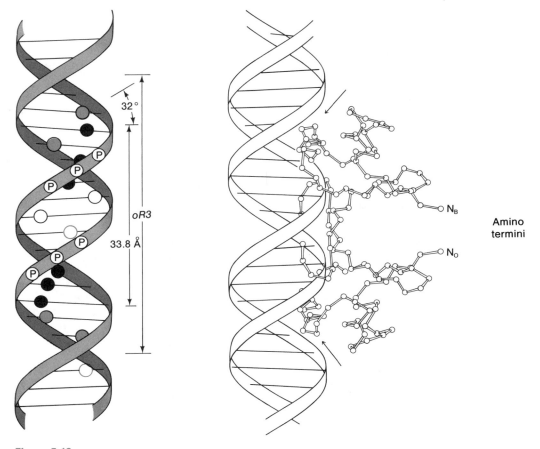

Amino
termini

Figure 5-12
Points at which the *oR3*
region of λ DNA makes
contact with Cro. Phos-
phates are labeled P;
guanines and adenines in
contact are solid black
and red circles; guanines
and adenines not in con-
tact are open black and
red circles, respectively.
Note that the bases in
contact are only in the
wider (major) groove of
the DNA. (After W. F. An-
derson et al., *Nature*,
1981. 290: 754.)

Figure 5-13
Presumed structure of the Cro-*oR3* complex. The DNA is rotated 90° relative to that in
Figure 5-12 so that the contact points are on the right side of the molecule. The amino
termini of each Cro monomer are labeled N_o and N_B. A pair of α helices, related by twofold symmetry, occupy successive major grooves of the DNA (arrows), and the two
extended chains run parallel to the axis of the DNA. (After W. F. Anderson et al., *Nature*,
1981. 290: 754.)

region. Figure 5–13 shows the Cro–DNA complex. The symmetry of
the Cro dimer is apparent; the relevant α helices are in the upper and
lower parts of the dimer and can be seen to fit nicely into the major
groove of the DNA. This kind of symmetric binding array has been
observed in repressor–operator binding (Chapters 14 and 15), in binding
of the *E. coli* CAP protein to the relevant sequence in the *lac* operon
(Chapter 14), and in chromatin.

   A close look at the complex (for which there is not enough detail
in the figures) would show that amino acids in each Cro monomer are

situated in a way that they can form hydrogen bonds with the DNA bases, which confers binding specificity. They also interact electrostatically with the phosphate groups, which confers binding strength.

Biological Membranes

Biological membranes are organized assemblies consisting mainly of proteins and lipids. The structures of all biological membranes have many common features but small differences exist to accommodate the varied functions of different membranes.

A basic function of membranes is to separate the contents of a cell from the environment. However, it is always necessary that cells take up nutrients from the surroundings; therefore, the enclosing membrane of a cell must be permeable. The permeability of the membrane must be selective though, so that the concentrations of compounds within a cell can be controlled; otherwise, intracellular concentrations would be inseparable from extracellular concentrations. Selective permeability is attained by means of restricting free passage of most intra- and extracellular substances and allowing transport of specific substances by molecular pumps, channels, and gates. External membranes also have a signal function. For instance, they often contain receptors for signal molecules such as hormones. The external membrane is also responsible for transmitting electrical impulses, as in nerve cells.

There are also numerous types of intracellular membranes. These membranes serve to compartmentalize various intracellular components as well as to provide surfaces on which certain molecules can be adsorbed prior to chemical reaction with other adsorbed molecules.

Figure 5-14 shows an electron micrograph in which the membrane enclosing a red blood cell is seen in cross-section. The most notable feature of this membrane is that it consists of two layers—it is called a **membrane bilayer.** This structure is a consequence of the chemical nature of the lipids that form the membrane. There are many membrane lipids, of which the most prevalent are the phospholipids, molecules consisting of a polar phosphate-containing group and two long nonpolar hydrocarbon chains. Figure 5-15 shows a space-filling model of one phospholipid and a schematic drawing. The essential feature of the molecule is a polar "head" group to which hydrocarbon "tails" are attached. Such a molecule, which has distinct polar and nonpolar segments, is called **amphipathic.** When placed in water, amphipathic molecules tend to spontaneously aggregate. This is because only the polar head is capable of interaction with the polar water molecules. The hydrocarbon tails are brought together by hydrophobic (water hating) interactions—that is, they cluster because they are unable to interact with water. If the length

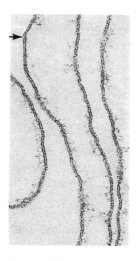

Figure 5-14
Electron micrograph of a preparation of plasma membranes from red blood cells. A single membrane is denoted by an arrow. (Courtesy of Vincent Marchesi.)

of the hydrocarbon tail satisfies various geometric constraints, a collection of amphipathic molecules forms a **micelle,** a spherical array of molecules in which the nonpolar tails form a hydrocarbon microdroplet enclosed in a shell composed of the polar heads (Figure 5-16(a)). The geometric constraints limit the size of a stable micelle, and as the size of the hydrocarbon tail increases, the array becomes ellipsoidal (Figure 5-16(b)) with the ratio of the lengths of the major axis to the minor axis of the ellipsoid increasing with length of the tail. Above a certain tail length, the only stable configuration is an ellipsoid whose major axis is "infinitely" long—that is, an extended lipid bilayer sheet, as shown in part (c) of the figure. The lipid bilayer can also be stabilized by van der Waals attractions between the hydrocarbon tails.

The ends of a lipid bilayer are unstable, as the hydrocarbon chains are exposed to water. Thus lipid bilayers tend to close upon themselves to form hollow bilayer spheres known as **vesicles** (Figure 5-17). If either natural membranes or surface films of individual lipid molecules are physically disrupted, many fragments will form vesicles. If such vesicles are formed from a solution of a small molecule, some will contain these molecules. Studies of synthetic vesicles formed from a single type of phospholipid and containing a variety of small molecules have shown that the lipid bilayer is impermeable to all ions and highly polar molecules, raising the question of how these molecules get in and out of cells. The answer is that naturally occurring biological membranes contain certain proteins that are responsible for transport of all molecules having polar regions and for most nonpolar molecules as well.

There are two classes of membrane proteins—the **integral membrane proteins,** which are contained wholly or in part within the membrane, and the **peripheral proteins,** which lie on the membrane surface and are bound to the integral proteins. An electron micrograph of the plasma membrane of a red blood cell in which integral membrane proteins can be seen is shown in Figure 5-18. The outer surface of the membrane is smooth, but the inner surface is covered with many globular protein molecules.

Numerous physical measurements have shown that the integral membrane proteins can freely diffuse laterally throughout the bilayer. The degree of motion is determined by the fluidity of the lipid layer, which is in turn a function of the van der Waals attractions between the hydrocarbon tails of the particular phospholipids in the membrane. The **fluid mosaic model** of membrane structure (Figure 5-19) recognizes these features. It was further shown that the proteins do not spontaneously rotate (flip-flop) from one side of the membrane to another. The integral proteins do not spontaneously rotate because they have polar and nonpolar regions as shown in Figure 5-19. Since the external region is polar and the internal region is nonpolar, rotation would require the passage of the polar region through the nonpolar center of the bilayer.

Since different proteins protrude from the two sides of the mem-

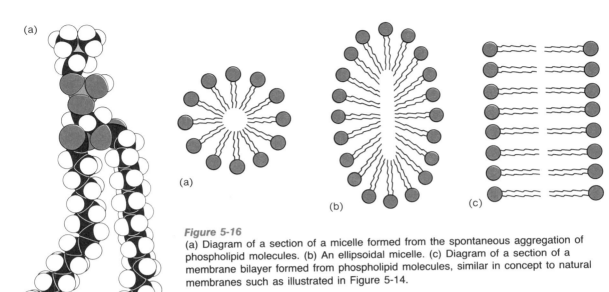

(a)

(a)

(b)

(c)

Figure 5-16
(a) Diagram of a section of a micelle formed from the spontaneous aggregation of phospholipid molecules. (b) An ellipsoidal micelle. (c) Diagram of a section of a membrane bilayer formed from phospholipid molecules, similar in concept to natural membranes such as illustrated in Figure 5-14.

(a)

(b)   Polar
             (hydrophilic)
             head group

             Hydrocarbon
             (hydrophobic)
             tails

Figure 5-15
(a) Space-filling model of phosphatidyl choline, a phospholipid. (b) The essential features of a phospholipid or glycolipid molecule, such as the one illustrated in (a).

brane, the membrane is an asymmetric structure having, in the case of a cell, both an inside and an outside. It is this asymmetry that determines the direction of movement of molecules entering and leaving a cell.

There are many modes of transport of molecules across the bilayer and only a few will be mentioned. One mode makes use of channels

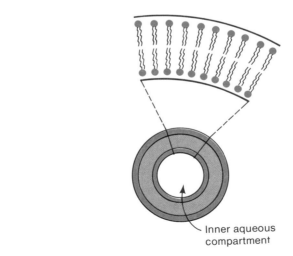

Inner aqueous
compartment

Figure 5-17
Schematic diagram of a lipid vesicle. The membrane is enlarged to show the double layer.

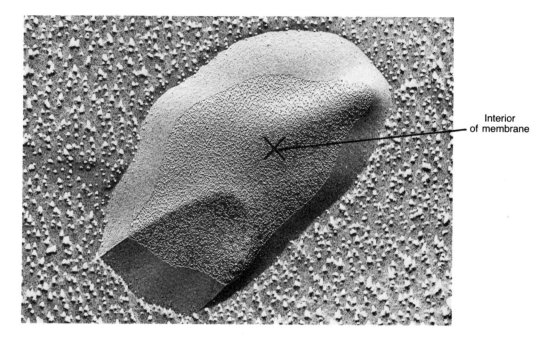

Interior
of membrane

Figure 5-18
An electron micrograph of the plasma membrane of red blood cells. The interior of the
membrane, which has been exposed by fracture of the membrane, contains numerous
globular particles having a diameter of about 7.5 nm. These particles are termed integral
membrane proteins. (Courtesy of Vincent Marchesi.)

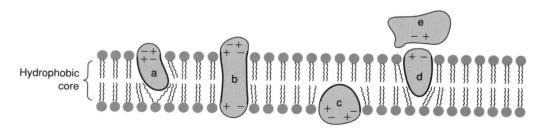

Figure 5-19
The structure of a membrane according to the fluid mosaic model. Four integral proteins
(a)–(d) are embedded to various degrees into a lipid bilayer such that the hydrophobic
surface of each protein (heavy lines) is in the membrane and the polar region (indicated
by + and −) is external. Peripheral proteins (e.g., (e)) are on the surface and bound to a
polar region of an integral protein. Integral proteins can drift laterally but cannot flip-flop.

through the membrane. These channels usually are passages through the integral membrane proteins; the passages will have an abundance of polar amino acids if their function is to allow transit of polar substances. Often the channels can be opened and closed by means of conformational changes of the membrane proteins. Such channels are found in nerve cell membranes. When open, they permit sodium ions to enter and thereby change the electrical properties of the cell. Other transport systems utilize chemical reactions that convert the substance to be transported into a molecule that can enter the membrane and then, after transit of the modified molecule, restore the original molecule at the other side of the membrane; these chemical mechanisms are usually very complex and consume a great deal of energy.

Since the types of proteins which are included in the lipid bilayer (Figure 5-19) to a large extent define the functional properties of membranes of different kinds of cells, methods for characterizing those proteins have been developed. Typically, detergents are employed to solubilize the membrane and thereby release intramembrane proteins. Those proteins are then initially characterized by using analytical methods such as gel electrophoresis (Chapter 2).

Summary

The *E. coli* chromosome (also called a nucleoid) consists of a single supercoiled, circular DNA molecule which is attached to a dense protein-containing scaffold at various points. Using the enzyme DNase, it has been demonstrated that the DNA is attached to the scaffold at approximately 45 points. The bacterial chromosome also has around 45 binding sites for DNA gyrase, an enzyme responsible for supercoiling DNA. The organization of the eukaryotic genome is quite different, in that more than one chromosome is present in the cell, and the degree of compaction is much higher. The DNA molecule in a eukaryotic chromosome is first complexed with proteins called histones. Histones are rich in lysine and arginine, and bind to DNA through electrostatic interactions with the phosphate groups. The DNA-histone complex (called chromatin) has the appearance of beads on a string. The beads represent the core particle—two molecules each of histones H2A, H2B, H3, and H4 that form an octameric disc wrapped by approximately 140 base-pairs of DNA. The DNA connecting the core particles is called linker DNA, and it is to this that histone H1 binds. The chromatin coils to form a solenoid with six nucleosomes per turn. Further folding and re-folding, and complexing with non-histone chromosomal proteins, results in a highly compact eukaryotic chromosome.

Certain proteins, namely polymerases, some nucleases, and regulatory proteins bind to DNA molecules at specific base sequences. Although different proteins bind to different sequences, these proteins share several characteristics: the protein and the DNA sequence to which it binds have two-fold symmetry, the region of contact is on one side of the DNA molecule, and α-helices of the protein lie within the major groove to allow amino acid-base interaction.

Biological membranes are bilayers composed of amphipathic phospholipids. These membranes are impermeable to most molecules in the environment. Associated with the membranes are many protein molecules. Integral membrane proteins are contained, at least in part, within the membrane, while peripheral proteins are found on the surface of the membrane, attached to integral proteins.

Drill Questions

1. What are the differences between a nucleosome, a core particle, and an octameric disc? Which directly involve the histone H1?

2. Why is the level of compaction of DNA higher in eukaryotes than in prokaryotes?

3. Why is it necessary for DNA to exist at various levels of compaction?

4. At what stage of the eukaryotic cell cycle would you expect DNA to be the most compact? At what stage would you expect it to be the least compact? Why?

5. Referring to the Cro protein, how far apart are the two α-helices which contain the amino acids that interact with the bases of the DNA molecule?

6. Where are the polar and the non-polar regions of a membrane-spanning integral protein?

7. What non-covalent interactions are involved in stabilizing a lipid bilayer?

Problems

1. If a DNA molecule is 2.4 cm long, how many nucleosomes could be present in its most highly compacted state? (Assume linker DNA has 60 base-pairs.)

2. When an isolated DNA-protein complex was treated with 2M NaCl the protein dissociated from the DNA. Digestion of the complex with a nuclease and a protease yields only free nu-

cleotides and amino acids. What is the main type of interaction involved in binding the protein to the DNA?

3. Another DNA-protein complex is isolated. When treated with 2M NaCl, the complex does not dissociate. The complex was treated with a nuclease and a protease and, in addition to free nucleotides and amino acids, a component was found that was neither a free nucleotide or amino acid. What is likely to be the identity of this component, and what interaction is responsible for protein-DNA binding?

4. A DNA-binding protein binds tightly to double-stranded DNA, but very poorly to single-stranded DNA. It can bind to all base sequences with equal affinity. Binding is poor in 1M NaCl. What part of the DNA molecule is likely to be the binding site?

5. Acridine molecules bind to DNA by intercalating between the base-pairs. If the Cro protein and DNA are mixed, the Cro-DNA complex forms. What would happen if acridine were added to the DNA before the Cro protein?

Conceptual Questions

1. What features advantageous to cell function accrue from the fact that histones bind to DNA without regard to base sequence?

2. What reasons might explain why it is beneficial to a cell that flip-flopping of membrane proteins occurs infrequently?

3. Speculate on why most functions carried out by a cell rely on macromolecular interactions and/or the use of complex aggregates of macromolecules instead of smaller, simpler molecules.

Robert L. Sinsheimer
Professor of Biological Sciences
Department of Biological Sciences

Birthday: **5 February 1920**

Birth Place: **Washington, D.C.**

Undergraduate Degree: **Massachusetts Institute of Technology**
Major: Biology 1941

Graduate Degree: **Massachusetts Institute of Technology, Ph.D. 1948**

Present Position: **University of California, Santa Barbara, Biological Sciences Department**

Address: **Santa Barbara, California**

MY CURRENT RESEARCH program is aimed at the development of techniques for the direct reading of nucleotide sequences in nucleic acids.

What advice can you offer to undergraduates who want to become scholars of molecular biology?

I would advise undergraduates who intend to become scholars in molecular biology to obtain a firm background in basic mathematics, physics, and chemistry (especially physical chemistry). They should also become sufficiently well versed in genetics so that they can utilize genetic approaches as a tool in whatever biological system they are studying. And I would strongly urge that they acquire research laboratory experience in off-hours, summers, etc. Experience with real research problems provides both excitement and motivation to pursue other studies.

6

The Genetic Material

The genetic material of any organism is the substance that carries the information determining the inherited properties of that organism. This substance must provide for the transfer of the genetic information, in largely unchanged form, from parent to progeny. In almost all organisms the genetic material is DNA; the exceptions are a few bacteriophages and numerous plant and animal viruses in which the genetic material is RNA. It was, in fact, two classic experiments that led to the identification of DNA as the genetic material and, in so doing, laid the foundation for molecular genetics. We will review these experiments in the following section.

Originally, however, it was believed that the genetic material consisted of proteins. Since inherited traits exhibit tremendous diversity, it was assumed genes would likewise be very diverse in composition and structure. As you learned in Chapter 4, proteins are very diverse in amino acid composition, size, and shape. Nucleic acids, which consist of merely four bases, were considered to be too simple in structure. Many scientists therefore believed that proteins were the logical candidates for storing and transferring genetic information. DNA was considered to function in a structural capacity—holding the "informational proteins" together in chromosomes. Now, in the modern era of biology, it is of course taken for granted that DNA is the genetic material.

We will learn about the landmark experiments, performed in some cases over 60 years ago, that paved the way for accepting the fact that genes consist of a simple macromolecule such as DNA.

In this chapter you will learn:

1. About the properties of DNA as the genetic material, including its storage, transmission, stability, and mutation.

Identification of DNA as the Genetic Material

THE TRANSFORMATION EXPERIMENTS

The development of our understanding that DNA is the genetic material began with an observation in 1928 by Fred Griffith who was studying the bacterium responsible for human pneumonia—i.e., *Streptococcus pneumoniae* or *Pneumococcus*. The virulence of this bacterium was known to be dependent on a surrounding polysaccharide capsule that protects it from the defense systems of the body. This capsule also causes the bacterium to produce smooth-edged (S) colonies on an agar surface. It was known that mice were normally killed if infected by S bacteria (Figure 6-1(a)). Griffith then isolated a rough-edged (R) colony mutant which proved to be both nonencapsulated and nonlethal (Figure 6-1(b)).

As expected, heating the S bacteria killed them and destroyed their infectivity (Figure 6-1(c)). But Griffith then made a very significant observation: a mixture of live R bacteria and heat-killed S bacteria was lethal, although neither component was lethal by itself. (Figure 6-1(d)).

Of even greater significance was that live bacteria isolated from a mouse which had died from such a mixed infection were only of the S type—i.e., the live R bacteria had somehow been replaced by or **transformed** into S bacteria, which then reproduced as S bacteria.

Several years later it was shown that a mouse was not needed to mediate this transformation. When a mixture of live R cells and heat-killed S cells was cultured in a nutritive medium, some live S cells were produced.

A possible explanation for this surprising result was that the R cells had somehow revived the heat-killed S cells. This possibility was eliminated, however, by the observation that live S cells could be obtained when the live R cells were mixed, not with heat-killed S cells, but with a cell extract prepared from broken S cells that had been freed from intact cells and capsular polysaccharides by centrifugation (Figure 6-2). Hence it was concluded that the S cell extract contained a **transforming principle,** the nature of which was unknown.

The next development occurred some 15 years later when Oswald Avery, Colin MacLeod, and Maclyn McCarty partially purified the transforming principle from the S cell extract and demonstrated that it was DNA. These workers modified known procedures for isolating DNA and prepared samples of DNA from the S bacteria. They added this DNA to a live R bacterial culture; after a period of time they placed a sample of this bacterial culture on an agar surface and allowed it to grow to form colonies. Some of the colonies (about 1 in 10^4) that grew were S type (Figure 6-2). To show that this was a permanent genetic change, they dispersed many of the newly formed S colonies and placed

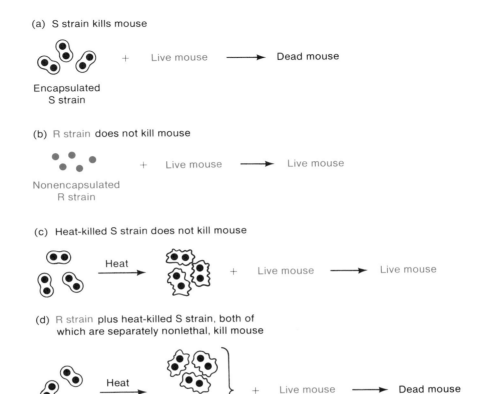

(a) S strain kills mouse

Encapsulated
S strain   +   Live mouse ⟶ Dead mouse

(b) R strain does not kill mouse

Nonencapsulated
R strain   +   Live mouse ⟶ Live mouse

(c) Heat-killed S strain does not kill mouse

Heat   +   Live mouse ⟶ Live mouse

(d) R strain plus heat-killed S strain, both of
which are separately nonlethal, kill mouse

Heat   +   Live mouse ⟶ Dead mouse

Figure 6-1
The Griffith experiment showing conversion of a nonlethal bacterial strain to a lethal form
by a cell extract.

them on a second agar surface. The resulting colonies were again S type.
When an R type colony arising from the original mixture was dispersed,
only R bacteria grew in subsequent generations. Hence the R colonies
retained the R character, whereas the transformed S colonies bred true
as S. Because S and R colonies differed by a polysaccharide coat around
each S bacterium, the ability of purified S polysaccharide to transform
was also tested, but no transformation was observed. Since the procedures for isolating DNA then in use produced DNA containing many
impurities, it was necessary to provide evidence that the transformation
was actually caused by the DNA alone.

This evidence was provided by the following four procedures:

1. Chemical analysis showed that the major component was a deoxyribose–containing nucleic acid.

Preparation of transforming principle from S strain

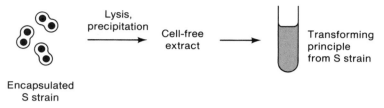

Addition of transforming principle to R strain

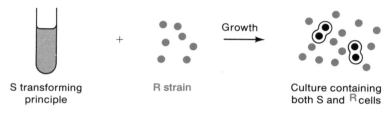

Figure 6-2
The transformation experiment.

2. Physical evidence showed that the sample contained a highly viscous substance having the properties of DNA.

3. Experiments demonstrated that transforming activity is not lost by reaction with either (a) purified proteolytic (protein-hydrolyzing) enzymes—trypsin, chymotrypsin, or a mixture of both—or (b) ribonuclease (an enzyme that hydrolyzes RNA).

4. It was demonstrated that treatment with materials known to contain DNA-hydrolyzing activity (DNase) inactivated the transforming principle.

The transformation experiment was not immediately accepted by the scientific community as proof that DNA is the genetic material, because it was widely believed that DNA was a simple tetranucleotide incapable of carrying the information required of a genetic substance. The tetranucleotide hypothesis was based on chemical analyses that indicated that DNA seemed to consist of equimolar amounts of the four bases; this conclusion was based on inadequate chemical procedures for analyzing base composition and on the use of higher organisms as sources of DNA. In these organisms the base composition is, in fact, not far from being equimolar. However, as techniques improved and a greater range of organisms was examined, it was found that the base composition of DNA varies widely from one species to another, and that DNA is neither a simple, small molecule nor a simple repeating structure. With these results, the idea that DNA is the genetic material *became* acceptable. The following experiment provided additional proof.

THE BLENDER EXPERIMENT

An elegant confirmation of the genetic nature of DNA came from an experiment with *E. coli* phage T2. This experiment, known as the **Blender Experiment** because a kitchen blender was used as a major piece of apparatus, was performed by Alfred Hershey and Martha Chase. They demonstrated that the DNA, injected by a phage particle into a bacterium, contained all the information required to achieve the synthesis of progeny phage particles.

A single particle of phage T2 consists of DNA (now known to be a single molecule) encased in a protein shell (Figure 6-3(a)). The DNA is the only phosphorus-containing substance in the phage particle; the proteins of the shell, which contain among others the amino acids methionine and cysteine, have the only sulfur atoms. Thus by growing phage in a nutrient medium in which the radioactive phosphate ($^{32}PO_4^{3-}$) is the sole source of phosphorus, phage containing radioactive DNA can be prepared. If instead the growth medium contains radioactive sulfur (as $^{35}SO_4^{2-}$), phage containing radioactive proteins are obtained. If these two kinds of labeled phage are used in the infection of a bacterial host, the phage DNA and the protein molecules can always be located by their specific radioactivity. Hershey and Chase used these phage to show that ^{32}P but not ^{35}S is injected into the bacterium.

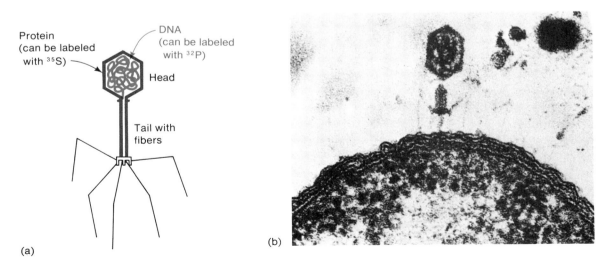

(a)

(b)

Figure 6-3
(a) A diagram of T2 phage. (b) Phage T2 adsorbed to the surface of *E. coli*. (Courtesy of Lee Simon and Thomas Anderson.)

Each phage T2 particle has a long tail by which it attaches to sensitive bacteria (Figure 6-3(b)). Hershey and Chase showed that an attached phage can be torn from a bacterial cell wall by violent agitation of the infected cells in a kitchen blender. Thus it was possible to separate an adsorbed phage from a bacterium and determine the component(s) of the phage that could not be shaken free by agitation—presumably those components that had been injected into the bacterium.

In the first experiment, ^{35}S-labeled phage particles were adsorbed to bacteria in a few minutes. The bacteria were separated from unadsorbed phage and phage fragments by centrifuging the mixture and collecting the sediment (the pellet) which consisted of the phage-bacterium complexes. These complexes were resuspended in liquid and blended. The suspension was again centrifuged, and the pellet, which now consisted almost entirely of bacteria, and the supernatant were collected separately. It was found that 80% of the ^{35}S label was in the supernatant, and 20% was in the pellet. The 20% of the ^{35}S that remains associated with the bacteria was shown some years later to consist mostly of phage tail-fragments that adhered too tightly to the bacterial surface to be removed by blending.

A very different result was observed when the phage population was labeled with ^{32}P. In this case 70% of the ^{32}P remained associated with the bacteria in the pellet after blending, and only 30% was in the supernatant. Of the radioactivity in the supernatant, roughly one-third could be accounted for by breakage of the bacteria during the blending; the remainder was shown some years later to be a result of defective phage particles that could not inject their DNA. Most important, when the pellet material was resuspended in growth medium and reincubated, it was found to be capable of phage production. Thus the ability of a bacterium to synthesize progeny phage is associated with the transfer of ^{32}P and hence of DNA, from parental phage to the bacteria (Figure 6-4).

This concept was definitively proven a few years later when it was shown that the naked, purified DNA of the small bacteriophage φ-x-

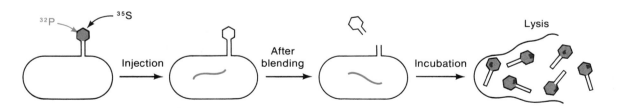

Figure 6-4

The blender experiment. The parental DNA molecule is drawn in red. During incubation the DNA replicates and progeny phage are produced. Some, but not all, of the parental DNA appears in these progeny. Although not known to Hershey and Chase, progeny DNA molecules engage in genetic recombination resulting in dispersal of parental DNA among the progeny phage, as shown.

174 could successfully infect bacterial protoplasts (cells lacking a portion of their cell wall) in the absence of any viral protein.

Properties of the Genetic Material

The genetic material must have the following properties:

1. Ability to store genetic information and express it in the cell as needed.

2. Ability to transfer information to daughter cells with minimal error.

3. Physical and chemical stability so information is not lost over long periods (years).

4. Potential for heritable change without major loss of parental information.

We will see in the following sections that DNA is particularly suited to be the genetic material.

STORAGE AND TRANSMISSION OF GENETIC INFORMATION

BY DNA

The information possessed and conveyed by the DNA in a cell is of several types:

1. The sequence of all RNA molecules synthesized by the cell.

2. The sequence of amino acids in every protein synthesized by the cell.

3. Start and stop signals for the synthesis of each RNA, the processing of each RNA, and the synthesis of each protein.

4. A set of signals that interacts with cellular components and determines whether and when a particular RNA or protein is to be made, and how many molecules are to be made per unit time.

5. Signals that serve as origins and terminators for replication of the DNA.

This information is contained in the sequence of DNA bases. DNA serves principally as the template for the synthesis of specific RNA molecules called transcript RNA. The base-sequence of each RNA is complementary to a region on one of the DNA strands. The complementary RNA is made by means of an enzymatic system that adds a ribonucleotide to the growing end of the RNA molecule only if that base can hydrogen-bond with the DNA base being copied. This restriction ensures a copy error rate of less than 1 in 10^4.

After synthesis each complementary strand of RNA is then "processed"—it may be trimmed or modified at the ends, cut into defined fragments, internal segments may be excised, or individual bases may be modified by components of the cellular machinery. Processing is much more extensive in eukaryotic cells than in prokaryotic cells. (See Chapters 9 and 16).

Some of the processed RNA is used for varied cellular functions. Much of the processed RNA—the **messenger RNA**—is used, however, to specify the amino acid sequences of cellular proteins. For this purpose the RNA nucleotide sequence is read on ribosomes in sequential groups of three bases called **triplets** or **codons.** Each group corresponds to a particular amino acid, or to a start or stop signal.

This two-stage process has an advantage—the genetic DNA molecule does not have to be used very often or enter the protein-synthesis mechanism.

Specific regions of the DNA serve, by interaction with cellular components, to modulate the transcription of adjacent messages. Other regions direct cellular components to initiate and terminate DNA replication.

The use of specific patterns of hydrogen-bonding enables the cell to use less genetic material to hold information than if van der Waals attractions had been selected in the course of evolution; this is because van der Waals attractions are so weak that the error frequency in transmitting information would be unacceptable unless more bases (or different molecules) were used.

TRANSMISSION OF INFORMATION FROM PARENT TO PROGENY

When a cell divides, each daughter cell must contain identical genetic information. That is, each DNA molecule must become two identical molecules, each carrying the information that was contained in the parent molecule. This duplication process is called **replication.** Once again, the importance of the hydrogen-bonding capability of nucleotide bases is apparent. As with transcription, the replication system only requires that the base being added to the growing end of a new DNA chain is capable of hydrogen-bonding to the base being copied.

However, the fidelity required for replication is much higher than for transcription. Occasional synthesis of an aberrant protein can be tolerated. Occasional error in the replication of a gene, however, would often be lethal.

In consequence, cells have developed additional mechanisms to insure the fidelity of gene duplication. The enzymes involved in DNA replication possess an editing function which re-checks the correctness of the nucleotides that have just been inserted into the growing chain, and excise them if an error has been made. This procedure removes 99.9% of the few errors made in the initial insertions.

In addition, the cells have a means to differentiate—for a period of time—the newly-synthesized DNA strand from the older strand. Special enzymes monitor the DNA at all times, looking for mismatched base-pairs. If such a mismatch is found on recently duplicated DNA strand, it is corrected using the older DNA strand as a template. This combination of error-correcting mechanisms reduces the copy error rate in DNA duplication in cells to between 1 in 10^9 and 1 in 10^{10} per base-pair.

WHY IS DNA USED AS THE GENETIC MATERIAL?

In long-lived organisms, a single molecule may have to last 100 years or more. Furthermore, the information contained in the molecule is passed on to succeeding generations over millions of years with only small changes. Thus DNA molecules must have great stability.

The sugar-phosphate backbone of DNA is extremely stable. The C—C bonds in the sugar are resistant to chemical attack under all conditions except strong acids at very high temperatures. The phosphodiester bond is a little less stable, and can be hydrolyzed at room temperature at pH 2, but this is not a physiological condition. In considering the stability of the phosphodiester bond, one can see immediately why 2'-deoxyribose rather than ribose is the constituent sugar in DNA. The phosphodiester bond in RNA is rapidly broken in alkali. The chemical mechanism that breaks the bond requires an OH group on the 2' carbon of the sugar. In DNA, because deoxyribose is used, there is no 2'-OH group, so the molecule is exceedingly resistant to alkaline hydrolysis. The N–glycosydic bond is also very stable, though it would not be so if the bases were not hydrophobic rings.

An alteration in the chemical structure of a base would mean a loss of genetic information. In a cell there are certainly a large number of chemical compounds that can attack a free base. In considering this problem, one can see immediately the value of the double helix. The molecule is redundant in the sense that identical information is contained in both strands; that is, the base-sequence of one strand is complementary to the other strand, and can be derived from it. In fact, there exist in cells elegant repair systems that can remove an altered base, read the sequence on the complementary strand, and then reinstate the correct base. (Repair will be discussed in Chapter 8.)

Another merit of the double-helical structure is that it provides the bases of the DNA with great protection against chemical attack. The bases are hydrophobic rings with charged groups that contain the genetic information. It is these charged groups that specifically need protection. The hydrophobic structure of the bases causes them to stack so tightly that water is almost completely excluded from the stacked array. As a

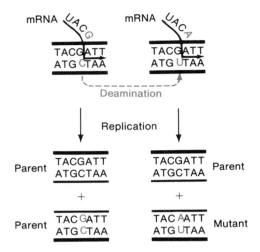

Figure 6-5
The effect of deamination of cytosine to form uracil in the base sequences of mRNA and of two daughter DNA molecules. The C→U transition is shown in red in the uppermost panel. "Parent" and "mutant" refer to base sequences before and after deamination. Newly-replicated DNA is indicated by a thin line.

result, water-soluble compounds are often unable to come into close contact with the "dry" stack of bases and the likelihood of their reaching the hydrogen-bonded charged groups is small.

The bases themselves, with the exception of cytosine, are very stable. At a very low rate, however, cytosine is deaminated to form uracil:

$$\text{Cytosine} + H_2O \longrightarrow \text{Uracil} + NH_3$$

Cytosine   **Uracil**

Deamination is a potentially disastrous change because the deamination product, uracil, pairs with adenine rather than with guanine. This could have two effects: (1) An incorrect base could appear in RNA; and (2) an adenine instead of a guanine would thus appear in newly-replicated DNA strands (Figure 6-5). To prevent these effects, there exists an intracellular system that removes uracil from DNA and replaces it with cytosine. It will be described more fully in Chapter 8.

The necessity of eliminating uracil formed by deamination of cytosine, explains why DNA uses thymine and not uracil as the complementary base for adenine. If uracil were a DNA base, there would be no way to distinguish between a correct uracil and a uracil produced by deamination of cytosine. Because DNA uses thymine, the cell can follow a simple rule:

Always remove uracil from DNA because it is unwanted.

It is not known why RNA uses uracil (and not thymine) nor why DNA evolved with cytosine rather than with a base that would not deaminate. This may have been an evolutionary accident. It is possible that the original RNA (and DNA?) molecules contained cytosine and uracil because these were the only pyrimidines available in the primordial sea. Cells may have later gained the ability to methylate uracil to form thymine because this would provide a cell with a criterion for eliminating the result of a cytosine-uracil conversion. It is suggestive that the final step in the synthesis of thymidylic acid is the methylation of a deoxy-uridylic acid.

THE ABILITY OF DNA TO CHANGE: MUTATION

mutation

All of the genetic information contained in a cell resides in its DNA. Thus, if a cell or an organism is to be able to evolve through time, the base-sequence of its DNA must be capable of change. Furthermore, the new sequence must persist so that progeny cells or organisms will have the new property. The process by which a base-sequence changes is called **mutation.** At the nucleotide level there are two principal mechanisms of mutation:

1. A **chemical alteration** of the base that gives it new hydrogen-bonding properties and thus causes a different base to be incorporated upon replication.

2. A **replication error** by which an incorrect base or an extra base is accidentally inserted or deleted in the daughter molecules.

On the average, mutational changes are deleterious and lead to cell death. Therefore, it is important that not too many mutations occur in a single DNA molecule at one time because otherwise the rare, advantageous alteration would always occur in a virus or cell destined to die by virtue of lethal mutations, or in some viruses. Keeping the mutation rate low is accomplished in two ways. First, the hydrophobic, water-free core of the DNA molecule reduces its accessibility to attacking molecules. Second, the cell has evolved several repair mechanisms for correcting alterations or replication errors. These repair systems are not completely efficient, so mutations occur at a rate that is very low, but useful in an evolutionary sense.

Mutations are, as we have said, usually deleterious, so it is important that when such evolutionary experiments take place, the original parental information not be discarded. Such loss is prevented in two ways: (1) The other members of the species retain the parental base sequence, and

(2) The double-stranded DNA molecule is redundant. Normally only one strand is altered. When DNA replicates, it does so in such a way that after cell division, the DNA molecule in each daughter cell contains only one of the parental single strands. Thus, it is possible for one of the DNA daughter molecules to be a copy of the parent, and the other to be mutant. If the mutant is better equipped to survive and multiply than the parent organisms or any other member of the species, then, after a great many generations, Darwin's principle of survival of the fittest will lead to ultimate replacement of the parental genotype by the mutant.

Although most replication errors are corrected by the cellular systems described in Chapters 7 and 8, a small fraction of such copy errors do persist, and these give rise to mutations.

In addition to mutations of nucleotide sequences, mutations may involve larger chromosomal modifications—duplications, inversions, or translocations—all with varied consequences. Any of these mutations may affect protein or RNA sequences or the control of the expression of a gene or gene-cluster.

RNA as the Genetic Material

In the preceding sections the properties of DNA which particularly suit it to be the genetic material have been described. However, it seems likely that RNA may have been the primordial genetic material. As a single-stranded molecule its hydrogen-bonding groups are available to perform catalytic functions and fulfill a genetic role. DNA presumably evolved as a more stable genetic material. RNA, nonetheless, has survived as the genetic material in some viruses, even though it lacks the beneficial features of base protection and redundancy associated with a double helix (although it should be noted that some viruses have double-stranded RNA).

In viruses the RNA is protected from the environment by a protein coat. Viral RNA molecules spend most of their time as inert particles, replicating infrequently in the cells of their host organisms. When they do replicate, they do so rapidly and a large number of progeny particles are produced in a short period of time. The sheer numbers compensate for the lesser chemical stability of RNA. Thus the RNA phages and viruses have evolved special compensatory features that enable them to survive despite their lesser stability, relative to DNA-containing organisms.

Summary

Two classic experiments led to the identification of DNA as the genetic material. The transformation experiment by Griffith using bacterial pneumonia led to the discovery that a bacterial species with a stable genotype (R strain) could acquire the genotype of a heat-killed variant of the same species (S strain) when the living and dead bacteria were put in contact with one another, for example, when both were injected into a mouse. Hershey and Chase showed that when T2 phage adsorbed to *E. coli*, only DNA was injected into the bacterium, and this DNA provided all of the genetic information necessary for development of progeny phage.

The central dogma of modern biology is: *DNA makes RNA, makes protein*. DNA has many features that make it useful as a genetic material: (1) It can store genetic information in its base-sequence. (2) It can transfer this information to the protein-synthesizing apparatus of a cell as needed, by synthesizing an RNA molecule whose sequence is complementary to the DNA. (3) It has great physical and chemical stability. (4) DNA can transfer genetic information to progeny cells by a replication with each strand serving as a template for synthesis of a complementary strand. (5) Genetic change can occur without loss of parental information via mutational alterations that result in the production of one daughter molecule identical in base-sequence to the parent molecule, and a second daughter molecule with a slightly different base-sequence.

Drill Questions

1. What is meant by bacterial transformation?

2. In which organisms is RNA the sole genetic material?

3. What term describes the base-sequence that corresponds to a single amino acid?

4. In the blender experiment what radioactive atoms were used to label DNA and protein?

5. What is the process whereby DNA is changed?

6. What is the product of deamination of cytosine?

7. How are uracil and thymine chemically related?

8. What are two main mechanisms of mutation?

Problems

1. Why are viruses able to use RNA as a genetic material?

2. What experimental result showed that transformation consists of a permanent genetic change?

3. Why is RNA (but not DNA) hydrolyzed at alkaline pH?

4. Why was the transformation experiment not accepted by the scientific community as proof that DNA is the genetic material?

5. How was the transforming principle in Griffith's transformation experiment later shown to be DNA?

6. How does a replication error cause mutations?

Conceptual Questions

1. What other experiments could be performed to show that DNA is the genetic material?

2. Is it beneficial to a species that most mutational changes are deleterious?

3. Are mutational changes necessary for the various species of living organisms to prosper?

Kenneth Marians
Department of Molecular Biology

Birthday: **23 November 1951**

Birth Place: **Brooklyn, New York**

Undergraduate Degree: **Polytechnic Institute of Brooklyn Major: Chemistry 1972**

Graduate Degree: **Cornell University, Ph.D. 1976, Biochemistry**

Postgraduate Training: **Albert Einstein College of Medicine, 1976–1978**

Present Position: **Memorial Sloan-Kettering Cancer Center, Molecular Biology Department**

Address: **New York, New York**

MY INTEREST IN molecular biology stems from a desire to understand how enzymes and DNA interact efficiently to accomplish such impressive macromolecular transactions as the complete and accurate duplication of the genetic content of the cell. The techniques of molecular biology have allowed us to identify the genes encoding the enzymes involved, to overexpress these gene products (thereby providing a consistent and reliable source of these nominally scarce enzymes), and to manipulate the DNA molecules so that we can construct essentially any template we need.

What advice can you offer to undergraduates who want to become scholars of molecular biology?

My advice to students entering the field of molecular biology is to be justifiably impressed, perhaps even awed, by the ability of modern molecular biological techniques to unlock nature's closely guarded secrets, but they should also remember that a complete understanding of the action of these gene products will only be revealed by biochemically determining their catalytic activities and how they fit into their respective metabolic pathways.

7

DNA Replication

Genetic information is transferred from parent to progeny organisms by a faithful replication of the parental DNA molecules. Usually the information resides in one or more double-stranded DNA molecules. Some bacteriophage species contain single-stranded instead of double-stranded DNA. In these systems replication consists of several stages in which single-stranded DNA is first converted to a double-stranded molecule, which then serves as a template for synthesis of single strands identical to the parent molecule.

Replication of double-stranded DNA is a complicated process that is not completely understood. This complexity is, at least in part, an acknowledgement of the importance of the following facts: (1) replication requires a supply of energy to unwind the helix; (2) single-stranded DNA tends to form intrastrand base-pairs; (3) a single enzyme can catalyze only a limited number of physical and chemical reactions; (4) several safeguards have evolved that are designed both to prevent replication errors and to eliminate the rare errors that do occur; and (5) both circularity and the enormous size of DNA molecules impose physical constraints on the replicative system. To add to the difficulty for the researcher, there is not a unique mode of replication common to all organisms having double-stranded DNA.

All genetically relevant information contained in a nucleic acid molecule resides in its base sequence, so the prime role of any mode of replication is to duplicate the base sequence of the parent molecule. The specificity of base pairing—adenine with thymine and guanine with cytosine—provides the mechanism used by all replication systems.

Furthermore,

1. Nucleotide monomers are added one by one to the end of a growing strand by an enzyme called a **DNA polymerase.**

2. The sequence of bases in each new or **daughter strand** is complementary to the base sequence in the old or **parent strand** being copied—that is, if there is an adenine in the parent strand, a thymine will be added to the end of the growing daughter strand when the adenine is being copied.

In the following section we consider how the two strands of a daughter molecule are physically related to the two strands of the parent molecule.

Semiconservative Replication of Double-stranded DNA

The purpose of DNA replication is to create daughter DNA molecules that are identical to the parent molecule. This is accomplished by **semi-**

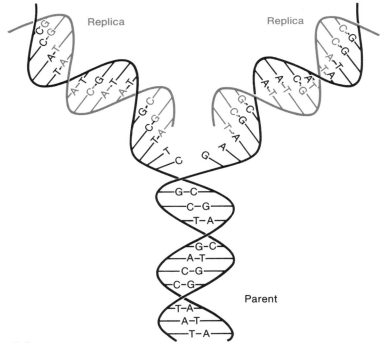

Figure 7-1
The replication of DNA according to the mechanism proposed by Watson and Crick. The two replicas consist of one parent strand (black) plus one daughter strand (red). Each base in a daughter strand is selected by the requirement that it form a base-pair with the parent base.

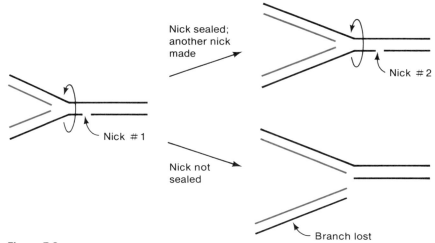

Figure 7-2
Mechanism by which a nick ahead of a replication fork allows rotation. If the nick is not sealed, a newly-formed branch is lost.

conservative replication. In this replication mode each parental single strand is a template for the synthesis of one new daughter strand, and as each new strand is formed it is hydrogen–bonded to its parental template (Figure 7-1). Thus, as replication proceeds, the parental double helix unwinds and then rewinds again into two new double helices, each of which contains one original parent strand and one newly–formed daughter strand.

Unwinding a double helix in semiconservative replication is a mechanical problem. Either the two daughter branches at the Y-fork shown in Figure 7-1 must revolve around one another or the unreplicated portion must rotate. If the molecule were fully extended in solution, there would be no problem and rotation of the unreplicated portion would be the simpler motion, as there would be less friction with the solvent. However, since the molecule is 600 times longer than the cell that contains it (so that it must be repeatedly folded, as was shown in Chapter 5), such rotation is unlikely. A simple solution would be to make a single-strand break in one parent strand ahead of the growing fork; this would enable a small segment of the unreplicated region to rotate, thereby eliminating the geometric problem (Figure 7-2). The only requirement would then be to re-form the broken bond. However, this repair would have to occur before the replication fork had passed the nick; otherwise a daughter strand would be lost, as shown in the figure.

Most DNA molecules replicate as circular structures (Figures 7-3 and 7-4). These circles resemble the Greek letter θ (theta), so the term **θ replication** is used to describe this replication mode. Replication of a circle introduces a topological problem that is more severe than that for a linear molecule. This problem is solved by an enzyme, DNA gyrase.

The unwinding problem in θ replication is formidable because lack

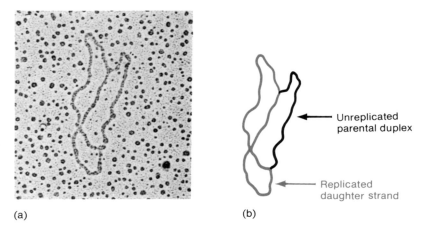

(a)   (b)

Figure 7-3
Electron micrograph of a ColE1 DNA molecule (molecular weight = 4.2×10^6) replicating by the θ mode. The parent and daughter segments are shown in the drawing. (Courtesy of Donald Helinski.)

of a free end makes rotation of the unreplicated portion impossible. An advancing nick, as in Figure 7-2, or a swivel at the replication origin (the point of initiation of replication) would solve the problem. These suggestions, though not quite correct, anticipate the correct mechanism.

As replication of the two daughter strands proceeds along the helix in the absence of some kind of swiveling, the nongrowing ends of the daughter strands would cause the entire unreplicated portion of the molecule to become overwound (Figure 7-4). This in turn would cause *positive* supercoiling (see Chapter 3) of the unreplicated portion. This

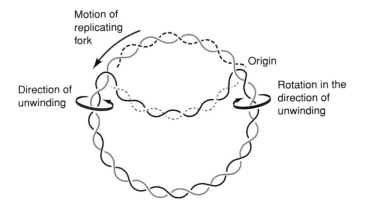

Figure 7-4
Drawing showing that the unwinding motion (curved arrows) of the daughter branches of a replicating circle lacking positions at which free rotation can occur causes overwinding of the unreplicated portion.

supercoiling obviously cannot increase indefinitely because, if it were to do so, the unreplicated portion would become coiled so tightly that no further advance of the replication fork would be possible. This topological constraint ought to be avoidable by the simple nicking-sealing cycle just described but, in general, the twists are removed in another way. As discussed in Chapter 3, most naturally occurring circular DNA molecules are *negatively* supercoiled. Thus, initially the overwinding motion is no problem because it can be taken up by the underwinding already present in the negative supercoil. However, after about 5% of the circle is replicated, the negative superhelicity is used up, and the topological problem arises.

topoisomerase

Most organisms contain one or more enzymes called **topoisomerases.** These enzymes can produce a variety of topological changes in DNA; the most common are production of negative superhelicity and the removal of superhelicity. In *E. coli* there is an enzyme called **DNA gyrase** (Eco topoisomerase II) which is able to produce negative superhelicity, and it is DNA gyrase that is responsible for removing the positive superhelicity generated during replication. The evidence for this point comes from experiments using antibiotics or drugs that inhibit DNA gyrase (e.g., coumermycin, nalidixic acid, oxolinic acid, and novobiocin) and from the study of gyrase mutants. Addition of any of these drugs to a growing bacterial culture inhibits DNA synthesis; in a phage-infected cell they prevent supercoiling of injected phage DNA molecules (that is, the molecules circularize but do not supercoil). Proof that these effects are due to inhibition of DNA gyrase comes from isolating a strain of *E. coli* that can grow normally in the presence of these antibiotics. With this strain the antibiotics have no effect on either DNA replication or supercoiling. If DNA gyrase is isolated from both wild-type *E. coli* and the mutant *E. coli* strains and each form is tested for binding of the antibiotics, it is found that only the wild-type enzyme binds the antibiotic. (This approach of probing for an activity with an inhibitor and comparing *in vitro* and *in vivo* results is used in many problems in molecular biology.)

DNA gyrase does have a DNA strand breakage and reunion activity, but this activity cannot by itself introduce negative superhelical turns. Twisting is a complex process that involves cutting the DNA, then folding it, and finally re-sealing it.

Enzymology of DNA Replication

The enzymatic synthesis of DNA is a complex process, primarily because of the need for high fidelity in copying the base sequence and for physical separation of the parent strands. The number of steps that must be completed is far too great to be accomplished by a single enzyme and, in fact, about twenty proteins are known to be necessary. Thus, in an effort to provide some understanding with a minimum of confusion, each step in the process will be treated separately. We will consider the basic chemistry of polymerization, the source of the precursors, the problems raised by the chemistry of polymerization, the means of initiating and terminating synthesis, and the mechanisms for eliminating replication errors.

THE POLYMERIZATION REACTION AND THE POLYMERASES

In 1957, Arthur Kornberg showed that in extracts of *E. coli* there exists a DNA polymerase (now called **polymerase I** or **pol I**). This enzyme is able to synthesize DNA from four precursor molecules—namely, the four deoxynucleoside 5'-triphosphates (dNTP), dATP, dGTP, dCTP, and dTTP—as long as a DNA molecule to be copied (a **template** DNA) is provided. Neither 5'-monophosphates nor 5'-diphosphates, nor 3'-(mono-, di-, or tri-) phosphates can be polymerized—only the 5'-triphosphates are substrates for the polymerization reaction; soon we will see why this is the case. Some years later, it was found that pol I, though playing an essential role in the replication process, is not the major polymerase in *E. coli;* instead, the enzyme responsible for advance of the replication fork is polymerase III or pol III.★ Pol III also exclusively uses 5'-triphosphates as precursors and requires a DNA template before polymerization can occur. Pol I and pol III have some features in common. The overall chemical reaction catalyzed by both DNA polymerases is:

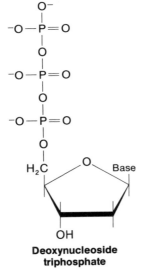

Deoxynucleoside triphosphate

$$\text{Poly(nucleotide)}_n\text{-3'-OH}^+ \text{dNTP} \rightarrow \text{Poly(nucleotide)}_{n+1}\text{-3'-OH}^+ \text{PP}$$

in which PP represents pyrophosphate cleaved from the dNTP.

Both pol I and pol III only polymerize deoxynucleoside 5'-triphosphates and can do so only while copying a template DNA. Furthermore, polymerization can only occur by addition to a **primer**—an oligonucleotide hydrogen-bonded to the template strand and whose terminal 3'-OH group is available for reaction (that is, a "free" 3'-OH group). The meaning of a primer is made clear in Figure 7-5, which depicts six potential template molecules; of these, only three can be said

---

★ The numbers I and III refer only to the order in which the enzymes were first isolated, not to their relative importance. Another polymerase, pol II, has also been isolated from *E. coli.* It plays no other role in DNA replication. Although mutant cells lacking pol II grow normally, recent studies indicate that the enzyme may be involved in certain DNA damage-induced DNA repair mechanisms. The recent cloning (see Chapter 13) and characterization of the pol II gene should help determine its actual role in the cell. Its biological function is unknown; mutant cells lacking pol II grow normally in the laboratory. For convenience, the abbreviation "pol" will be used in this chapter to refer to specifically named polymerases.

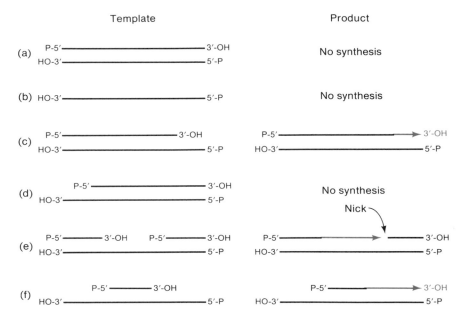

Figure 7-5
The effect of various templates used in DNA polymerization reactions. A free 3'-OH on a hydrogen-bonded nucleotide at the strand terminus and a non-hydrogen-bonded nucleotide at the adjacent position on the template strand is needed for strand growth. Newly synthesized DNA is red.

to be active—(c), (e), and (f)—each of which has a free 3'-OH group. The lack of activity with (d) and the direction of synthesis with (e) and (f) indicate that nucleotides do not add to a free 5'-P group. The lack of any synthesis with (a) or (b) indicates that addition to a 3'-OH group cannot occur if there is nothing to copy. Thus, we draw two conclusions:

1. Both a primer with a free 3'-OH group and a template are needed.

2. Polymerization consists of a reaction between a 3'-OH group at the end of the growing strand and an incoming nucleoside 5'-triphosphate. When the nucleotide is added, it supplies another free 3'-OH group. Since each DNA strand has a 5'-P terminus and a 3'-OH terminus, strand growth is said to proceed in the 5' → 3' (5'-to-3') direction.

These two points are summarized in Figure 7-6.

Occasionally polymerases add a nucleotide terminus that cannot hydrogen-bond to the corresponding base in the template strand. This may be purely a mistake, or may result from the tautomerization of adenine and thymine discussed in Chapter 10. In any case, it is important that the unpaired base be removable while its incorrectness is recogniz-

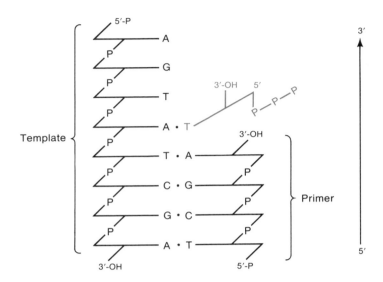

Figure 7-6
Schematic diagram of a replicating DNA molecule showing the distinction between template and primer and the meaning of 5' → 3' synthesis.

able—namely, as an unpaired base at the 3'-OH terminus of a growing strand.

Pol I responds to an unpaired terminal base by terminating polymerizing activity because the enzyme requires a primer that is correctly hydrogen-bonded. When such an impasse is encountered, a 3' → 5' exonuclease activity, which is a distinct catalytic activity of the enzyme located at a site distant from the active site for nucleotide polymerization, is stimulated and the unpaired base is removed (Figure 7-7). After removal of this base, the exonuclease activity stops, polymerizing activity is restored, and chain growth begins again. This exonuclease activity is called the **proofreading** or **editing function** of pol I.

Another function of polymerase I is that of a 5' → 3' exonuclease. This activity has the following features:

1. Nucleotides are removed from the 5'-P terminus only, one by one.
2. More than one nucleotide can be removed by successive cutting.
3. The nucleotide removed must have been base-paired.
4. The nucleotide removed can be either of the deoxy- or the ribotype.
5. Activity can be at a nick as long as there is a 5'-P group.

We will see shortly that the main function of the 5' → 3' exonuclease activity is to remove ribonucleotide primers.

The 5' → 3' exonuclease activity at a single-strand break (nick) can

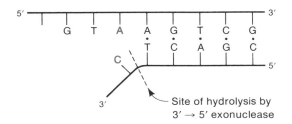

Figure 7-7
The 3′ → 5′ exonuclease activity of DNA polymerase I showing the site of hydrolysis. The C that is removed (red) does not base-pair with the A being copied (red).

occur simultaneously with polymerization. That is, as a 5′-P nucleotide is removed, a replacement can be made by the polymerizing activity (Figure 7-8). Since pol I cannot form a bond between a 3′-OH group and a 5′-monophosphate, the nick moves along the DNA molecule in the direction of synthesis. This movement is called **nick translation.**

nick translation

Experimental conditions can be chosen so that polymerization will occur at a single-strand break without concomitant 5′ → 3′ exonuclease activity. The growing strand then displaces the parental strand (Figure 7-8). This is thought to be an important step in the mechanism of genetic recombination. Of all *E. coli* polymerases known to date, polymerase I is the only one capable of carrying out an unaided displacement reaction. In other strand displacement reactions, auxiliary proteins are required and ATP is cleaved to fuel the unwinding of the helix; this will be discussed when the events at a replication fork are described.

Pol III is a very complex enzyme. In its most active form it is associated with nine other proteins to form the **pol III holoenzyme,**

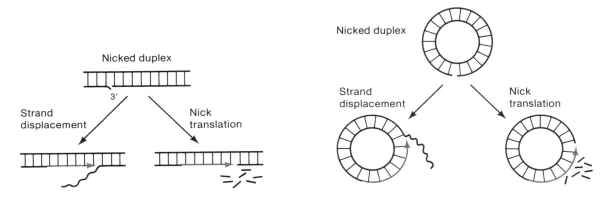

Figure 7-8
Strand displacement and nick translation on linear and circular molecules. In nick translation a nucleotide is exonucleolytically removed for each nucleotide added. The growing strand is shown in red.

occasionally termed pol III. The term **holoenzyme** refers to an enzyme that contains several different subunits and retains some activity even when one or more subunits is missing. The smallest aggregate having enzymatic activity is called the **core enzyme.** The activities of the core enzyme and the holoenzyme are usually very different. Genes encoding five of the subunits have been identified; these are called *dnaE, dnaN, dnaQ, dnaX,* and *dnaZ.* The *dnaE* protein possesses the major polymerizing activity but each of the subunits, except for the *dnaQ* protein, is essential for replication. Pol III shares with pol I a requirement for a template and a primer but its substrate specificity is much more limited. For instance, neither the core enzyme or the holoenzyme can act at a nick. The core enzyme is most active *in vitro* on gapped DNA, where the gap is less than 100 nucleotides long. On the other hand, the holoenzyme is extraordinarily active on long single-stranded DNA templates primed with either a DNA or RNA primer. This reflects the action of the holoenzyme at the replication fork, where as the parental duplex is unwound, the template is presented essentially as one very long single-strand. Pol III cannot carry out strand displacement either, and another system is needed to unwind the helix in order that a replication fork will be able to proceed (this will be discussed in detail in a later section). The enzyme, like pol I, possesses a $3' \rightarrow 5'$ exonuclease activity which performs the major editing function in DNA replication. This function resides in the *dnaQ* subunit. The activity of this $3' \rightarrow 5'$ editing exonuclease is stimulated greatly when the *dnaQ* subunit is complexed together with the *dnaE* subunit in either the core enzyme or the holoenzyme. A role for this activity in providing fidelity to the replication process is clearly indicated, since strains carrying mutations in the *dnaQ* gene have a higher mutation frequency. Unlike pol I, pol III does not possess a $5' \rightarrow 3'$ exonuclease activity.

Although pol III holoenzyme is the major replicating enzyme in *E. coli,* much less is known about it than about pol I, because it is a more complex enzyme. Study of pol III is currently an active field of research.

Later in this chapter, when the events at the growing fork catalyzed by the *E. coli* replication system are examined, it will be seen that pol I and pol III holoenzyme are both essential for *E. coli* replication. However, a requirement for two polymerases is not common to all organisms: for instance, *E. coli* phage T4 synthesizes its own polymerase, and this enzyme is capable of carrying out all necessary polymerization functions.

All known polymerases (for both DNA and RNA) are capable of chain growth in only the $5' \rightarrow 3'$ direction; that is, the growing end of the polymer must have a free 3'-OH group. It is possible for the following reasons that the enzymes evolved in this way to facilitate editing. If $3' \rightarrow 5'$ growth were to occur, the growing strand would also be terminated with a 5'-triphosphate and the 3'-OH group of the incoming nucleotide would react with it. Chemically this is certainly acceptable but since the bonds formed contain only a single phosphate, an editing function would leave a free 5'-monophosphate. In order for chain growth to proceed, an enzymatic system would be needed to enter the replication fork and convert the monophosphate to a triphosphate. There is already a great deal going on in the replication fork, so that it would seem more economical for the cell to require 5' → 3' growth exclusively. However, the observation that chain growth proceeds in only one direction introduces what is probably the greatest complication in the entire replication process; this will be described shortly.

DNA Ligase

Neither replication from a primed circular single strand nor gap filling results in a continuous daughter strand. Discontinuity results because no known polymerase can join a 3'-OH and a 5'-monophosphate group. The joining of these groups is accomplished by the enzyme **DNA ligase,** which functions in replication and other important processes.

E. coli DNA ligase can join a 3'-OH group to a 5'-P group as long as both are termini of adjacent base-paired deoxynucleotides—the enzyme cannot bridge a gap (Figure 7-9).

In the usual polymerization reaction, the activation energy for phosphodiester bond formation comes from cleaving the triphosphate. Since DNA ligase has only a monophosphate to work with, it needs another source of energy. It obtains this energy by hydrolyzing either ATP or nicotine adenine dinucleotide (NAD); the energy source depends upon the organism from which the DNA ligase is obtained. The E. coli DNA ligase uses NAD.

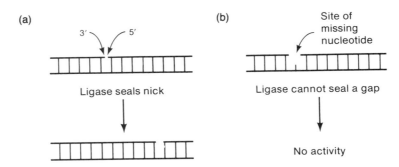

Figure 7-9
The action of DNA ligase. (a) A nick having a 3'-OH and a 5'-P terminus is sealed. (b) If one or more nucleotides are absent, the gap cannot be sealed.

Discontinuous Replication

In the model of replication shown in Figure 7-1 both daughter strands are drawn as if replicating continuously. However, no known DNA molecule replicates in this way—instead,

> One of the daughter strands is made in short fragments, which are then joined together.

Polymerase I and polymerase III can add nucleotides only to a 3'-OH group. Examination of the growing fork indicates that if both daughter strands grew in the same overall direction, only one of these strands would have a free 3'-OH group; the other strand would have a free 5'-P group because the two strands of DNA are antiparallel (Figure 7-10). Thus one of the following must be true:

1. There is another polymerase that can add a nucleotide to the 5' end; that is, it would catalyze strand growth in the 3' → 5' direction.

2. The two strands both grow in the 5' → 3' direction, but from opposite ends of the parental molecule, as shown in Figure 7-11. If this were correct, a significant fraction of the unreplicated molecule would have to be single-stranded.

3. The two strands both grow in the 5' → 3' direction at a single growing fork, and hence do not grow in the same direction along the parent molecule. One way to accomplish this is shown in Figure 7-12. This mode of synthesis is called **discontinuous synthesis.** The **leading strand** is synthesized continuously, and the **lagging strand** is synthesized discontinuously in the form of short fragments, each growing in the correct 5' → 3' direction. In time, the fragments can be joined together to form a continuous strand.

leading strand (DNA)
lagging strand (DNA)

A great deal of evidence eliminates the first two models and supports the idea of discontinuous synthesis. No DNA polymerase has been discovered that adds nucleotides to the 5'-P terminus, and at most about 0.05% of intracellular DNA is ever in single-stranded form. The most critical evidence is the following series of experiments.

In 1968 R. Okazaki demonstrated that in *E. coli* newly synthesized DNA consists of fragments that later attach to one another to generate continuous strands. The presence of fragments supported the discontinuous replication model of Figure 7-12. Okazaki did two experiments to demonstrate this. In the first experiment, [^{3}H]dT was added to a growing bacterial culture in order to label the new DNA strands with radioactivity, and 30 seconds later the cells were collected and all of their DNA was isolated. This is called a **pulse-labeling experiment.** The DNA was then sedimented in alkali, which causes strand separation. The type of data obtained (Figure 7-13(a)) showed that the most recently made

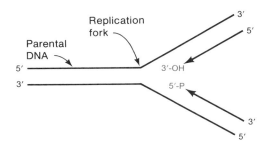

Figure 7-10
The termini (red) that would be present in a replication fork if both strands were to grow in the same overall direction.

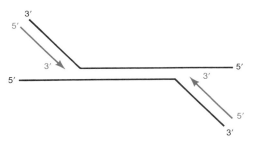

Figure 7-11
One way to replicate an antiparallel DNA molecule by means of 5′ → 3′ chain growth. Daughter strands are shown in red.

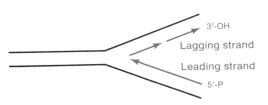

Figure 7-12
A growing fork showing the direction of growth of the leading and lagging strands (both in red).

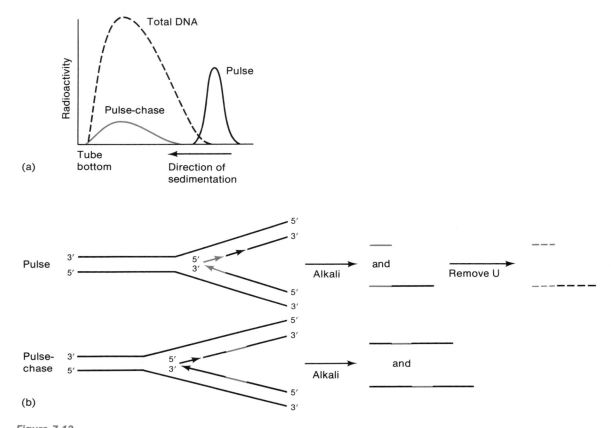

Figure 7-13
(a) The type of data obtained by alkaline sedimentation of pulse-labeled DNA (black) and pulse-chased DNA (red). Total DNA is the sum of the nonradioactive and radioactive DNA, as might be indicated by the optical absorbance. The s value of the sedimenting material increases from right to left. (b) The location of radioactive DNA (red) at the time of pulse-labeling and after a chase. The radioactive molecules present in alkali are shown. The fragmentation resulting from removal of uracil (see later in text) accounts for the fact that all pulse-labeled DNA sediments slowly before the chase.

(pulse-labeled) DNA sediments very slowly in comparison with single strands obtained from parent DNA (even though these strands usually break in the course of isolation); from the s value★, it was estimated that the fragments of pulse-labeled DNA range in size from 1000 to 2000 nucleotides, whereas the isolated parent DNA is usually 20 to 50 times as large. In the second experiment, the bacteria were pulse-labeled for 30 seconds; then, the [³H]dT was replaced with nonradioactive dT, and the bacteria were allowed to grow for several minutes. This is called

★ The s value, or sedimentation coefficient, is a measure of the speed with which a particle moves under the influence of centrifugal force. For DNA, it can be related simply to molecular weight.

a **pulse-chase experiment** and it allows one to examine the current state of molecules synthesized at an earlier time. Okazaki observed that the *s* value of the radioactive material increased with time of growth in the nonradioactive medium. These experiments are presented and interpreted in Figure 7-13 in terms of the discontinuous replication model shown in Figure 7-12. Apparently the joining of the pulse-labeled fragments to the growing daughter strands caused the label to sediment with the bulk of the DNA.

Okazaki fragments

These fragments, called **precursor fragments** or **Okazaki fragments,** have the properties predicted by the discontinuous model: they are initially small and then become large as they are attached to previously made DNA. However, the model predicts that only half of the radioactivity should be found in small fragments whereas the data shown in Figure 7-13(a) indicate that all of the newly synthesized DNA consists of fragments. This result should be surprising because there is no reason why the DNA of the 3′-OH-terminated strand should be synthesized discontinuously. In fact, it is not. Occasionally DNA polymerase adds a uracil nucleotide instead of a thymine nucleotide to the end of the growing strand when copying an adenine in the template strand. When this occurs, a repair system excises the uracil, replacing it with thymine. This repair system does not act at the terminus (in contrast with the proofreading 3′–5′ exonuclease) but instead when the uracil is well within the growing strand. The excision process produces a transient single-strand break, which is sealed by DNA ligase after the correct nucleotide is inserted. Thus, the leading strand is made continuously but, after synthesis, is fragmented at the rare uracil sites.

Additional evidence for the discontinuous-synthesis model comes from high-resolution electron micrographs of replicating DNA molecules, showing a short single-stranded region on one side of the replication fork (Figure 7-14). This region results from the fact that synthesis of the discontinuous strand is initiated only periodically; perhaps a particular base sequence or some other signal is required for initiation. In fact, the 3′-OH terminus of the continuously replicating strand is always ahead of the discontinuous strand. This is the origin of the terms **leading strand** and **lagging strand** for the continuously and discontinuously replicating strands, respectively.

Pol III cannot initiate chain growth because like all DNA polymerases, it needs a primer. This problem applies to initiating both the leading strand and the precursor fragments. Either a ribo- or deoxynucleotide, hydrogen-bonded to the template strand, could serve as a primer, because all known DNA polymerases can add nucleotides to the 3′-OH termini of both types. In every case examined, the primer for both leading and lagging strand synthesis is a short RNA oligonucleotide that consists of 1 to 60 bases; the exact number depends on the particular organism. This RNA primer is synthesized by copying a particular base sequence from one DNA strand and differs from a typical RNA molecule in that after its synthesis the *primer remains hydrogen-bonded to the DNA*

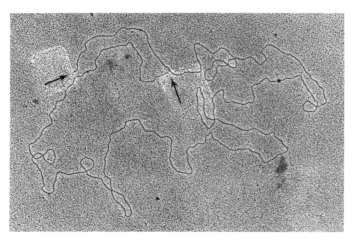

Figure 7-14
A replicating θ molecule of phage λ DNA. The arrows show the two replicating forks. The segment between each pair of thick lines at the arrows is single-stranded DNA; note that it appears thinner and lighter. (Courtesy of Manuel Valenzuela.)

RNA polymerase

template. In bacteria two different enzymes are known to synthesize primer RNA molecules—**RNA polymerase,** which is the same enzyme that is used for synthesis of most RNA molecules such as messenger RNA, and **primase** (the product of the *dnaG* gene). Experimentally, these enzymes can be distinguished *in vivo* by their differential sensitivities to the antibiotic rifampicin—RNA polymerase is inhibited by the antibiotic, but primase is not. In *E. coli*, initiation of leading-strand synthesis is rifampicin-sensitive, presumably because RNA polymerase is used. Initiation of precursor-fragment synthesis is resistant to the drug, as it uses primase. A precursor fragment has the following structure, while it is being synthesized:

PPP-5' ———— ———————————— 3'-OH
         RNA   DNA

Precursor fragments are ultimately joined to yield a continuous strand. This strand contains no ribonucleotides so that assembly of the lagging strand must require removal of the primer ribonucleotides, replacement with deoxynucleotides, and then joining. In *E. coli* the first two processes are accomplished by DNA pol I and joining is catalyzed by DNA ligase. How this is done is shown in Figure 7-15. Pol III extends the growing strand until the RNA nucleotide of the primer of the previously synthesized precursor fragment is reached. Pol III can go no further since its 5' → 3' exonuclease is inactive on base-paired DNA; it cannot join a 5'-triphosphate at the terminus of a polymer (i.e., on the primer) to a 3'-OH group on the growing strand, and it cannot carry out strand

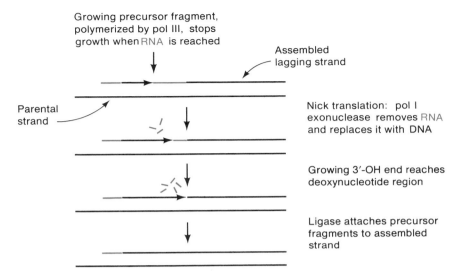

Figure 7-15
Sequence of events in assembly of precursor fragments. RNA is indicated in red. The replication fork (not shown) is at the left.

displacement. Thus pol III dissociates from the DNA, leaving a nick. *E. coli* DNA ligase cannot seal the nick because a triphosphate is present, even if an additional enzyme could cleave the triphosphate to a monophosphate. DNA ligase would be inactive when one of the nucleotides is in the ribo form. However, pol I works efficiently at a nick as long as there is a 3'-OH terminus. In this case the enzyme carries out nick translation, probably proceeding into the deoxy section, but there DNA ligase can compete with pol I and seal the nick. Thus, the precursor fragment is assimilated into the lagging strand. By this time, the next precursor fragment has reached the RNA primer of the fragment just joined and the sequence begins anew. Another enzyme, RNase H, a riboendonuclease specific for RNA in the form of a RNA–DNA hybrid, can also participate in the removal of primer ribonucleotides.

Synthesis of precursor fragments follows synthesis of the leading strand. In the next section, how the leading strand advances into the parent double helix is described. Once this is understood, we can return to the question of how RNA primer synthesis gets started—a surprisingly complicated process.

Events in the Replication Fork

DNA replication requires not only an enzymatic mechanism for adding nucleotides to the growing chains, but also a means of unwinding the

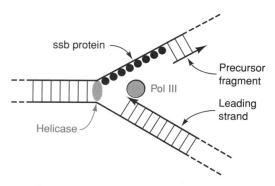

Figure 7-16
The unwinding events in a replication fork.

parent double helix. In this section we will see that these are distinct processes and that the unwinding mechanism is closely related to the initiation of synthesis of precursor fragments.

Polymerase III cannot carry out strand displacement (as has already been discussed), because it is unable to unwind the helix. In order for a helix to be unwound, hydrogen-bonds and hydrophobic interactions must be eliminated, and this requires energy. Pol I is able to utilize both the free energy of hydrolysis of the triphosphate group and the binding energy of forming a new hydrogen-bond to unwind the parent molecule as it synthesizes the leading strand. No other polymerase can do that; usually a helix must first be unwound in order for a DNA polymerase to advance. Helix-unwinding is accomplished by enzymes called **helicases.** The helicase active in *E. coli* DNA replication hydrolyzes ATP and utilizes the free energy of hydrolysis in an unknown way to unwind the helix. The ratio of ATP molecules hydrolyzed per base-pair broken at the replication fork is unknown.

In *E. coli* the pol III molecule synthesizing the leading strand is most likely immediately behind the advancing helicase (Figure 7-16). However, the helicase leaves in its wake one single-stranded region to be copied by precursor fragments. In order to prevent the single-stranded regions from forming intrastrand hydrogen-bonds, the single-stranded DNA is coated with *E. coli* single-strand binding protein (**ssb protein**). This is one of a class of binding proteins that bind tightly to both single-stranded DNA and to one another. As pol III advances, it displaces the ssb protein so that base pairing of the nucleotide being added can occur.

In some replication systems the binding of the ssb protein can also unwind the helix. This is true of *E. coli* phage T4, which synthesizes a protein called the gene-32 protein. The 32-protein binds very tightly to single-stranded DNA and exceedingly tightly to itself. Its binding energy is great enough to unwind the helix. This should not be confused with the action of a DNA helicase, which utilizes the energy of ATP hydrolysis to unwind the helix.

helicase

ssb protein

The mechanism for initiation of replication of precursor fragments is closely connected to the unwinding of the helicase. A direct study of initiation of synthesis of precursor fragments has not yet been carried out because experimentally the problem is very complex. Instead, attention has been directed to several *E. coli* phages having single-stranded DNA, in which the first step in replication is the conversion of single-stranded circular DNA to a double-stranded covalent circle. Since some of the phages studied use *E. coli* enzymes exclusively, it is thought that they can serve as a model for understanding initiation of precursor fragment synthesis.

The most complicated system studied so far and the one that is most closely related to *E. coli* DNA synthesis is that of the circular single-stranded DNA of phage ϕX174. In this system primer synthesis requires seven proteins. Primer synthesis begins by the formation of a complex called a **preprimosome,** which contains six proteins: the PriA, PriB, PriC (Pri = *primo*some), DnaT, DnaB, and DnaC proteins. This preprimosome is capable of moving along the DNA. Occasionally, the primase (the DnaG protein) will join the preprimosome forming the complete primosome. Both assembly and movement of the primosome requires ATP. Interestingly, two of the proteins complexed together in the primosome (DnaB and PriA) are DNA helicases and each of these proteins is capable of moving the primosome along the DNA. What is most startling about this phenomenon is that the DnaB and PriA proteins actually act to drive the primosome in opposite directions along single-stranded DNA. It is thought that this causes the generation of a loop in the template for the lagging strand of single-stranded DNA anchored by the primosome (see Figure 7-17). The significance of this loop is not clear. Current evidence suggests that it is the DnaB protein that is at the base of the replication fork unwinding the parent helix. It is also not clear what signal causes the primase to initiate synthesis of an RNA primer, although the priming sites seem to be chosen at random. Replication then proceeds by pol III, pol I, and ligase, in sequence. The important characteristics of the prepriming reaction are: the lack of a specific sequence at which initiation occurs; the use of a complex of six proteins; the requirement for ATP; and synthesis of RNA by primase.

A Summary of Events at the Replication Fork

Figure 7-17 summarizes the proposed events that occur in or near a replication fork in *E. coli*. The DnaB helicase, complexed within the primosome and driven by ATP hydrolysis, unwinds the helix. The unpaired bases on the template for the lagging strand are coated with ssb protein. The leading strand advances along one parent strand by nucleotide addition catalyzed by the pol III holoenzyme. Periodically,

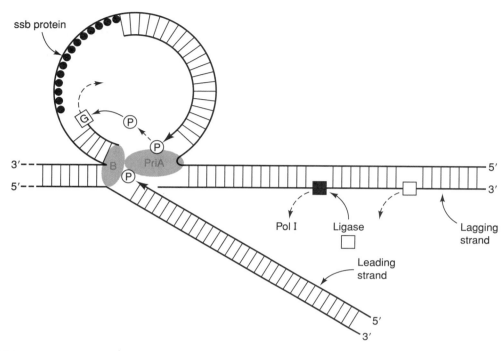

Figure 7-17
Summary of the proposed events in a replication fork of *E. coli* DNA. A solid arrow from an enzyme symbol indicates that the enzyme joins the DNA at the site shown. A dashed line indicates that the enzyme leaves the DNA. The loop-like structure exists to facilitate movement of the primosome in opposite directions on the leading and lagging strands. Ⓟ = pol III

primase, acting within the primosome, synthesizes a primer RNA. The pol III holoenzyme adds nucleotides to this primer, thereby synthesizing a precursor fragment. Synthesis of this fragment may occur within a loop generated by the combined action of the DnaB and PriA proteins. This may allow the pol III holoenzyme to find the new primer very quickly. This synthesis continues up to the primer of the preceding precursor fragment; at this point pol I replaces pol III. Nick-translation removes the RNA and replaces it with DNA. Once the RNA is gone, DNA ligase seals the nick, thereby joining the precursor fragment to the lagging strand. The advance of the replication fork continues until replication is completed.

The reader must surely be impressed with the complexity of this process. However, in some systems replication is somewhat simpler. For instance, with *E. coli* phage T7, priming is accomplished by a phage RNA polymerase and all replication is performed by a single DNA polymerase encoded in the phage DNA; none of the known *E. coli* enzymes are needed. Phages T4 and λ replicate in a somewhat more complicated manner than T7 but still the mechanism is far simpler than

E. coli replication. Why should the process in *E. coli* be so complex? This may be a consequence of the requirement for fidelity. The larger the DNA molecule, the greater the probability that a replication error will be made during a round of replication. Furthermore, a bacterial DNA is replicated only once per generation; a phage replicates many times in an infected cell, so that one defective replica would not significantly affect the successful outcome of the infection. Thus, in order to reduce the error frequency to a tolerable level, organisms with a larger DNA molecule and a single replication event need more replication proteins, since each protein can be designed to minimize or correct a particular type of error.

We now leave our analysis of the replication fork and examine how the fork itself is created at the outset; what is the process by which synthesis of double-stranded DNA is initiated? This is one of the least understood subjects of replication, especially because it may occur in several ways; it is described in the following section.

Initiation of Synthesis of the Leading Strand

ori

All known double-stranded DNA molecules initiate a round of replication at a unique base sequence, called the **replication origin** or *ori.* The sequence is specific to each organism, though there is one example in which two organisms have the same sequence.

Initiation can occur in two ways—*de novo* **initiation,** in which the leading strand is started afresh, and **covalent extension,** in which the leading strand is covalently attached to a parental strand. We begin with *de novo* initiation.

DE NOVO INITIATION

Most known DNA molecules replicate as circles and hence initiate within the helix. Even most molecules that replicate as linear molecules initiate within the helix rather than at one end. Since we know that all DNA synthesis catalyzed by DNA polymerases occurs on a single-stranded template, *de novo* initiation requires a localized unwinding or denaturation event of the parent helix at the specific origin of replication. In the systems that have been examined to date, this can occur by one of two mechanisms: either a DNA sequence-specific origin-binding protein acts to locally unwind the helix; or leading strand synthesis is initiated before lagging strand synthesis with the action of RNA polymerase providing a primer. The initial extension of this primer before synthesis of the first precursor fragment begins, generates a replication bubble that con-

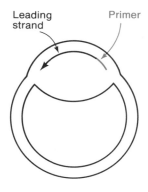

Leading strand   Primer

Figure 7-18
A circular DNA molecule
with a D loop.

sists of one double-stranded branch, made up of one parental strand paired with the leading strand, and one single-stranded branch, which is the unreplicated, second parental strand (Figure 7-18). Since the leading strand displaces the unreplicated parent strand, the bubble is called a **displacement loop** or **D loop.** Such a configuration is ordinarily a transient one, existing only until synthesis of precursor fragments begins—which may require that the leading strand release a specific sequence (in single-stranded form) that can be used for prepriming. However, in certain circumstances, namely in a replication system that does not employ DNA gyrase to relieve topological constraints, a D loop may be long lived.

In the initial stages of replication of a naturally occurring circular DNA molecule, advance of the replication fork does not require the presence of DNA gyrase, because such a circular DNA molecule initially is negatively supercoiled and the negative twists compensate for the positive turns introduced by movement of the fork. However, once the fork has advanced sufficiently that the negative twists are used up, a topological constraint-relieving system such as gyrase is needed. The superhelix density of all naturally occurring DNA molecules is 0.05 twists per turn of the helix. This means that, if positive superhelicity is forbidden, the leading strand could move roughly 5% of the distance along a circular, negatively superhelical molecule before gyrase would be needed. A replication complex cannot in fact move along a double helix if the movement would produce more than one or two turns of overwinding. Thus, if neither gyrase nor any other topoisomerase system were present, replication would cease when the unreplicated portion had lost its supercoiling. The molecular configuration of the bubble would depend upon whether the first precursor fragment had been initiated; if it had not, the structure in Figure 7-18 would result.

Such a structure is seen in animal cell mitochondrial DNA (Figure 7-19). The molecular weights of these molecules in most species is about 10^7 and the molecular weight of a precursor fragment is typically about 10^6, so the lack of double-stranded DNA in both branches is consistent with the fact that only 4% of the DNA (or 2×10^5 molecular weight units of the leading strand) is replicated. The fact that 70% of all replicating mitochondrial DNA molecules are in the D-loop configuration indicates that indeed replication has ceased at this point.

A more common mechanism of *de novo* initiation is shared by DNA molecules as diverse as the *E. coli* chromosome, phage λ DNA, and the DNA of the eukaryotic simian virus 40 (SV40). During the initiation of replication of the *E. coli* chromosome (shown in Figure 7-20), the DnaA protein binds to a 9 nucleotide sequence that is repeated 4 times within a 250 nucleotide sequence that has been defined as *ori*C (C for chromosomal). In the presence of ATP, the bound 50-kDa DnaA protein monomers interact to form a complex protein-DNA structure consisting of 15–20 DnaA monomers and about 150 base-pairs of DNA. Formation of this structure acts to locally denature an A + T-rich region directly

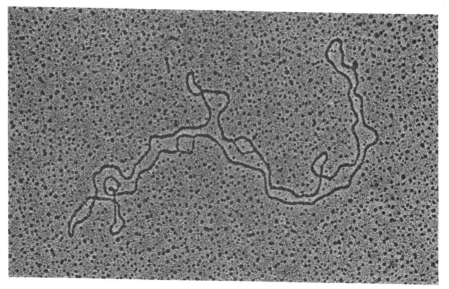

Figure 7-19
An electron micrograph of a dimer of mouse mitochondrial DNA showing diametrically opposing D loops. The single strand in each D loop appears thinner than the double-stranded DNA. The total length of the molecule is 10 μm. (Courtesy of David Clayton.)

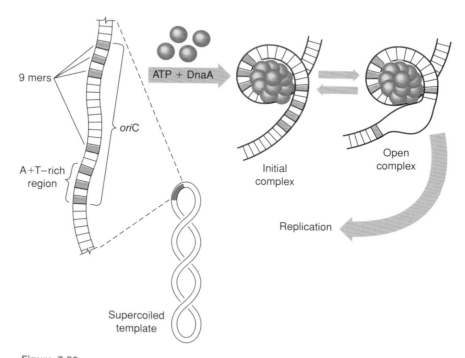

Figure 7-20
Initiation at *ori*C occurs after DnaA protein binds the four 9 mers. The A + T-rich region is then denatured, and this open complex serves as a replication site. (Adapted and redrawn from D. Brainhill and A. Kornberg. *Cell,* 1988:54; 915–918.)

adjacent to it. Then, because of specific interactions between the DnaA protein and the other replication proteins (like DnaB), the replication fork machinery is recruited to the unwound site on the DNA. Once all the requisite enzymes are present and primers have been synthesized, the pol III holoenzyme can begin to synthesize nascent DNA. In the case of *ori*C, two replication forks then proceed in opposite directions away from the origin (see Bidirectional Replication), although the exact mechanism for initiation of the leading and lagging strands is unclear and currently under investigation.

During phage λ replication in *E. coli*, as well as the replication of the animal virus SV40, a very similar mechanism is used, and a virally-encoded protein binds repeated sequences at the *ori* to form the initial complicated nucleoprotein structure. Thus, this mechanism seems to be general.

INITIATION BY COVALENT EXTENSION: ROLLING CIRCLE REPLICATION

There are numerous instances in which, in the course of replication, a circular phage DNA molecule gives rise to linear daughter molecules in which the base sequence of the DNA present in the phage is repeated numerous times, forming a concatemer (Figure 7-21). These concatemers are usually an essential intermediate in phage production. Likewise, in bacterial mating, a linear DNA molecule is transferred by a replicative process from a donor cell to a recipient cell. Both phenomena are consequences of initiation by covalent extension, an event that gives rise to a replication mode known as **rolling circle replication.**

rolling circle replication

Consider a duplex circle in which, by some initiation event, a nick is made having 3'-OH and 5'-P termini (Figure 7-22). Under the influence of a helicase and ssb protein, a replication fork can be generated. Synthesis of a primer for the leading strand is unnecessary because of the 3'-OH group. Leading-strand synthesis proceeds by elongation from this terminus. At the same time, the parental template for lagging-strand synthesis is displaced. Primers are, however, synthesized for the lagging strand in the same manner as illustrated previously for conventional DNA replication (for example, Figure 7-17). The polymerase used for this synthesis is apparently polymerase III. (In this circumstance we can ignore the fact that the enzyme cannot ordinarily carry out a displacement reaction, because in this case, displacement is a result of a coupling of

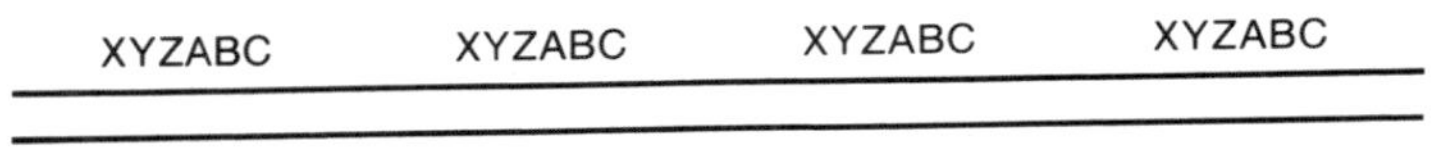

| XYZABC | XYZABC | XYZABC | XYZABC |

Figure 7-21
A concatemer consisting of the repeating unit ABC . . . XYZ. Note that the definition of concatemer does not make any requirements about the terminal sequences.

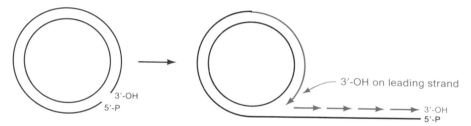

Figure 7-22
Rolling circle or σ replication. Newly synthesized DNA is shown in red.

helicase, ssb protein, and polymerase III.) The displaced parent strand is replicated in the usual way by means of precursor fragments. The result of this mode of replication is a circle with a linear branch; it resembles the Greek letter *sigma* and is called σ replication, or rolling circle replication.

There are four significant features of rolling circle replication:

1. The leading strand is covalently linked to the parental template for the lagging strand.

2. Before precursor fragment synthesis begins, the linear branch has a free 5'-P terminus.

3. Rolling circle replication continues unabated, generating a con- catemeric branch.

4. The circular template for leading-strand synthesis never leaves the circular part of the molecule.

Bidirectional Replication

In this section we examine the events following a D-looplike initiation step in which chain growth occurs continuously. It will be seen that whether a DNA molecule replicates unidirectionally or bidirectionally depends on whether (because of the mechanism by which the replication fork machinery is assembled) DNA helicases are capable of moving in opposite directions away from the *ori*.

In Figure 7-23(a) and (b), the counterclockwise moving replication fork has just been initiated. The helicase/primase component of this fork would be positioned as shown; moving counterclockwise around the circle on the template for the lagging strand unwinding the helix and priming the synthesis of precursor fragments. It was noted previously that the synthesis of each precursor fragment is terminated when the growing-end reaches the primer of the previously synthesized fragment. However, in the case of the first precursor fragment, there can be no previously made fragment. As also noted previously (Figure 7-15), only the DNA polymerase is actively engaged at the end of this fragment.

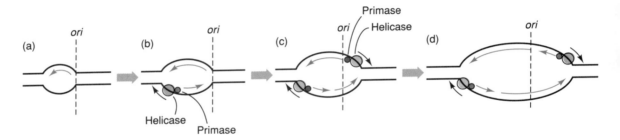

Figure 7-23
The formation of a bidirectionally enlarging replication bubble. (a) The leftward-leading strand starts at *ori*. (b) The leading strand has progressed far enough that the first rightward precursor (lagging strand) fragment begins. (c) The leftward-leading strand has progressed far enough that the second rightward precursor (lagging strand) fragment has begun. The first rightward precursor fragment has passed *ori* and has become the rightward-leading strand. (d) The rightward-leading strand has moved far enough that the first leftward (lagging strand) precursor fragment has begun. There are now two complete replication forks. The helicase/primase complex shows how *E. coli* replicates DNA simultaneously in both clockwise and counterclockwise directions. The black illustrates the direction of movement the helicase (unwinding) protein.

Thus, since no DNA polymerase involved in DNA replication can strand displace, replication will terminate in Figure 7-23(b) when the nonreplicated parent duplex is reached. Under these circumstances, replication will be unidirectional and counterclockwise. However, if (as in the case for *E. coli oriC*) a second helicase/primase complex is positioned as shown in Figure 7-23(c) so that it can move clockwise around the circle, replication can continue. The first leftward-moving precursor fragment becomes the rightward-moving leading strand, and the second helicase/primase complex acts to unwind the helix in a clockwise direction and prime the synthesis of the rightward-moving precursor fragments. The result of these events is that the DNA molecule will have two replication forks moving in opposite directions around the circle. This is called **bidirectional replication.**

The method used to study the direction of replication was developed by Ross Inman and Maria Schnös and is called **denaturation mapping.** It is done as follows. If a DNA molecule is heated to a temperature at which melting is just detected, single-stranded bubbles form in the regions having very high A + T content. If formaldehyde is added and the DNA is cooled, the bubbles persist. DNA treated in this way can be observed by electron microscopy and the position of the bubbles can be noted (Figure 7-24). When the technique is applied to a replicating molecule, the bubbles serve as fixed reference points against which the positions of a branch-point can be plotted. Figure 7-25 shows the kinds of molecules that would be expected for unidirectional and bidirectional replication of a hypothetical molecule. Note that:

1. The relation between the positions of the bubbles and the branch points differ for the two modes.

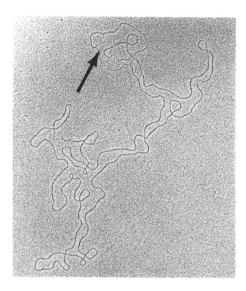

Figure 7-24
An electron micrograph of a partially denatured circular phage λ DNA molecule. With this method of sample preparation the single strands are very thin and faint compared to the double-stranded DNA. An arrow indicates the position of a replication fork. (Courtesy of Manuel Valenzuela.)

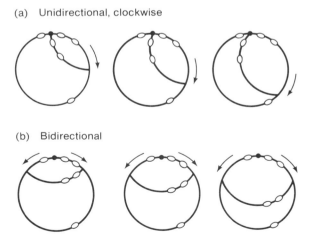

(a) Unidirectional, clockwise

(b) Bidirectional

Figure 7-25
A diagram showing the relative positions of branch points and a replication bubble for a DNA molecule replicating (a) unidirectionally and clockwise or (b) bidirectionally. The arrows indicate the direction of movement of the replication fork, and the dots indicate the replication origin.

2. A replication fork is identified by its changing distance from a bubble.

In unidirectional replication, one branch point remains at a fixed position with respect to the bubbles; this position defines the replication origin. In bidirectional replication both branch points move with respect to the bubbles so that each branch point is a replication fork. If both replication forks move at the same rate, the origin is always at the midpoint of each branch of the replication loop.

Bidirectional replication has been widely observed with phage, bacterial, and plasmid DNA. A small number of phages and plasmids use the unidirectional mode exclusively. One plasmid uses a surprising variant of the bidirectional mode in that initiation of movement of one fork occurs at a much later time than the primary initiation event. Bidirectional replication is the major mode of chain growth with eukaryotes.

Replication of Eukaryotic Chromosomes

The replication of eukaryotic chromosomes presents many problems not found in the prokaryotes because of the enormous size of eukaryotic chromosomes and the geometric complexity imposed by the organization of the DNA into nucleosomes. How these problems are handled is described in this section.

The rate of movement of a replication fork in *E. coli* is $\approx 10^5$ base-pairs per minute. In eukaryotes the polymerases are much less active, and the rate ranges from 500 to 5000 base-pairs per minute. Since a typical animal cell contains about fifty times as much DNA as a bacterium, the replication time of an animal cell should be about 1000 times as great as that of *E. coli* or about 30 days. However, the duration of the replication cycle is usually several hours and this is accomplished by having multiple initiation sites. For instance, the DNA of the fruit fly *Drosophila* has about 5000 initiation sites, each separated by about 30,000 bases, and each site replicates bidirectionally. The number of sites is regulated in a way that is not understood. For example, in the round of replication following fertilization of *Drosophila* eggs, the number of initiation points reaches 50,000, and it takes only 3 minutes to replicate all of the DNA. An example of a fragment of this rapidly replicating *Drosophila* DNA is shown in Figure 7-26.

The enormous number of growing forks in eukaryotic cells is reflected in the number of polymerase molecules. In *E. coli* there are between 10 and 20 molecules of pol III holoenzyme. However, a typical animal cell has 20,000 to 60,000 molecules of polymerase α, which is one of the two DNA polymerases thought to be involved in replication. Replication of double-stranded DNA proceeds through both a poly-

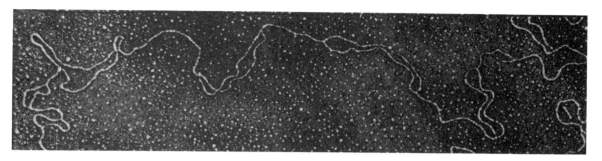

Figure 7-26
Replicating DNA of *Drosophila melanogaster* showing many replicating eyes. The molecular weight of the segment shown is roughly 20×10^6. (Courtesy of David Hogness.)

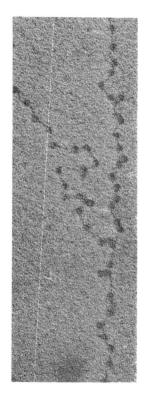

Figure 7-27
A replicating fork showing nucleosomes on both branches. The diameter of each particle is about 110 Å. (Courtesy of Harold Weintraub.)

merization step (nucleotide addition) and a dissociation step (strand separation). The replication of chromatin, which is the form that DNA has in eukaryotes, proceeds through an additional dissociation step—namely, dissociation of DNA and histone octamers—and a histone-DNA reassociation step (see Chapter 5 for a discussion of the structure of the octamers contained in nucleosomes). In chromatin, DNA is wrapped around a histone octamer to form a nucleosome and if the DNA were never unwrapped from the histone spool, severe geometric problems would arise at the growing fork. Moreover, after DNA dissociates from the histones, newly formed DNA must rejoin with the nucleosomal octamers so that each daughter molecule will be organized into nucleosomes, just as the parent was.

Examination of replication forks in DNA that has not been deproteinized during isolation indicates that nucleosomes form very rapidly after replication. For example, Figure 7-27 shows that all portions of a replication eye have the beadlike appearance characteristic of nucleosomes.

The synthesis of histones occurs simultaneously with DNA replication—that is, histones are made in the cell as they are needed, so that the cell does not contain an appreciable amount of unassociated histone molecules. In light of this, we would like to know whether newly synthesized histones mix with parent histones in the octamers associated with daughter DNA molecules.

Unfortunately, even after many years of investigation, the answer is not clear. In addition, the answer to the other important question about the behavior of histones during replication is unclear—how are the histone octamers resident on the parent duplex distributed on the daughter molecules? (Do they all go to only one daughter molecule, or are they dispersed between both daughter molecules?) Sufficient experimental evidence exists to support either claim. Very recently, histone assembly has been coupled to replication *in vitro*. Studies with this type of system should provide definitive answers.

ENZYMOLOGY OF EUKARYOTIC REPLICATION

In prokaryotes, studies *in vitro* on the replication of infecting phages were the key to unlocking the secrets of the enzymological mechanisms and the identification and purification of the cellular replication proteins. Recently, a similar approach has been used to identify human replication proteins. In particular, these studies have exploited the ability of both adenovirus and the previously mentioned SV40 virus to replicate in human cell cultures.

Adenovirus is a long (about 30,000 nucleotides) linear virus that replicates from the ends. The obvious priming problem (if it replicates from the very last nucleotide, how is a 3'-OH primer end provided?) is solved because the primer is actually a virally-encoded protein. The first nucleotide inserted is covalently attached to the primer protein. This protein (called the terminal protein) remains attached to the newly synthesized viral DNA and is encapsulated into the viral particle.

Studies on SV40 replication have revealed two different polymerases, each apparently involved in the synthesis of only one strand. Polymerase α synthesizes the lagging strand, while polymerase δ synthesizes the leading strand. Most eukaryotic α polymerases also have an associated primase activity—thus its involvement in lagging-strand synthesis, which calls for the repeated generation of primers, seems appropriate. It is interesting to note the functional similarities between development of the eukaryotic and prokaryotic replication forks. One can think of the combination of polymerases α and δ, along with several other proteins that bind to these polymerases, as being equivalent to the prokaryotic polymerase III holoenzyme. In the next few years, our knowledge of the enzymology of the eukaryotic replication should begin to rival that of the prokaryotic system.

Summary

DNA replication is accomplished by an enzymatic polymerization reaction in which a DNA strand is used as a template to synthesize a copy with a base sequence that is complementary to the template. Replication is semiconservative; each parent single strand is present in one of the double-stranded progeny molecules. The growth of a DNA strand is catalyzed by a DNA polymerase enzyme. Strand growth always moves in the 5' to 3' direction. Since double-stranded DNA is antiparallel, only one strand (the leading strand) grows in the direction of movement of the replication fork. The second daughter strand (the lagging strand) is synthesized in the opposite direction as short, precursor fragments that are subsequently joined together. DNA polymerase cannot initiate synthesis, so a primer is always needed. In most replication systems the primer is a short RNA oligonucleotide synthesized either by RNA polymerase or by DNA primase.

In *E. coli* two enzymes, polymerase I and polymerase III, carry out nucleotide addition. Pol III is responsible for overall chain growth; pol I is used in the completion of precursor fragments and also in repair processes. The primer is removed at later stages of replication by the 5'-3' exonuclease activity of DNA polymerase I. Pol I and pol III both possess a 3'-5' exonuclease activity that corrects incorporation errors.

Replication of both linear and circular DNA molecules is usually bidirectional; the replication loops enlarge by movement of replication forks at both ends of the loop. A DNA molecule of a prokaryote is usually initiated at a single site, and has a single replication loop; in contrast, eukaryotic DNA molecules have multiple initiation sites, and hence many loops.

Drill Questions

1. Is DNA replication conservative or semiconservative?

2. Fill in the blank. A parent strand serves as a ____ for synthesis of a daughter strand.

3. In semiconservative replication, what fraction of the DNA consists of one original parent strand and one daughter strand after 1, 2, and 3 rounds of replication?

4. In which mode of replication does a parent circular DNA molecule yield two daughter circles?

5. In which mode of replication does a parent circle generate a circle with a linear branch?

6. What kind of supercoiling is produced by replicative movement of the growing fork in a circle?

7. What kind of supercoiling is produced by DNA gyrase?

8. Name three enzymatic activities of DNA polymerase I.

9. DNA polymerization occurs by addition of a deoxynucleotide to which chemical group?

10. What two reactions are coupled when nick translation occurs?

Problems

1. Will a ^{15}N-labeled circle replicating in ^{14}N medium using the rolling circle mode ever achieve the density of ^{14}N^{14}N DNA?

2. What are the roles of the various exonuclease activities of the DNA polymerases in DNA replication?

3. How do pol I and pol III differ with respect to their ability to unwind the parent DNA in a replication fork?

4. What is the chemical difference between the groups joined by a DNA polymerase and by DNA ligase?

5. How do organisms allow DNA polymerases to move in the same direction along a template strand, even though double-stranded DNA is antiparallel?

6. What must be done to two precursor fragments before they can be joined together?

Conceptual Questions

1. Why isn't there a single mode of replication common to all organisms?

2. What specific constraint to DNA replication occurs at the ends of linear DNA molecules? How does replication occur at the ends of linear DNA molecules?

3. Why has DNA replication been called "the most complex process in a cell?"

Philip C. Hanawalt
Professor of Biological Sciences
Department of Biological Sciences

Birthday: **25 August 1931**

Birth Place: **Akron, Ohio**

Undergraduate Degree: **Oberlin College**
Major: Physics 1954

Graduate Degree: **Yale University, M.S.**
Physics 1955; Ph.D. 1959, Biophysics

Postdoctoral Training: **University of**
Copenhagen, Denmark, 1958–1960;
California Institute of Technology,
Pasadena, 1960–1961

Present Position: **Stanford University,**
Biological Sciences Department

Address: **Stanford, California**

OUR LONG TERM goal is to understand how the processing of damaged DNA in mammalian cells relates to mutagenesis, carcinogenesis, aging, and human disease. Cellular protooncogenes have been found to be mutated, translocated, or amplified in a variety of transformed cells. In several inherited human diseases the effected individuals are cancer-prone, and their cells are abnormally sensitive to DNA damaging agents, apparently as a consequence of DNA repair deficiencies. We have developed many of the biochemical techniques now used routinely to study DNA repair. Recently we pioneered development of methods to analyze the fine structure of DNA damage processing in mammalian chromatin. We have shown that UV-induced pyrimidine dimers are efficiently removed from the active dihydrofolate reductase gene, but not from silent upstream sequences in hamster cells. We have also determined the comparative *rates* of repair in a number of different genes and have shown that repair in an active gene in human cells is much faster than in the nontranscribed alpha DNA or the overall genome. We have demonstrated differences in the repair of pyrimidine dimers in transcriptionally active and inactive c-*abl* and c-*mos* protooncogenes in mouse cells. Recently we discovered that selective repair in active genes is due to preferential repair in the transcribed DNA strand. Our current research focus is the elucidation of the features of DNA sequence, chromosome structure, or expression that determine if a particular region of the genome is accessible to repair enzymes, if preferential repair differs with the type of DNA damage present, and if damage tolerance processes differ among DNA sequences of different functions.

What advice can you offer to undergraduates who want to become scholars of molecular biology?

Obtain a strong grounding in basic physics and chemistry, then turn to genetics and biochemistry. Otherwise, start to work part-time in a molecular biology research laboratory at your earliest opportunity. That will be the most important and rewarding experience.

8

DNA Repair

There is no single molecule whose integrity is as vital to the cell as DNA. Indeed, survival of the species depends upon maintaining the nucleotide sequence intact. The genetic material is, however, subject to constant challenge. During DNA replication, RNA transcription, and even while in an inactive, resting state, damage can occur. Copying errors, breaks, and damage to nucleotide bases if not corrected lead to permanent change—mutation. Most mutations are potentially dangerous, since they lead to either blocks in DNA replication or the production of defective proteins. Thus, in the course of hundreds of millions of years there have evolved efficient systems for correcting occasional replication errors and for eliminating damage to DNA caused by environmental agents and intracellular chemicals. A few of these systems are examined in this chapter. We begin by describing the principal kinds of damage that may occur in a DNA molecule. Those damage control systems recognize errors in base pairing, breaks, and altered bases, and promptly repair them.

Needless to say, DNA repair systems are very complex. In fact, they are probably just as complicated as DNA replication and RNA transcription. In this chapter we shall examine a few of the repair systems. We will learn how most of the repair systems "read" the undamaged or error-free complementary strand of the double helix. It is usually the bar sequence from that other strand that is employed by the repair enzymes to rectify errors.

In this chapter you will learn:

1. How the mismatch repair system corrects mutated base pairs.

2. Several mechanisms for thymine dimer repair.

Alterations of DNA Molecules

There are a number of distinct mechanisms for altering the structure of DNA. These include base substitutions during replication, base changes resulting from the inherent chemical instability of the bases or of the N-glycosylic bond, and alterations resulting from the action of other chemicals and environmental agents. These mechanisms are responsible for the occurrence of the following types of defects:

1. *An incorrect base in one strand that cannot form hydrogen-bonds with the corresponding base in the other strand.* This defect can result from a replication error that by chance is not corrected by the editing function of the DNA polymerase, or by spontaneous loss of an amino group (**deamination**), converting cytosine to uracil or adenine to hypoxanthine.

2. *Missing bases.* The N-glycosylic bond of a purine nucleotide is spontaneously broken at physiological temperatures, though at a very low rate. This process is called **depurination** because the purine is lost from the DNA leaving a non–informational site, but without breaking the backbone. The rate of spontaneous depurination is about 10^4 purines per day in a mammalian cell.

3. *Altered bases.* Bases can be changed into strikingly different compounds by a variety of chemical and physical agents. For instance, ionizing radiation (such as the β particles emitted by naturally occurring

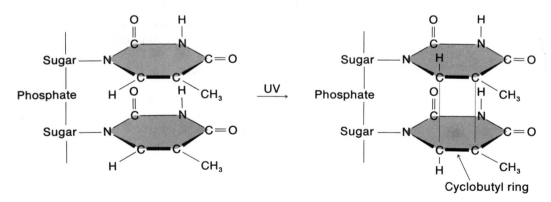

Figure 8-1

Structure of a cyclobutylthymine dimer. Following ultraviolet (UV) irradiation, adjacent thymine residues in a DNA strand are joined by formation of the bond shown in red. Although not drawn to scale, these bonds are considerably shorter than the spacing between the planes of adjacent thymines, so that the double-stranded structure becomes distorted. The shape of the thymine ring also changes as the C=C double bond of each thymine is converted to a C—C single bond in each cyclobutyl ring.

radioisotopes or x-rays) can break purine and pyrimidine rings. Ultraviolet radiation causes the formation of several types of dimers between adjacent pyrimidines in a strand of DNA. The most prominent of these dimers is the thymine dimer shown in Figure 8-1. The significant effects of thymine dimers are the following: the DNA helix becomes distorted as the thymines are pulled toward one another (Figure 8-2); and as a result of the distortion, hydrogen–bonding to adenines in the opposing strand, although possible (because the hydrogen–bonding groups are still present), is significantly weakened. Some chemical carcinogens (such as those found in cigarette smoke) form covalent bonds with bases, also resulting in substantial distortion of the DNA helix. Others simply add a methyl group at an inappropriate position, such as the O^6 position in guanine. Although such an addition does not interfere with replication, during replication it causes the altered guanine to behave as though it were an adenine.

4. *Single-strand breaks.* A variety of agents can break phosphodiester bonds. Among the more common chemicals are peroxides, and metal ions such as Fe^{2+} and Cu^{2+}. Ionizing radiation produces strand breaks directly and indirectly by production of highly reactive free radicals in water surrounding the DNA.

5. *Double-strand breaks.* If a DNA molecule receives a sufficiently large number of randomly located single-strand breaks, two breaks may be situated opposite one another, resulting in breakage of the double helix.

6. *Cross-linking.* Some antibiotics (for example, mitomycin C) and some reagents (the nitrite ion) can form covalent linkages between a base in one strand and another base in the complementary DNA strand. This prevents strand separation during DNA replication, and also causes a local distortion of the helix. DNA crosslinking agents such as cis-platinum and psoralen (activated by ultraviolet light) have been used effectively in chemotherapy for cancer and hyperproliferative skin diseases like psoriasis.

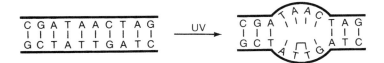

Figure 8-2
Distortion of the DNA helix caused by two thymines joined in a dimer. The dimer is shown as two joined lines.

Repair of Incorrect Bases

As has already been described, DNA polymerases occasionally catalyze incorporation of an incorrect base which cannot form a hydrogen-bond with the template base in the parent strand; such errors are usually corrected by the editing function of these enzymes. However, the integrity of the base sequence of DNA is so important that a second system exists for correcting the occasional error missed by the editing function. This correction system is called **mismatch repair.** In mismatch repair a pair of non-hydrogen-bonded bases is recognized as incorrect, and a polynucleotide segment is excised from one strand, thereby removing one member of the unmatched pair. The resulting gap is filled in by DNA polymerase, using this "second chance to get it right"; then the final seal is made by DNA ligase.

If it is to eliminate errors, the mismatch repair system must be able to distinguish the correct base in the parental strand from the incorrect base in the daughter strand. In *E. coli* the critical information is provided by the location of particular adenines in the sequence G-A-T-C. These bases carry methyl groups not found elsewhere in the DNA. Methylation of the adenines in this sequence is associated with replication, occurring in the *daughter* DNA strands. However, it does not take place in the replication fork, but is somewhat delayed. The effect is that while parental strands are fully methylated, the newly-synthesized daughter strands near the replication fork are not yet methylated. The mismatch repair system recognizes the degree of methylation of each strand, and when a mismatch is found, it preferentially excises nucleotides from the undermethylated strand—that is, from the daughter strand. Thus, the parent strand is always the template, enabling the repair system to correct misincorporation errors. Many types of cells use schemes other than adenine methylation to designate parent strands for mismatch repair.

When cytosine loses an amino group by *deamination*, it becomes uracil. After one round of replication this would lead to replacement of a G · C pair with an A · U pair, which would then become an A · T pair after another round of replication. Since this would be mutagenic, cells have evolved a mechanism for replacing the unwanted U by a C. The first step of this repair cycle (Figure 8-3) is removal of the uracil by the enzyme **uracil N-glycosylase.** This enzyme cleaves the *N*-glycosylic bond and leaves the deoxyribose in the backbone. A second enzyme, **AP endonuclease** (acting on apurinic or "baseless" sites in general), makes a single cut, freeing one end of the deoxyribose. This is followed by removal of the deoxyribose and several adjacent nucleotides (possibly through exonuclease activity of pol I), after which pol I fills the gap with the correct nucleotides. (This sequence, endonuclease–exonuclease–polymerase, an example of a general repair mechanism called excision-repair, is described below). A specific glycosylase is

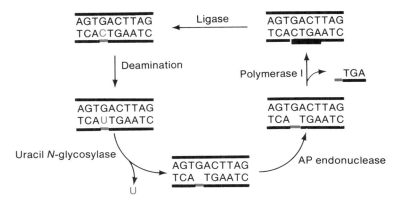

Figure 8-3
Scheme for repair of cytosine deamination. The same mechanism could remove a uracil that is accidentally incorporated.

also available to remove hypoxanthine, the consequence of adenine deamination.

The latter part of the repair sequence, namely that beginning with AP endonuclease (shown in Figure 8-3), is also used in the repair of missing bases arising by depurination.

A curious situation arises when the base 5-methyl cytosine (formed in many eukaryotes) spontaneously deaminates. It becomes the normal base thymine, and will mispair with guanine. How would a mismatch repair system determine which base is the correct one? This particular mismatch is recognized by an enzyme that *always* removes the mispaired thymine rather than the guanine, so that the correct base-pairing sequence is restored. The new cytosine can be re-methylated at a later time.

Repair by Direct Reversal

The simplest mechanism for repair of altered bases is to restore them to normal without major surgery on the DNA molecule. In the case of the pyrimidine dimers formed in DNA by ultraviolet light this reversal can be accomplished very simply by an enzyme that recognizes and binds to these dimers. That enzyme, called **photolyase,** is activated by visible light and cleaves the dimers to yield the intact pyrimidines. Photolyase is found in many types of cells, but it operates only on pyrimidine dimers and only when activated by light (300–600 nm).

Another example of direct reversal is the methyltransferase that recognizes O^6 methyl guanine in DNA and removes the offending methyl group by attaching it instead to one of its own cysteines. The exchange reaction restores the normal guanine in the DNA, but inactivates the enzyme. The importance of this particular repair scheme to the cell is underscored by the fact that an entire protein molecule is expended for each O^6 methyl guanine repaired.

Excision-Repair

excision-repair

The most ubiquitous repair scheme, and one that can deal with a large variety of structural defects in DNA, is called excision-repair. This multistep enzymatic process is best understood in *E. coli* (in which it was originally discovered); it is a mechanism by which pyrimidine dimers can be repaired in the DNA without photolyase. The essential four steps are shown in Figure 8-4, although the sequence of the **excision** and **repair** synthesis steps may be reversed in some cellular systems. The sequence is referred to colloquially as "cut and patch." In the first step, **incision,** a repair endonuclease recognizes the distortion produced by a thymine dimer and makes a cut in the sugar-phosphate backbone near the dimer. In *E. coli* the product of three genes (*uvrA*, *uvrB*, and *uvrC*)

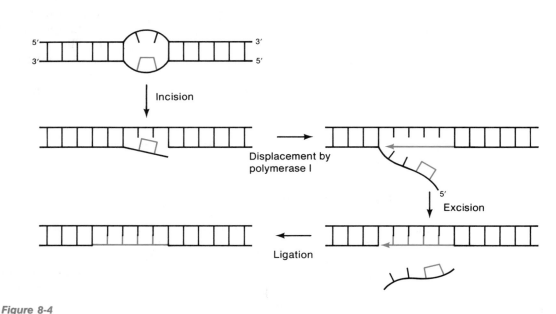

Figure 8-4
Scheme for excision-repair of a thymine dimer by the "cut and patch" mechanism. The thymine dimer and the repair patch are both shown in red.

act in concert to produce two incisions 12–13 nucleotides apart that bracket the dimer or other lesion. However, one incision is adequate to initiate the repair sequence. At the incision site illustrated in Figure 8-4 there is a 5'-P group on the side of the cut containing the dimer and a 3'-OH group on the other side. The 3'-OH group is recognized by polymerase I, which can then synthesize a new strand while displacing the DNA segment carrying the thymine dimer. This segment is then excised by the 5' → 3' exonuclease activity of pol I. The final step is joining the newly-synthesized segment to the original strand by DNA ligase. The excision repair system has been shown to preferentially repair pyrimidine dimers that are in certain essential regions of the genome, notably within expressed genes. In genes that are being actively transcribed, the repair is somehow targeted selectively to the strand that is being used as a template for messenger RNA synthesis. This special treatment makes sense because a pyrimidine dimer poses an absolute block to the process of transcription, and could kill the cell quite directly if it prevented expression of an essential gene.

In mammalian cells, chromatin structure is much more complex than in simple *E. coli* bacterium; and the process of excision repair requires many more enzymes. In fact, at least 9 or 10 genes are involved in the incision step. Xeroderma pigmentosum, a human genetic disease in which the victims are extremely sensitive to sunlight and develop skin cancer after very short exposures to the sun, is known to involve a deficiency in one or another of the genes controlling the incision step of excision repair. Cells from xeroderma pigmentosum patients are sensitive to ultraviolet light and have a reduced ability to remove thymine dimers from their DNA. Another known genetic disease, Cockayne's syndrome, also is characterized by a sensitivity to sunlight, but without the tendency to develop cancer. Interestingly, the defect in Cockayne's syndrome affects the preferential repair of expressed genes discussed above.

RECOMBINATIONAL REPAIR AND TRANSLESION SYNTHESIS

The excision repair systems just described are responsible for the removal of many of thymine dimers and other lesions. However, before sufficient time has elapsed for their repair, many dimers have interfered with various cellular processes. The deleterious effects of some remaining dimers are eliminated by recombinational repair, which is carried out by the system responsible for genetic exchange.

In order to discuss the mechanism of recombinational repair, it is necessary to know the effect of a thymine dimer on DNA replication. When polymerase III reaches a thymine dimer, the replication fork fails to advance. A thymine dimer is still capable of forming hydrogen-bonds with two adenines because the chemical change in dimerization does not alter the groups that engage in hydrogen-bonding. However, the dimer introduces a distortion into the helix and when an adenine is added to the growing chain, polymerase III reacts to the distorted region as if a mispaired base had been added; the editing function then removes the adenine. The cycle begins again—an adenine is added and then it is removed; the net result is that the polymerase is stalled at the site of the dimer. (The same effect would occur if, instead of a dimer, radiation or chemical damage resulted in formation of a base to which no nucleoside triphosphate could base-pair.) Evidence that such a phenomenon occurs after ultraviolet irradiation is the existence of an ultraviolet-light-induced idling process—that is, rapid cleavage of deoxynucleoside triphosphates to monophosphates without any net DNA synthesis (i.e., without advance of the replication fork). A cell in which DNA synthesis is permanently stalled cannot complete a round of replication, and does not divide.

There are basically two different ways in which DNA synthesis can get going again—**postdimer initiation** and **transdimer synthesis.**

One way a cell could deal with a thymine dimer block is to pass it by and initiate chain growth beyond the block perhaps at the starting point for the next Okazaki fragment (Figure 8-5). The result of this process is that the daughter strands have large gaps, one for each unexcised thymine dimer. There is no way to produce viable daughter cells by continued replication alone, because the strands having the thymine dimer will continue to turn out gapped daughter strands, and the first set of gapped daughter strands would be fragmented when the growing fork enters a gap. However, by a recombination mechanism called **sister-strand exchange** proper double-stranded molecules can be made.

The essential idea in sister-strand exchange is that a single-stranded segment free of any defects is excised from a "good" strand on the homologous DNA segment near the replication fork and somehow inserted into the gap created by excision of a thymine dimer (Figure 8-6). The combined action of polymerase I and DNA ligase joins this inserted piece to adjacent regions, thus filling in the gap. The gap formed in the donor molecule by excision is also filled in completely by polymerase I and ligase. If this exchange and gap filling are done for each thymine dimer, two complete daughter single-strands can be formed,

Figure 8-5
Blockage of replication by thymine dimers (represented by joined lines) followed by re-starts several bases beyond the dimer. The black region is a segment of ultraviolet-light-irradiated parent DNA. The red region represents synthesis of a daughter molecule from right to left. The daughter strand contains gaps.

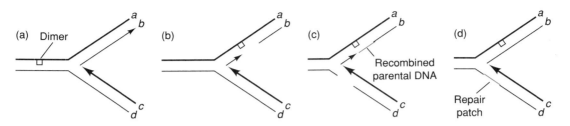

Figure 8-6
Recombinational bypass. (a) A molecule containing a thymine dimer (red box) in strand *a* is being replicated.
(b) Replication is blocked at the site of the dimer, leaving a gap because an Okazaki fragment can not be completed.
(c) That gap is filled by a sister-strand recombination event in which a segment of the parent strand *d* is utilized. That
now leaves a gap in strand *d*. (d) The daughter strand *c* can serve as a template for repair DNA synthesis to fill the
gap in strand *d*. DNA synthesized after irradiation is shown in red. Heavy and thin lines are used only to identify
strands of the same polarity.

and each can serve in the *next* round of replication as a template for
synthesis of normal DNA molecules. Note that the system fails if two
dimers in opposite strands are very near one another because then no
undamaged sister-strand segments are available to be excised. The molecular details of recombinational repair are not known with precision,
so that the model shown in Figure 8-6 must be considered to be a
working hypothesis that is at present consistent with the facts.

Recombinational repair is an important mechanism because it eliminates the necessity for delaying replication for the many hours that
would be needed for excision repair to remove all thymine dimers. It
may also be the case that some kinds of damage cannot be eliminated
by excision repair—for example, alterations that do not cause helix
distortion but do stop DNA synthesis.

Since recombinational repair occurs after DNA replication, in contrast with excision repair, it is often called **postreplicational repair.**
However, it should be noted that what is repaired is a daughter strand
gap, rather than a thymine dimer.

In spite of the evidence that thymine dimers and other helix-dis-
torting lesions block the progression of DNA polymerase it is clear that
most types of cells, including mammalian cells, can tolerate large numbers of persisting lesions without resorting to recombinational strand
exchange. Somehow replication is able to overcome the lesions, albeit
with an expected high error rate at those non-coding sites.

Inducible Responses

While most of the DNA damage responses are constitutive (i.e., active
all the time) a few are activated in response to some signal such as a
blocked replication fork. Most notable is the **SOS system,** characterized

in *E. coli* as a complex regulatory scheme in which the product of the *recA* gene governs the expression of a dozen other genes when it is activated by binding to single-stranded DNA, at blocked replication forks. Some of these genes, which include *uvrA* and *urvB*, are involved in excision repair, while others like *umuC* and *umuD* seem to be required to help replication bypass the offending lesion, as illustrated in Figure 8-7. Since most lesions provide misinformation (or no information) to the replicative polymerase this process has been called **error-prone replication,** and results in a higher than normal level of mutagenesis. The tradeoff is that at least the cell is able to replicate its DNA, and that gives it a better chance for survival in an otherwise disastrous situation. The enhanced mutagenesis may be seen as a benefit in that it may result in progeny that through mutation are better adapted to live in the noxious environment that led to induction of the SOS system. A general observation is that cells can control and alter the level of mutagenesis that operates during their proliferation.

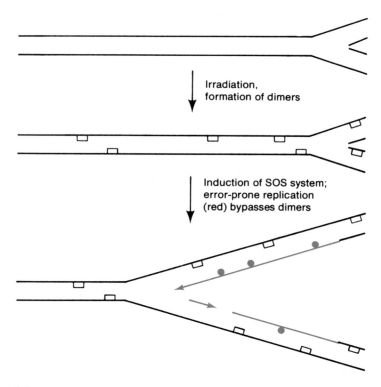

Figure 8-7
SOS repair. A DNA molecule in an early stage of replication is irradiated with ultraviolet light and thymine dimers are formed. The SOS system is induced and all subsequent replication (shown in red) has a higher than usual number of misincorporated bases (red dots). Some of these bases can be removed by the mismatch repair system.

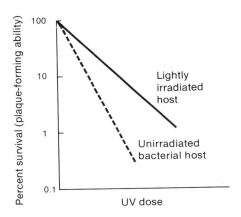

Figure 8-8
UV reactivation of ultraviolet-light-irradiated phage λ. The dashed line shows the survival curve (for plaque-forming ability) obtained when λ phage irradiated with various doses of ultraviolet light are plated on unirradiated bacteria. The solid line represents survival of plaque-forming ability, when ultraviolet-light-irradiated λ are plated on lightly irradiated bacteria.

The first evidence for the SOS system was obtained 25 years before the mechanism began to be understood. The experiment, illustrated in Figure 8-8, involved an analysis of the survival of UV-irradiated bacteriophage λ plated on *E. coli*. The survival of the UV-damaged phage was markedly enhanced if the bacterial host had also been irradiated with a low UV dose. However, in that case a much higher percentage of mutants were obtained among the surviving phages. This phenomenon (termed **UV reactivation**) is now understood in terms of the SOS system. The low UV dose to the bacteria activates the SOS error-prone replication of the damaged phage.

There are examples of inducible cellular responses to other types of environmental stress, such as heat shock, oxidative DNA damage, or agents that produce O^6 methyl guanine. In the latter case, there is an *adaptive response* in *E. coli* that increases the level of the methyltransferase enzyme 100-fold when the cells are "conditioned" by growth in the presence of agents which produce this lesion.

Summary

There are several ways in which a DNA molecule can be altered. One way is to have an incorrect base in one strand. This usually results from uncorrected errors in the replication process. It is possible for bases to be missing from the DNA molecule (so that the molecule resembles a ladder missing half of a rung). Physical and chemical agents, including ultraviolet radiation (UV), can alter existing bases. UV radiation can cause two adjacent pyrimidines to become covalently linked; these dimers distort the helix and affect the replication process. Another type of damage involves one or more breaks in the sugar-phosphate backbone of the helix. A final alteration discussed in this chapter is the covalent linking of one strand of a double-stranded molecule with its complement. Processes exist to repair these types of alterations. Mismatch repair is used to replace an incorrect base. This repair system recognizes a pair of non-hydrogen-bonded bases, excises the one that is incorrect, and replaces it with the proper base. A pathway involving the enzyme uracil N-glycosylase is responsible for repair of cytosine deamination. Four mechanisms exist for the repair of thymine dimers. Photolyase is an enzyme that cleaves the cyclobutyl ring of the dimer when activated by visible light. The excision repair system makes a cut in the backbone on either side of the dimer, allowing the subsequent displacement and subsequent replacement of the strand containing the dimer. Recombinational repair involves postdimer re-initiation of DNA synthesis and sister-strand exchange, and prevents long delays of replication. The SOS "repair" system also prevents delay of replication; this system allows for transdimer synthesis—at the expense of a higher rate of base misincorporation by the replication enzymes.

Drill Questions

1. Two DNA alterations occur so frequently that they are considered to be the weak points of a DNA molecule. What are these two changes?

2. Which repair system involves an enzyme that catalyzes the cleavage of:
   a. N-glycosidic bonds?
   b. carbon-carbon bonds?
   c. phosphodiester bonds?

3. What are two features of the SOS repair system that distinguish it from other repair systems?

4. Uvr$^+$ bacteria possess the excision-repair system. The ability of ultraviolet-irradiated T4 phage to form plaques is the same on both Uvr$^+$ and Uvr$^-$ bacteria. How might you explain this fact?

5. Which of the enzymes listed below is involved in repair of *both* thymine dimers and deaminate cytosine in *E. coli*?
   a. DNA polymerase III
   b. Photolyase
   c. Uracil N-glycosylase
   d. DNA polymerase I
   e. AP endonuclease

Problems

1. The cell has a pathway for repair of deaminated cytosine in DNA, but not for RNA. What are some possible explanations for this?

2. A bacterial repair system called X removes thymine dimers. You have in your bacterial collection the wild-type (X$^+$) and an X$^-$ mutant. Phage, when ultraviolet-irradiated and then plated, gives a larger number of plaques on X$^+$ than on X$^-$ bacteria. It has been proposed on the basis of survival curve analysis that the X enzyme is inducible. To test this proposal, UV-irradiated phage are adsorbed to both X$^+$ and X$^-$ bacteria in the presence of the antibiotic chloramphenicol (which inhibits protein synthesis). No thymine dimers are removed in the X$^+$ cell and 50% are removed

in the X^- cell. The same results were obtained in the absence of chloramphenicol.

a. Is X an inducible system?

b. Suppose 5% of the thymine dimers had been removed in the presence of chloramphenicol and 50% in its absence; how would your conclusion be changed?

3. An unexcised thymine dimer produces a partial block to DNA replication. It is observed that if there are unexcised thymine dimers in parent strands, the daughter strands contain large gaps which are frequently several thousand nucleotides long. Would you expect to find gaps in both daughter strands if there were only a single thymine dimer in the parent molecule? If not, in which strand? Explain.

Conceptual Questions

1. Would different types of organisms, e.g., bacteria, fruit flies, frogs, and humans be expected to sustain different types of DNA damage?

2. Mitochondria and chloroplasts also contain DNA. Do you suppose they would employ the same or different repair mechanisms as chromosomal DNA?

3. How might it be established whether DNA damage-repair mechanisms evolved first to repair replication errors, and only secondarily, to repair damage from environmental influences; or vice versa?

John Richardson
Professor of Chemistry
Department of Chemistry

Birthday: **27 June 1938**

Birth Place: **Pittsfield, Massachusetts**

Undergraduate Degree: **Amherst College**
Major: Biophysics 1960

Graduate Degree: **Harvard University,**
Ph.D. 1966

Postdoctoral Training: **Institut de Biologie**
physico-chimique, Paris, 1965–1967

Present Position: **Indiana University,**
Chemistry Department

Address: **Bloomington, Indiana**

MY RESEARCH IS on the mechanism of transcription of RNA from DNA templates and on the regulation of RNA metabolism in bacteria and higher organisms.

A major aim of my research is to determine how RNA synthesis is stopped at the ends of genes and of gene groups (operons) in bacteria. It is known that termination at the ends of many genes or operons requires the action of a protein factor known as rho. My work has shown that rho has to interact with the nascent RNA transcript and that it uses energy from the hydrolysis of ATP to break up the transcription complex of nascent RNA, RNA polymerase, and DNA.

In my group we are currently interested in characterizing the structural organization of rho protein, finding out what sequences or structures on RNA molecules are recognized by rho factor, and establishing how the energy of ATP hydrolysis affects the interactions between rho and the various parts of the transcription complex. We are also investigating how NusA protein, a transcriptional elongation factor, interacts with RNA polymerase and the nascent RNA.

We solve many of our problems by a combination of biochemical and genetic approaches. To identify sequences on an RNA molecule that are recognized by rho protein, for instance, our genetic approach involves making alterations in the DNA coding for the RNA and analyzing how these mutations affect the efficiency of termination at the end of the gene. Our biochemical approach involves the use of photochemical RNA-protein cross-linking techniques and selective modifications of the RNA to identify the nucleotides that make direct contact with rho protein.

What were the major influences or driving forces that led you to a career in molecular biology?

The major deciding factor that led me to a career in molecular biology was reading the review article, "Genetic Regulatory Mechanisms in the Synthesis of Proteins" by Francois Jacob and Jacques Monod in one of the spring 1961 issues of the *Journal of Molecular Biology*. The other factor was the exciting work being done in Jim Watson's laboratory at Harvard University that spring on the discovery of messenger RNA. Molecular biology was very young and very exciting then and I decided I wanted to be on the bandwagon, so on I jumped!

9

Transcription

Gene expression is accomplished by the transfer of genetic information from DNA to RNA molecules and then from RNA to protein molecules. RNA molecules are synthesized by using the base sequence of one strand of DNA as a template in a polymerization reaction that is catalyzed by enzymes called DNA-dependent RNA polymerases or simply **RNA polymerases.** The process by which RNA molecules are initiated, elongated, and terminated is called **transcription.**

In this chapter we will quickly realize that although complementary base pairing is common to both RNA and DNA synthesis, several features of RNA synthesis differ from DNA synthesis. For example, the RNA product is single stranded, unlike the double helix from which it is transcribed. It is also relatively short in length, since only a small portion of the total DNA is represented in any single RNA product.

Initially, we will learn about transcription in prokaryotes. Then we will review the more complicated eukaryotic gene expression system. Because of their relative simplicity, the pioneering experiments on transcription were carried out in prokaryotes. Indeed, it was a series of elegant genetic experiments with *E. coli* that led to the discovery of what is now known as messenger RNA (mRNA). Through mRNA, the information stored in DNA is converted to amino acid sequences of the ultimate gene product—protein.

Three aspects of transcription will be considered in this chapter: the enzymology of RNA synthesis, the signals that determine at what points on a DNA molecule transcription starts and stops, and the types of transcription products and how they are converted to the RNA molecules needed by the cell.

In this chapter you will learn:

1. The sequence of events that take place in transcription of DNA to RNA, and how those events differ in prokaryotic and eukaryotic systems.

2. About the structures and functions of different classes of RNA.

Enzymatic Synthesis of RNA

In this section we describe the basic features of the polymerization of RNA, the identity of the precursors, the nature of the template, the properties of the polymerizing enzyme, and the mechanisms of initiation, elongation, and termination of synthesis of an RNA chain.

The essential chemical characteristics of the synthesis of RNA are the following:

1. The precursors in the synthesis of RNA are the four ribonucleoside 5'-triphosphates (rNTP) ATP, GTP, CTP, and UTP. On the ribose portion of each NTP there are two OH groups—one each on the 2'- and 3'-carbon atoms (Figure 2-5).

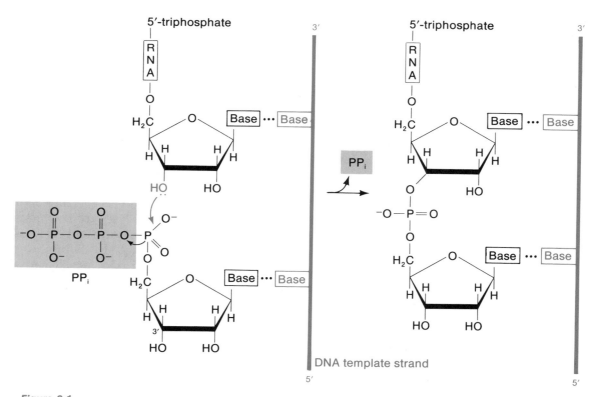

Figure 9-1
Mechanism of the chain-elongation reaction catalyzed by RNA polymerase. The red arrow joins the reacting groups. The pyrophosphate group (shaded in red) and the red hydrogen atom do not appear in the RNA strand. The DNA template and the RNA strands are antiparallel, as in double-stranded DNA.

2. In the polymerization reaction the 3'-OH terminus of the RNA being synthesized reacts with the 5'-triphosphate of a precursor rNTP; a pyrophosphate is released, and a phosphodiester bond results (Figure 9-1). This is the same reaction that occurs in the synthesis of DNA.

3. The sequence of bases in an RNA molecule is determined by the base sequence of the DNA. Each base added to the growing end of the RNA chain is chosen by its ability to base-pair with the DNA strand used as a template; thus, the bases C, T, G, and A in a DNA strand cause G, A, C, and U, respectively, to appear in the newly-synthesized RNA molecule.

4. The DNA molecule being transcribed is double-stranded, yet in any particular region only one strand serves as a template. This comes about because the two DNA strands are separated temporarily to allow one to be used as a template. The meaning of this statement is shown in Figure 9-2.

5. The RNA chain grows in the 5' → 3' direction: that is, nucleotides are added only to the 3'-OH end of the growing chain—this is the same as the direction of chain growth in DNA synthesis. Furthermore, the RNA strand and the DNA template strand are antiparallel to one another.

6. RNA polymerases, in contrast to DNA polymerases, are able to initiate chain growth; that is, no primer is needed.

7. Only ribonucleoside 5'-triphosphates participate in RNA synthesis, and the first base to be laid down in the initiation event retains

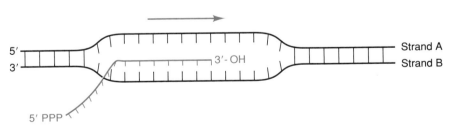

Figure 9-2
An RNA strand (shown in red) is copied only from strand B of a segment of a DNA molecule. No RNA is copied from strand A in that region of the DNA molecule. However, elsewhere, for example in a different gene, strand A might be copied; in that case, strand B would not be copied in that region of the DNA. The RNA molecule is antiparallel to the DNA strand being copied and is terminated by a 5'-*tri*phosphate at the nongrowing end. The red arrow shows the direction of RNA chain growth.

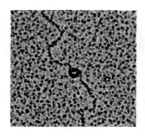

Figure 9-3
E. coli RNA polymerase
molecule bound to DNA.
(× 160,000). (Courtesy of
Robley Williams.)

its triphosphate. Its 3'-OH group is the point of attachment for the
subsequent nucleotide. Thus, the 5' end of a growing RNA molecule
terminates with a triphosphate (Figure 9-2).

The overall polymerization reaction may be written as

$$nNTP + XTP \xrightarrow[Mg^{2+}]{DNA, RNA\text{-}P} XTP\text{---}(NMP)_n + nPP_i$$

in which XTP represents the first nucleotide at the 5' terminus of the
RNA chain, NMP is a mononucleotide in the RNA chain, RNA-P is
RNA polymerase, and PP_1 is the pyrophosphate released each time a
nucleotide is added to the growing chain. The Mg^{2+} ion is required for
all nucleic acid polymerization reactions.

E. coli RNA polymerase consists of five subunits—two identical α
subunits and one each of types β, β', and σ. The σ subunit dissociates
from the enzyme easily and in fact does so shortly after polymerization
is initiated. The term **holoenzyme** is used to describe the complete
enzyme and **core enzyme** for the enzyme which lacks the σ subunit.
We use the name RNA polymerase when the holoenzyme is meant.
RNA polymerase is one of the largest enzymes known and can easily
be seen by electron microscopy (Figure 9-3).

The synthesis of RNA consists of four discrete stages: (1) binding
of RNA polymerase to a template at a specific site, (2) initiation, (3)
chain elongation, and (4) chain termination and release. A discussion of
these stages follows.

Transcription Signals

promoter

The first step in transcription is binding of RNA polymerase to a DNA
molecule. Binding occurs at particular sites called **promoters,** which
are specific sequences of about 40 base-pairs at which several interactions
occur. The most crucial interactions for positioning of *E. coli* RNA
polymerase at a promoter occur at two short-sequence patches on the
DNA. These patches are located about 10 and 35 base-pairs before the
first base which is copied into RNA. That first base copied is called the
start point of transcription, and is assigned as position $+1$. The two
patches therefore are at positions -10 and -35, with respect to the start
point. The σ subunit of RNA polymerase appears to have two separate
α-helical segments with amino acid residues that can make specific contacts with the exposed portions of the base-pairs in the major groove
of the DNA. The enzyme becomes positioned correctly when these

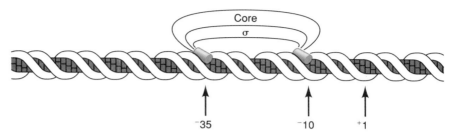

Figure 9-4
A diagram showing RNA polymerase bound to DNA at a promoter. The arrows indicate the position of the transcription start point (⁺1) and the ⁻10 and ⁻35 regions. The sigma (σ) factor is shown in red with two cylinders representing α-helical segments that make contact with the DNA in its major groove at the ⁻10 and ⁻35 regions.

separate α-helical segments contact the base pairs in the two patches simultaneously. Since a DNA molecule makes one helical turn every 10 to 11 base-pairs, these contacts would be nearly on the same side of DNA two turns away, or 78 Å along the DNA. Figure 9-4 shows a diagram of how an extended RNA polymerase could contact those two patches. *E. coli* RNA polymerase is a large protein that covers as much as 70 to 75 base-pairs of DNA from position −55 to position +20 when it is bound to a promoter. Thus, it can easily make the two separate contacts at the same time.

The sequences of many promoters have been identified by researchers. Figure 9-5 shows portions of the sequences along one strand for a few *E. coli* promoters with the start points and the −10 and −35 regions shown in red. The strand that is shown has its 5′ end at the left. One striking feature is that no two sequences are identical or even closely similar, except in the two key locations. The sequences shown in red from −7 to −12 are similar, as are the sequences from about −30 to about −35. The sequences around −10 are considered to be variants of a basic sequence, TATAAT, called the −10 consensus sequence, while the sequence around −35 is a variant of TTGACA, called the −35

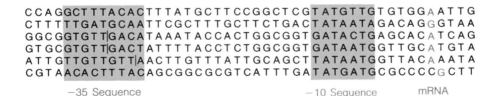

−35 Sequence   −10 Sequence   mRNA start

Figure 9-5
Base sequences in the non-template strand of six different *E. coli* promoters, showing the three important regions.

consensus sequence. A consensus sequence is a pattern of bases from which actual sequences observed in many different systems differ by usually no more than one or two bases. Although the figure only shows the sequences in one strand, remember that the DNA is double-stranded, and that the other complementary strand is part of the recognition site in the -10 and -35 regions.

With some promoters, the sequence at one of the two contact regions differs greatly from the consensus. In this case, RNA polymerase recognizes the sequence very poorly, and the promoter does not function very well. For some genes, that might be fine because the cell may not need to express that gene very well. For other genes, however, there is often an accessory protein that helps RNA polymerase bind to the promoter. This kind of accessory protein is called a **gene activator protein** and is basically a positive effector. For example, the λ p_{re} promoter is active only when the λ cll protein is present. These gene-activator proteins usually bind to specific sequences very near or even within the promoter sequence and they appear to have surfaces to which RNA polymerase can attach, when correctly positioned on the promoter. Thus a gene-activator protein helps the RNA polymerase bind to a promoter with a poor recognition signal. An important effector protein, which will be discussed further in Chapter 14, is the CAP protein; this protein is needed for the activation of many promoters for genes required for sugar metabolism, and regulation of the binding ability of CAP is a major means of regulating the expression of these genes.

After RNA polymerase binds tightly at the promoter region, it manages to unwind a small part of the DNA, from base-pairs $+3$ to -10 to form the **open-promoter complex.** This unwinding is necessary for pairing of the incoming ribonucleoside triphosphates, the building blocks for synthesis of the RNA. The base composition between the start site through the -10 region is generally rich in $A + T$, which renders the DNA especially susceptible to unpairing.

Once an open-promoter complex has formed, RNA polymerase is ready to initiate synthesis. RNA polymerase contains two nucleotide binding sites, called the initiation site and the elongation site, respectively. The initiation site prefers to bind purine nucleoside triphosphates, namely ATP and GTP, and one of these is usually the first nucleotide in the chain. Thus, the first base that is transcribed is usually a thymine or a cytosine. The initiating nucleoside triphosphate binds to the enzyme in the open-promoter complex and forms a hydrogen bond with the complementary DNA base (Figure 9-6). The elongation site is then filled with a nucleoside triphosphate that is selected strictly by its ability to form a hydrogen-bond with the next base in the DNA strand. The two nucleotides are then joined together by the reaction shown in Figure 9-1. During the joining process, RNA polymerase moves along the DNA to the next base-pair. As it moves, it separates the base-pair at position

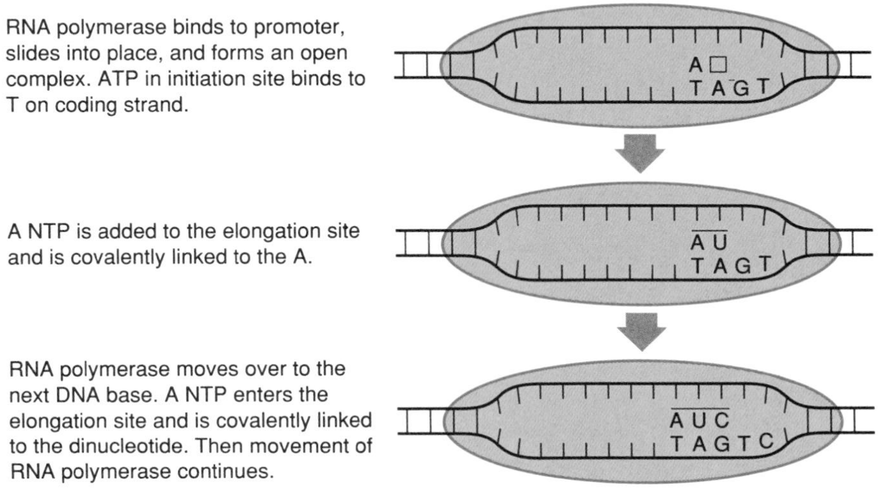

RNA polymerase binds to promoter, slides into place, and forms an open complex. ATP in initiation site binds to T on coding strand.

A NTP is added to the elongation site and is covalently linked to the A.

RNA polymerase moves over to the next DNA base. A NTP enters the elongation site and is covalently linked to the dinucleotide. Then movement of RNA polymerase continues.

Figure 9-6
A scheme for initiation of RNA synthesis.

+4, and allows the base-pair at position -10 to come together again, thus maintaining the same stretch of 13 base-pairs unwound, but now centered one base-pair further along the DNA. The two RNA nucleotides that were linked together stay paired to their complementary DNA bases, but now a new unpaired base (the one at position $^{+}3$) is positioned in the elongation site. Again, the elongation site is filled with a nucleoside triphosphate that is selected by its ability to form a hydrogen-bond with the DNA base at position 3. This nucleotide is then joined to the two that had been linked previously, and RNA polymerase moves along to the next base-pair. Repetition of this cycle of nucleoside triphosphate selection and joining, RNA polymerase motion, DNA base unpairing and DNA base repairing gradually elongates the RNA chain.

Note that the selection is made by pairing of the nucleoside triphosphate with only one of the two strands of the DNA. The DNA strands were purposely unwound to allow this selection to occur. The strand that has bases that pair with the incoming RNA nucleotide is the **template strand.** Since the sequence of the RNA is complementary to the template strand, it will have the same sequence as the other strand of DNA.

After several nucleotides (between four and eight) are added to the growing RNA chain, three important changes occur. First, the nucleotide at the 5′ end of the RNA chain (the first one to be incorporated) becomes unpaired from the base on the template strand to allow that base to pair

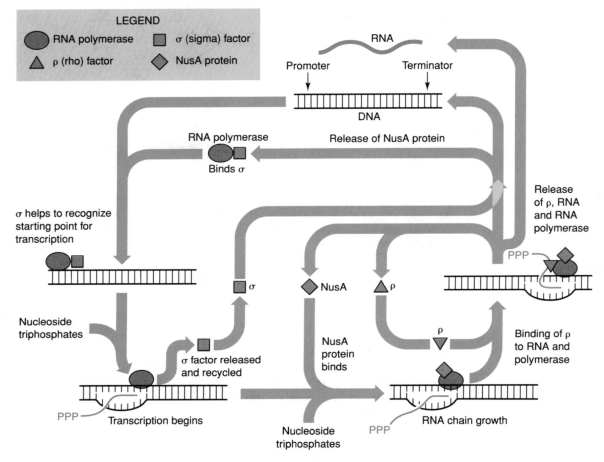

Figure 9-7
Stages in RNA synthesis. Synthesis starts at a promoter on DNA and ends at a terminator. Sigma (σ) factor finds the promoter sequence. NusA protein is an elongation factor, and rho (ρ) is a termination-RNA release factor. (Adapted from Watson et al., *Molecular Biology of the Gene,* Fourth Edition, Menlo Park, CA: Benjamin/Cummings, 1987.)

again with its complement on the other DNA strand. Second, RNA polymerase changes its structure and loses the σ subunit. Third, a protein elongation which aids (called NusA protein), binds. The rest of the elongation process is carried out by the core enzyme NusA complex (Figure 9-7). Each new nucleotide that is added at the 3' end of the growing RNA allows another nucleotide to be released from its binding to the template. Hence, as elongation continues a constant section of 4 to 8 nucleotides at the 3' end of the newly-made RNA is paired to the DNA template strand, while the rest of the RNA emerges from RNA polymerase as a single strand.

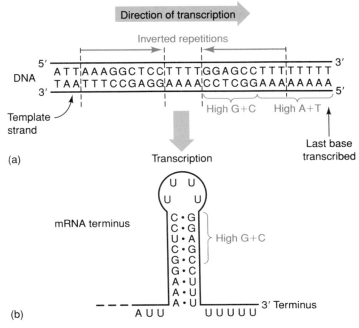

Figure 9-8
Base sequence of (a) the DNA of the *E. coli trp* operon at which transcription termination occurs and of (b) the 3′ terminus of the mRNA molecule. The inverted repeat sequence is indicated by reversed red arrows. The mRNA molecule is folded to form a stem-and-loop structure thought to exist. The relevant regions are labeled in red; the terminal sequence of U's in the mRNA is shaded in red.

Termination of RNA synthesis occurs at specific base-sequences in the DNA molecule, called **terminators.** Some termination sequences allow RNA polymerase to terminate elongation spontaneously. These are called intrinsic terminators. Other terminators require the action of a protein called Rho; they are called rho–dependent terminators. Intrinsic terminators have three characteristic features (Figure 9-8):

1. First, there is an inverted-repeat base-sequence containing a central nonrepeating segment; that is, the sequence in one DNA strand would read ABCDEF-XYZ-F′E′D′C′B′A′ in which A and A′, B and B′, and so on, are complementary bases. Thus, this sequence is capable of intrastrand base pairing, forming a stem-and-loop configuration in the RNA transcript and possibly in the DNA strands.

2. The second region is near the loop end of the putative stem (sometimes totally within the stem) and is a sequence having a high G + C content.

3. A third region is a sequence of A · T pairs (which may begin in

the putative stem) that yields in the RNA a sequence of six to eight uracils often followed by an adenine.

Rho-dependent terminators lack sequences with a stretch of adenine in the template strand. Rho acts by binding to a special sequence on the nascent RNA and forcibly pulls the RNA away from its contact with the DNA in the transcription-elongation complex with RNA polymerase. Rho derives its force from the hydrolysis of ATP molecules.

The final step in the termination process is dissociation of RNA polymerase from the DNA. This comes about by the rebinding of a σ subunit to core after the RNA is released. The released holoenzyme is free to search for a new promoter where it can start synthesis of a new RNA (Figure 9-7).

Classes of RNA Molecules

There are three major classes of RNA molecules—messenger RNA (mRNA), ribosomal RNA (rRNA), and transfer RNA (tRNA). All are synthesized from DNA base sequences. They are all involved in protein synthesis, but each class of RNA has a different function, as will be seen in this and the next chapter. There are also significant differences between the structures and modes of synthesis of the RNA molecules of prokaryotes and eukaryotes, though the basic mechanisms of their functions are nearly the same. The greatest amount of information has been obtained from studies with bacteria and bacterial cell extracts, so that it is here that we begin. Transcription in eukaryotes and the structure and synthesis of eukaryotic RNA molecules are discussed in a later section.

MESSENGER RNA

The base-sequence of a DNA molecule determines the amino acid sequence of every polypeptide chain in a cell, though amino acids have no special affinity for DNA. Thus, instead of a direct pairing between amino acids and DNA, a multistep process is used and the information contained in the DNA is converted to a form in which amino acids can be arranged in an order determined by the DNA base-sequence. This process begins with the transcription of the base-sequence of one of the DNA strands (the **template strand**) into the base sequence of an RNA molecule and it is from this molecule—messenger RNA—that the amino acid sequence is obtained by the protein-synthesizing machinery of the cell. As we will see in Chapter 10, the base sequence of the mRNA is read in groups of three bases (a group of three is called a **codon**) from a start codon to a stop point, with each codon corresponding either to one amino acid or a stop signal.

A DNA segment corresponding to one polypeptide chain plus the start and stop signals is called a **cistron** and a mRNA encoding a single polypeptide is called monocistronic mRNA. It is very common for a bacterial mRNA molecule to encode several different polypeptide chains; in this case it is called a **polycistronic mRNA** molecule.

polycistronic mRNA

In addition to cistrons and start and stop sequences for translation, other regions in mRNA are significant. For example, translation of a mRNA molecule (that is, protein synthesis) seldom starts exactly at one end of the mRNA and proceeds to the other end; instead, initiation of synthesis of the first polypeptide chain of a polycistronic mRNA may begin hundreds of nucleotides from the 5′ terminus of the RNA. The section of nontranslated RNA before the coding regions is called a **leader sequence;** in some cases, the leader contains a regulatory region (called an **attenuator**) that determines the level of gene expression. Untranslated sequences are found at both the 5′ and the 3′ termini, and a polycistronic mRNA molecule may contain intercistronic sequences (**spacers**) hundreds of bases long.

leader (sequence)

An important characteristic of prokaryotic mRNA is its short lifetime; usually within a few minutes after an mRNA molecule is synthesized, nuclease degradation occurs. Although this means that continuous synthesis of a particular protein requires ongoing synthesis of the corresponding mRNA molecule, rapid degradation of mRNA is nonetheless advantageous to bacteria, whose environment and needs often fluctuate widely. Synthesis of a necessary protein can be regulated as needed simply by controlling transcription. When a particular protein is needed by a cell, the appropriate mRNA molecule will be made. However when the protein is no longer needed, inhibition of synthesis of the mRNA is sufficient to prevent wasted synthesis, because the previously made mRNA molecules will soon all be degraded.

STABLE RNA: RIBOSOMAL RNA AND TRANSFER RNA

During the synthesis of proteins, genetic information is supplied by messenger RNA. RNA also plays other roles in protein synthesis. For example, proteins are synthesized on the surface of an RNA-containing particle called a **ribosome;** these particles consist of three classes of **ribosomal RNA** (**rRNA**), which are stable molecules. Also, amino acids do not line up against the mRNA template independently during protein synthesis but are aligned by means of a set of about fifty adaptor RNA molecules called **transfer RNA** (**tRNA**), also a stable species. Each tRNA molecule is capable of "reading" three adjacent mRNA bases (a codon) and placing the corresponding amino acid at a site on the ribosome at which a peptide bond is formed with an adjacent amino acid. Neither rRNA nor tRNA is used as a template. The roles of ribosomes and of tRNA molecules will be explained in the following chapter; here we are concerned only with their synthesis, which involves transcription from DNA and some transcriptional modification.

The synthesis of both rRNA and tRNA molecules is initiated at a promoter and completed at a terminator sequence, and in this respect, their synthesis is no different from that of mRNA. However, the following three properties of these molecules indicate that neither rRNA nor tRNA molecules are the **primary transcripts** (immediate products of transcription):

primary transcript

1. The molecules are terminated by a 5′-monophosphate rather than the expected triphosphate found at the ends of all primary transcripts.

2. Both rRNA and tRNA molecules are much smaller than the primary transcripts (the transcription units).

3. All tRNA molecules contain bases other than A, G, C, and U, and these "unusual" bases (as they are called) are not present in the original transcript.

All of these molecular changes are made after transcription by a process called **posttranscriptional modification** or, more commonly, **RNA processing.**

Both rRNA and tRNA molecules are excised from large primary transcripts. Often a single transcript contains the sequences for several different molecules—for example, different tRNA molecules, or both tRNA and rRNA molecules. Formation of rRNA molecules is a result of fairly straightforward excision of a single continuous sequence. However, tRNA molecules (which contain bases other than A, U, G, and C) are not only cut by several enzymes acting in a particular order, but also chemical modification of various bases when the primary transcript occurs. An example of the production of a particular tRNA molecule is given in Figure 9-9, to illustrate the complexity of the process.

Transcription in Eukaryotes

The basic features of the transcription and the structure of mRNA in eukaryotes are similar to those in bacteria. However, there are five notable differences:

1. Eukaryotic cells contain three classes of nuclear RNA polymerase and these are responsible for the synthesis of different classes of RNA.

2. Many mRNA molecules are very long lived.

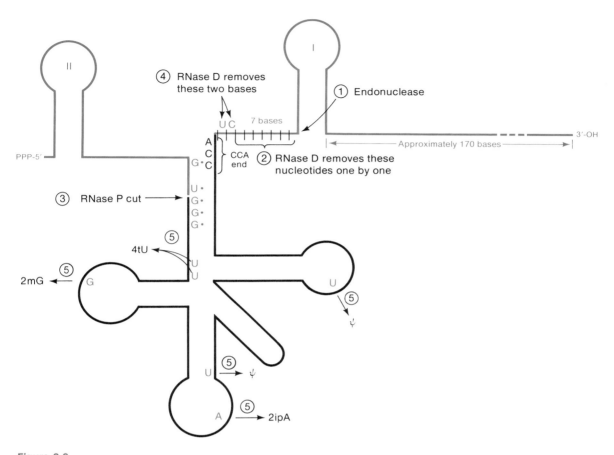

Figure 9-9
The stages in processing of the *E. coli* tRNA$_1^{Tyr}$ gene transcript. The five stages are given arabic numbers. Step 3 generates the 5'-P end. This step is catalyzed by RNase P which is a ribozyme, an unusual RNA enzyme (see Chapter 3). Step 4 generates the 3'-OH end (the CCA end). In step 5, six bases, all in or near the loops of the tRNA molecule, are modified to form pseudouridine (ψ), 2-isopentenyladenosine (2ipA), 2-o-methylguanosine (2mG), and 4-thiouridine (4tU). The continuous sequence that forms the final tRNA molecule is given in black.

cap

3. Both the 5' and 3' termini are modified; a complex structure called the **cap** is found at the 5' end and a long (up to 250 nucleotides) sequence of polyadenylic acid—poly(A)—is found at the 3' end.

intron

4. The mRNA molecule that is used as a template for protein synthesis is usually about one-tenth the size of the primary transcript. During mRNA processing intervening sequences called **introns** are excised and the fragments are rejoined.

5. All eukaryotic mRNA molecules are monocistronic.

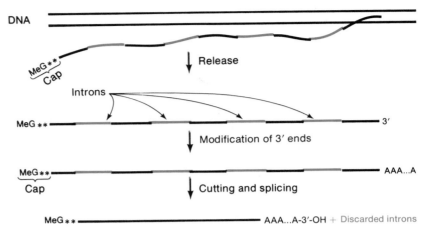

Figure 9-10
Schematic drawing showing production of eukaryotic mRNA. The primary transcript is capped before it is released. Then, its 3'-OH end is modified, and finally the intervening regions are excised. MeG denotes 7-methylguanosine and the two asterisks indicate the two nucleotides whose riboses are methylated.

These points are illustrated in Figure 9-10, which shows a schematic diagram of a typical eukaryotic mRNA molecule and how it is produced. There is clearly more to the production of mRNA in eukaryotes than just transcription of a DNA.

Initiation of transcription in eukaryotes is not well understood. A sequence—TATAAAA—analogous to the −10 consensus sequence in prokaryotes but centered at −29 is a part of many promoters. Other sequences called upstream activation sites and enhancers, which will be discussed in Chapter 16, affect the efficiency of initiation. These are usually sites where certain proteins bind. These proteins, called transcription factors, generally increase the transcription from promoters that are within the vicinity of the sequences to which they bind. In some cases, promoters as far as 2,000 and 3,000 bases away from an enhancer site (along the DNA) are activated.

In eukaryotic cells, DNA is bound tightly to histone proteins in a structure called chromatin. The structure of chromatin makes transcription more complicated in eukaryotes than in prokaryotes, and evidence exists suggesting that chromatin structure is altered in regions being transcribed; though how it is altered is far from clear. There is also a mechanism for terminating transcription in eukaryotic cells, but very little is known about the general properties of the sequences or the factors involved.

The three major classes of eukaryotic RNA polymerases are denoted I, II, and III; they can be distinguished by the ions required for their activity, their optimal ionic strength, and their sensitivity to various inhibitory compounds. All are found in the eukaryotic nucleus. RNA polymerases are found in mitochondria and chloroplasts. The locations and products of the nuclear RNA polymerases are listed below.

	Class I	Class II	Class III
Location:	Nucleolus	Nucleoplasm	Nucleoplasm
Product:	rRNA	mRNA	tRNA, 5S RNA

Note that RNA polymerase II is the enzyme responsible for all mRNA synthesis.

The biochemical reaction catalyzed by the eukaryotic RNA polymerases is the same as that catalyzed by *E. coli* RNA polymerase.

The 5′ terminus of a eukaryotic mRNA molecule carries a methylated guanosine derivative, 7–methylguanosine (7–MeG) in an unusual 5′-5′ linkage to the 5′-terminal nucleotide of the primary transcript. Occasionally the sugars of the adjacent nucleotides are also methylated. The unit

(7-MeG)-5′-PPP-5′-(G or A, with possibly methylated ribose)-3′-P-

in which P and PPP refer to mono- and triphosphate groups respectively, is called a **cap.**

Capping occurs shortly after initiation of synthesis of the mRNA, possibly before RNA polymerase II leaves the initiation site, and precedes all excision and splicing events. The biological significance of capping has not yet been unambiguously established, but it is believed that it is required for efficient protein synthesis. Capping may function to protect the mRNA from degradation by nucleases and to provide a feature for recognition by the protein–synthesizing machinery.

Most, but not all, animal mRNA molecules are terminated at the 3′ end with a poly(A) tract, which is added to the primary mRNA by a nuclear enzyme, poly(A) polymerase. The adenylate residues are not added to the 3′ terminus of the primary transcript. Transcription normally passes the site of addition of poly(A), so that an endonucleolytic cleavage must occur before the poly(A) is added. A base sequence— AAUAAA—10 to 25 bases upstream from the poly(A) site is a component of the system for recognizing the site of poly(A) addition. Interestingly, some primary transcripts contain two or more sites at which poly(A) can be added. The differentially terminated mRNA molecules usually play different roles in the life cycle of the particular organism. The length of the poly(A) segment can be from 50 to 250 nucleotides. The significance of the poly(A) terminus is unknown at present, but it is believed that it increases the stability of mRNA; a possible role in enhancement of mRNA translation has also been proposed. Some cellular

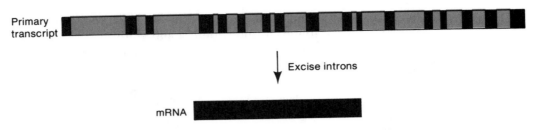

Figure 9-11
A diagram of the conalbumin primary transcript and the processed mRNA. The seventeen introns, which are excised from the primary transcript, are shown in red.

mRNA molecules lack poly(A), so its presence is not obligatory for successful translation.

Most of the primary transcripts of higher eukaryotes contain untranslated intervening sequences (**introns**) that interrupt the coding sequence and are excised in the conversion of the primary transcript to mRNA (Figures 9–10 and 9–11). The amount of discarded RNA ranges from 50% to nearly 90% of the primary transcript. The remaining segments (**exons**) are joined together to form the finished mRNA molecules. The excision of the introns and the formation of the final mRNA molecule by joining of the exons is called **RNA splicing.**

The number of introns per gene varies considerably (Table 9-1) and is not the same in all organisms for a given protein. Furthermore, within a particular gene, introns have many different sizes and are usually larger than exons.

Splicing occurs in the nucleus after capping and addition of poly(A). The existence of splicing explains how the nucleus can contain an enor-

exon

Table 9-1

Translated Eukaryotic Genes in Which Introns Have Been Demonstrated

Gene	Number of Introns
α-Globin	2
Immunoglobulin L chain	2
Immunoglobulin H chain	4
Yeast mitochondria cytochrome *b*	6
Ovomucoid	6
Ovalbumin	7
Ovotransferrin	16
Conalbumin	17
α-Collagen	52

Note: At present the genes for histones and interferon are the only known translated genes in the higher organisms that do not contain introns.

mous number of different RNA molecules whose size distribution is very great. All animal mRNA molecules are monocistronic, so that one might expect the mRNA to have a small range of sizes. However, the amount of RNA that is discarded as a result of excision and splicing varies widely, so the sizes of the primary transcripts cover a range of molecular weights from about 6×10^4 to 3×10^6. Furthermore, introns are excised one by one, and apparently ligation occurs before the next intron is excised, so a huge number of different nuclear RNA molecules is present at any instant. These precursor and partially processed molecules are called **heterogeneous nuclear RNA** (HnRNA). After processing is completed, the mature mRNA is transported to the cytoplasm to be translated; it is not known why the primary transcript and partially processed molecules are not transported.

Removing an intron without altering the coding sequence of the resulting mRNA molecule requires great fidelity in the cleavage process. For example, a cutting error that displaces the cutting site by a single base would completely destroy the reading frame of the bases in the mRNA. Fidelity is provided by the base-sequence itself. In all genes observed, in which splicing occurs, the splice sites of the primary transcript are marked by a sequence resembling

$$5'—^A_CAGG\underline{U}^A_GAGU \overset{intron}{\ldots\ldots} (Py)_6 XC\underline{AGG}^G_U—3'$$

in which Py is any pyrimidine, X is any base, and the arrows are the splice points. The underlined bases are the same for all observed introns (that is, they are conserved), whereas the remainder is a consensus sequence in which there is some variation from one splice site to the next.

In the nucleus the precursors to mRNA molecules are complexed with specific proteins, forming a **ribonucleoprotein particle** (**RNP**). In the course of the splicing reaction primary transcripts are assembled into particles called HnRNP.

The process of RNA splicing occurs on a separate structure called a **spliceosome.** Spliceosomes actually form on the splice sites through interactions with a number of small nuclear ribonucleoproteins called **snRNPs.** A key component of an snRNP is at least one small nuclear RNA (snRNA). Each snRNP has a separate function. The one with the 165-nucleotide U1 snRNA binds to the consensus sequence at the 5' end of an intron, while that with the U5 snRNA binds to the consensus sequence at the 3' end of the intron. These snRNPs bring the two junctions into close proximity at the spliceosome, which has enzymatic activities that perform the intricate process of cutting out the exon sequences and joining the two intron segments.

The existence of introns seems to be fairly wasteful, especially in view of the relative amounts of intron and exon RNA. In a few cases, the presence of introns increases the coding capacity of a particular stretch of genes in that alternative splicing patterns yield different mRNA molecules. This is common in DNA viruses, but not particularly frequent in cellular DNA, though several examples are known.

Several theories have been proposed to explain why genes from eukaryotic cells are interrupted by introns. Since introns are almost never found in the genes from bacteria, which are considered to be simpler forms of life, it is reasonable to suppose that gene interruptions were added during the course of evolution. Various types of recombination allow segments of DNA to be exchanged between different places on a chromosome. Thus, an interruption of a gene could appear by this type of recombination, and as long as the cell had acquired the means of splicing RNA, the right kinds of interruptions would yield a functional gene.

An alternative idea is that interrupted genes were present even in the earliest organisms, and during the course of evolution were lost by the bacteria in order to simplify their genomes and allow them to duplicate as rapidly as possible. This alternative provides a plausible explanation of how genes from complex protein might have evolved. Each exon could represent the coding sequence of an ancestral small protein and rearrangements that occurred during the course of evolution brought the sequences together from various parts of the genome, to form new

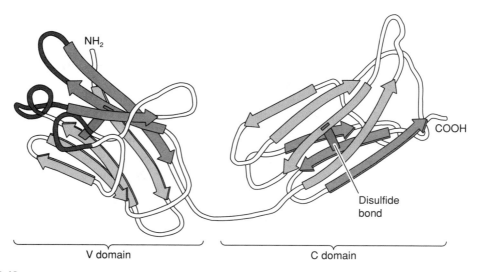

Figure 9-12
A diagram of a subunit of an antibody molecule, a protein with two domains, one black and one red. (Adapted from Watson et al., *Molecular Biology of the Gene,* Fourth Edition, Menlo Park, CA: Benjamin/Cummings, 1987.)

genes. The use of an RNA splicing mechanism that could work to remove all introns would allow a gene to be assembled and tried in new combinations without requiring precise recombination events to fuse parts and form a gene lacking interruptions.

Examination of the structure of proteins has revealed that many consist of several independently folded regions each separated by a short polypeptide segment. Each individually folded region is called a domain. A protein with two very distinct domains is shown in Figure 9-12. This particular protein, a part of an antibody molecule, is encoded by a gene consisting of two large and two small exons. The two large exons code for the two separated domains, suggesting that each domain is derived from an ancestral protein.

Means of Studying Intracellular RNA

Several procedures are used to study RNA metabolism *in vivo*. In most of the techniques, radioactive RNA is prepared by adding ^{3}H-labeled or ^{14}C-labeled uridine to growth medium. However, this also produces radioactive DNA, because uridine can be metabolized to cytosine and thymine; this is significant when studying RNA metabolism because when RNA is isolated from cells, it is usually contaminated with DNA. The contaminating DNA can be removed by treatment of the sample with pancreatic DNase, for this enzyme degrades the DNA to mononucleotides and small oligonucleotides, which can be easily separated from the RNA.

Most investigations of RNA *in vivo* are concerned with either the presence, synthesis, or degradation of specific mRNA species—for example, the mRNA from a particular gene. In order to carry out such analyses, it is necessary to distinguish the particular RNA molecule from all other RNA molecules. This is usually done by **DNA-RNA hybridization.** With this technique, DNA with a sequence complementary to the RNA molecule of interest is permanently fixed to a nitrocellulose filter, as described earlier in Chapter 3 (Figure 3-12). The filter is subsequently incubated with an extract containing radioactive RNA, using conditions leading to base-pairing between the RNA and DNA strands, and then washed repeatedly. RNA that is not present in DNA-RNA hybrids is removed; the amount of radioactivity remaining on the filter is a measure of the amount of the particular RNA present in the extract.

A variety of procedures are used to obtain the DNA used in a hybridization experiment. In the most useful experiments, a DNA segment containing a sequence complementary to the RNA of interest is inserted (by genetic engineering techniques described in Chapter 13) into

clone

a piece of foreign DNA that has no sequences in common with the RNA and that is easily purified. In this case, the DNA sequence of interest is said to be **cloned.** Typical foreign DNA molecules are plasmids, phage, and viral DNA. A DNA molecule carrying a cloned DNA sequence to be used in a hybridization experiment is called a **probe.**

The technique most commonly used is the **Southern transfer,** or **Southern blotting,** procedure (named after E. Southern, its developer), a method in which hybridization is performed with a large number of distinct DNA segments simultaneously. In this procedure the total DNA of an organism is broken into discrete fragments by restriction enzymes (Chapter 13), which make cuts in unique base-sequences. Fragments are separated by gel electrophoresis (as in the di-deoxy base-sequencing technique, Chapter 3) and hybridization is carried out on all of the fragments. Various techniques enable the positions of particular fragments to be identified, and the location of hybridized radioactive RNA can be matched with these fragments. The technique is performed as follows (Figure 9-13).

DNA is enzymatically fragmented and then electrophoresed through an agarose gel. Following electrophoresis the gel is soaked in a denaturing solution (usually NaOH), so that all the DNA in the gel is converted to single-stranded DNA, which is needed for hybridization. After de-

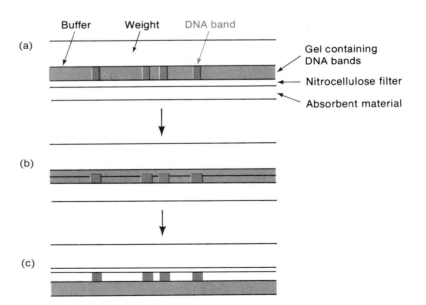

Figure 9-13
The Southern transfer technique. (a) A stack consisting of a weight, a gel, a filter, and absorbent material at the time the weight is applied. (b) A later time—the weight has forced the buffer (shaded area), which carries the DNA, into the nitrocellulose. (c) The lowest layer has absorbed the buffer, the DNA remains bound to the nitrocellulose.

naturing the gel is filtered through a nitrocellulose filter placed on top of several sheets of ordinary filter paper. The gel (which is typically in the form of a broad flat slab) is placed on the nitrocellulose and covered with a glass plate to prevent drying. A weight is placed on the top of the stack, and the liquid is squeezed out of the gel. The liquid passes downward through the nitrocellulose and the denatured DNA binds tightly to the nitrocellulose; the remaining liquid passes through and is absorbed by the filter paper. DNA molecules do not diffuse very much, so if the gel and the nitrocellulose filter are in firm contact, the positions of the DNA molecules on the filter will be identical to their positions in the gel. The nitrocellulose filter is dried in vacuum, to insure that the DNA remains on the filter during the hybridization step. The dried filter is then moistened with a very small volume of a solution of ^{32}P-labeled RNA, placed in a tight-fitting plastic bag to prevent drying, and held at a temperature suitable for renaturation (usually 65°C) for 16–24 hours. The filter is then removed, washed to remove unbound radioactive molecules, dried, and autoradiographed with x-ray film. The blackened positions on the film indicate the locations of the DNA molecules whose DNA base-sequences are complementary to the sequences of the added radioactive molecules. Since the genes contained in each fragment usually are known, specific mRNA molecules can be identified. The degree of blackening on the film is easily measured quantitatively and is usually proportional to the amount of RNA that has hybridized. Thus, the amount of mRNA transcribed from each region of a DNA molecule can be measured. Note that by this technique many different mRNA molecules can be studied simultaneously. The Southern transfer method also has many other uses, particularly in DNA-DNA hybridization analysis.

Summary

RNA polymerase initiates transcription by first binding to a promoter sequence on DNA. The bacterial enzyme does this by contacts between the σ subunit of the holoenzyme and sequences at 10 and 35 base–pairs upstream from the start point. Eukaryotic RNA polymerases rely on recognition of the promoter by separate initiation factors. RNA synthesis is initiated by a reaction between two nucleoside triphosphates paired to residues $+1$ and $+2$ of the template strand. Chains are elongated by addition of nucleotides at the 3′ end of the growing RNA at a rate of about 50 per second. Termination occurs spontaneously at certain DNA sequences, called intrinsic terminators or by action of a protein factor at a "factor-dependent" terminator sequence. In *E. coli* that factor is called rho, which acts at rho-dependent terminators. The three major classes of RNA are messenger RNA (mRNA), ribosomal RNA (rRNA) and transfer RNA (tRNA). rRNA and tRNA are derived from primary transcripts, synthesized as described above, by various processing reactions that include cutting, trimming, and nucleotide modification. In prokaryotes, mRNAs are usually primary transcripts while in eukaryotes mRNA are processed from precursor transcripts by addition of 5′ caps and poly(A) tails, and usually by splicing together exon sequences.

Drill Questions

1. a. From what substrates is RNA made?
   b. On what template?
   c. With what enzyme?
   d. Is a primer required?

2. Describe the differences, if any, between the chemical reactions catalyzed by DNA polymerase and RNA polymerase.

3. What chemical groups are present at the origin and terminus of a molecule of mRNA that has just been synthesized?

4. a. What is mRNA?
   b. How does mRNA sometimes differ from a primary transcript?
   c. Define cistron and polycistronic mRNA.
   d. Which parts of a mRNA molecule are not translated?

5. How many subunits are in E. coli RNA polymerase, and which one is responsible for correctly positioning the enzyme on a promoter?

6. a. What two regions are common to most prokaryotic promoters?
   b. What consensus sequence is present in a large number of eukaryotic promoters?

7. Answer these questions about eukaryotic RNA.
   a. What is a cap?
   b. At which end of the mRNA is the poly (A)?
   c. Are there eukaryotic mRNA molecules that do not contain either feature?

8. a. What are intervening sequences or introns?
   b. What is meant by mRNA "splicing"?

Problems

1. The ⁻10 region of an E. coli promoter is an example of what is termed a consensus sequence—namely, a base sequence from which other sequences having similar functions can be obtained by changing only one or two bases. Answer these questions about consensus sequences.
   a. What is the evolutionary significance of such a sequence?
   b. What is the biochemical significance of a conserved base, such as the conserved T in the ⁻10 region?
   c. What biochemical differences might you expect between sequences that differ only slightly in the nonconserved bases?

2. An RNA molecule is isolated, which has a 3′-OH terminus and a 5′-P terminus. What information does this fact provide?

3. A chromatographic column in which oligo-dT is linked to an inert substance is useful in separating eukaryotic mRNA from other RNA molecules. On what principle does this column operate?

4. A particular sequence containing six base-pairs is located in ten different organisms. The observed sequences are ACGCAC, ATACAC, GTGCAC, ACGCAC, ATACAC, ATGTAT, ATGCGC, ACGCAT, GTGCAT, and ATGCGC. What is the consensus sequence?

5. Write down the two RNA sequences which could conceivably result from complete transcription of a DNA molecule that has the following sequence in one of the strands—5′-AGCTGCAATG-3′. Indicate the 5′ and 3′ and of each transcript.

6. Genetic engineering techniques allow the movement of genes from one organism to another. When eukaryotic genes are placed into E. coli it has been observed that these genes are not always transcribed, while transcription occurs much more frequently when eukaryotic genes are transferred to yeast. Why?

Conceptual Questions

1. How might the expression of genes that are developmentally controlled be regulated at the level of their messenger RNA?

2. What could be the consequence(s) of mutations in various regions of a promoter?

3. How might an organism compensate for severe promoter mutations in order to survive?

Joseph Ilan
Professor of Anatomy
Department of Anatomy

Birthday: **14 March 1930**

Birth Place: **Tel Aviv, Israel**

Undergraduate Degree: **The Hebrew University, Jerusalem**
Major: Biology 1960 B.S., 1962 M.S.

Graduate Degree: **McGill University, Canada, Ph.D. 1965, Biochemistry**

Postdoctoral Training: **The Rockefeller University, New York, 1965–1967**

Present Position: **Case Western Reserve University, Anatomy Department**

Address: **Cleveland, Ohio**

IN 1965, WHEN I joined Dr. Fritz Lipmann at the Rockefeller University as a postdoctoral fellow, the laboratory was establishing cell-free systems to study the mechanisms of protein biosynthesis in bacteria. Almost nothing was known about the mechanisms of protein synthesis in eukaryotes. We decided to develop a cell-free system from insects during metamorphosis in order to follow developmental processes connected to protein synthesis. To our surprise the mRNA for the adult cuticular protein was stable for seven days of pupation, but was translated only at the last day. mRNA of first day pupae could be induced to translate adult cuticular protein in a cell-free system by an addition of enzyme and tRNA fractions from seventh day pupae, the stage at which adult cuticular protein is synthesized in vivo. This indicates that mRNA may be stable and under translational control. At that time the current dogma stated that mRNA must be a short-lived intermediate as shown by Jacob and Monod for the lactose operon in E. coli (chapter 14). Our finding led us to study translational regulation and the general mechanism of mRNA translation in eukaryotic organisms. Soon we found out that AUG-methionine is the initiator codon in eukaryotes and that initiation may be under translational regulation. Later we showed that elongation of the nascent polypeptide chain may also be under translational control. We also explored the mechanism of termination of the nascent polypeptide chain. It is now established that many genes are under translational control and the level of the final gene product, the protein, can be greatly changed many fold via translational regulation. The understanding of the mechanisms of translational regulation may lead to the designing of new drugs. Of special importance to future work is the employment of antisense RNA for regulating the translation of specific mRNA.

10

Translation

The synthesis of every protein molecule in a cell is directed by intracellular DNA. There are two aspects to understanding how this is accomplished—the **information** or **coding problem** and the **chemical problem.** The information problem is the mechanism by which a base-sequence in a DNA molecule is translated into an amino acid sequence of a polypeptide chain. The chemical problem includes the actual process of synthesizing a protein: the means of initiating synthesis; linking together the amino acids in the correct order; terminating the chain; releasing the finished chain from the synthetic apparatus; folding the chain; and, often, postsynthetic modification of the newly-synthesized chain. The overall process is called **translation.**

This chapter presents an outline of the process, in order to introduce the terminology, and present the major features of the coding and decoding systems, and the mechanism for polypeptide synthesis.

We will learn that once the necessary information is coded (in a triplet format) into mRNA, other types of RNAs as well as several enzymes join up to form the macromolecular complex that carries out protein synthesis. Eventually, we will begin to appreciate that many features of protein synthesis are much more complicated than nucleic acid synthesis. That complexity derives from the fact that amino acids do not have a direct affinity for the nucleotide bases of mRNA. Special molecules are therefore required to hold the mRNA in position while amino acids are lined up and covalently linked to one another to form a protein.

In this chapter you will learn:

1. The nature of codons, anticodons, and the "universal" genetic code.

2. The mechanisms by which polypeptides are synthesized.

3. The structure of transfer RNA and the function of aminoacyl tRNA synthetases.

4. The general structure and function of ribosomes.

Outline of Translation

ribosome

Protein synthesis occurs on intracellular particles called **ribosomes.** In prokaryotes, these particles consist of three RNA molecules and about 55 different protein molecules. Those proteins include the enzymes needed to form a peptide bond between amino acids, a site for binding the mRNA, and sites for bringing in and aligning the amino acids in preparation for assembly into the finished polypeptide chain. Amino acids themselves are unable to interact with the ribosome and cannot recognize bases in the mRNA molecule. Thus, there exists a collection of carrier molecules mentioned in the previous chapter, **transfer RNA (tRNA).** These molecules contain a site for amino acid attachment, and

anticodon

a region called the **anticodon** that recognizes the appropriate base sequence (the **codon**) in the mRNA. Proper selection of the amino acids for assembly is determined by the positioning of the tRNA molecules, which in turn is determined by hydrogen-bonding between the anticodon of each tRNA molecule and the corresponding codon of the mRNA.

A schematic diagram showing the events that occur on the ribosome is given in Figure 10-1. The scheme shown applies equally to prokaryotes and eukaryotes, with a single exception. In eukaryotes, since transcription occurs in the nucleus and protein synthesis occurs in the cytoplasm, the mRNA is not attached to the DNA during protein synthesis, in contrast with what is shown in the figure.

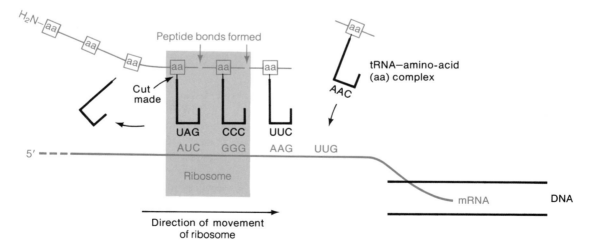

Figure 10-1
A diagram showing how a protein molecule is synthesized. Note that the relative directions of polypeptide and RNA synthesis are such that polypeptide synthesis can occur before the mRNA is completed. This is, in fact, the case in prokaryotes.

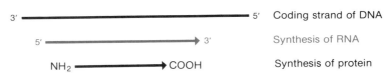

Figure 10-2
Directions of synthesis of RNA and protein with respect to the coding strand of DNA.

Clearly there must be many different tRNA molecules, because each amino acid must be brought in to the ribosome in a way that insures that it corresponds to the base-sequence of the mRNA. There are specific tRNA molecules that correspond to each amino acid. Furthermore, the linkage of each amino acid to its tRNA molecule is catalyzed by a specific enzyme, which insures that the appropriate amino acid will be attached to the correct tRNA molecule.

In earlier chapters several examples of directional synthesis of macromolecules were seen, for example, DNA and RNA. This is also the case for polypeptide synthesis, which begins at the amino terminus. Furthermore, translation of an mRNA molecule occurs in only one direction, namely, 5' to 3'. Figure 10-2 summarizes the polarity of synthesis of mRNA and protein with respect to the DNA coding strand. These directions are also shown in Figure 10-1.

The Genetic Code

genetic code

The **genetic code** is the collection of base-sequences (codons) that correspond to each amino acid and to translation signals.

Since there are 20 amino acids, there must be more than 20 codons to include signals for starting and stopping the synthesis of particular protein molecules. If all codons have the same number of bases, then each codon must contain at least three bases. The argument for this conclusion is the following: a single base cannot be a codon because there are 20 amino acids and only four bases. Pairs of bases also cannot serve as codons because there are only 4^2 or 16 possible pairs of the four bases. Triplets of bases are possible because there are 4^3 or 64 triplets, which is more than adequate. In fact, the genetic code is a triplet code, and all 64 possible codons carry information of some sort. Furthermore in translating mRNA molecules the codons do not overlap, but are read sequentially (Figure 10-3).

The general properties of the code—for example, that each codon contains three bases, and that codons do not overlap—were deduced from genetic experiments. The sequence of each codon was determined

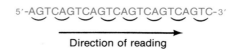

Direction of reading

Figure 10-3

Bases in mRNA are read sequentially in the 5' to 3' direction, in groups of three.

from *in vitro* protein-synthesizing experiments in which synthetic mRNA molecules of known sequence were used; for example, when polyuridylic acid was used as an mRNA, the peptide polyphenylalanine was made, indicating that UUU is the codon for phenylalanine. Use of poly(U) terminated with a 3' guanine yielded polyphenylalanine with a carboxyl-terminal leucine, indicating that UUG is a leucine codon. These and other experiments identified all of the codons. The code is shown in Table 10-1.

The following features of the code should be observed:

1. Most amino acids have more than one codon. In fact, only methionine and tryptophan have a single codon. Furthermore, multiple codons corresponding to a single amino acid usually differ only by the

Table 10-1

The "Universal" Genetic Code

First Position (5' end)	Second Position				Third Position (3' end)
	U	C	A	G	
U	Phe	Ser	Tyr	Cys	U
	Phe	Ser	Tyr	Cys	C
	Leu	Ser	Stop	Stop	A
	Leu	Ser	Stop	Trp	G
C	Leu	Pro	His	Arg	U
	Leu	Pro	His	Arg	C
	Leu	Pro	Gin	Arg	A
	Leu	Pro	Gin	Arg	G
A	Ile	Thr	Asn	Ser	U
	Ile	Thr	Asn	Ser	C
	Ile	Thr	Lys	Arg	A
	Met	Thr	Lys	Arg	G
G	Val	Ala	Asp	Gly	U
	Val	Ala	Asp	Gly	C
	Val	Ala	Glu	Gly	A
	Val	Ala	Glu	Gly	G

Note: The boxed codons are used for initiation. GUG is very rare.

third base. For example, GGU, GGC, GGA, and GGG all code for glycine. Thus, the code is said to be redundant (the term degenerate is also used).

2. Three codons signal termination of polypeptide synthesis—the stop codons UAA, UAG, and UGA.

3. One codon signals initiation of polypeptide synthesis—the start codon AUG, which codes for methionine. An important question is how a particular AUG sequence is designated as a start codon, whereas others located in a coding sequence act only as an internal codon for methionine. In prokaryotes, special base-sequences serve this function; a different mechanism is used for eukaryotes. In some organisms GUG is also used as a start codon for some proteins.

To date, the same codon-amino acid relations seem to exist for all organisms—viruses, prokaryotes, and eukaryotes—and the code is said to be universal. An exception are the codes of mitochondria, the energy-generating organelles of eukaryotes; and chloroplasts, the photosynthetic organelles of plants, in which there are several deviations. Furthermore, the particular deviations appear to be species-specific. The evolutionary significance of these differences is a widely discussed topic.

TRANSFER RNA AND THE AMINOACYL SYNTHETASES

aminoacyl-tRNA synthetases

The decoding operation by which the base-sequence within an mRNA molecule becomes translated to an amino acid sequence of a protein is accomplished by the tRNA molecules and a set of enzymes called the **aminoacyl tRNA synthetases.**

The tRNA molecules are small, single-stranded nucleic acids ranging in size from 73 to 93 nucleotides. Like all RNA molecules, they have a 3'-OH terminus, but the opposite end terminates with a 5'-monophosphate rather than a 5'-triphosphate, because tRNA molecules are cut from a large primary transcript. Due to pairing of complementary base-sequences short double-stranded regions form, causing the molecule to fold into a structure in which open loops are connected to one another by double-stranded stems (Figure 10-4). In two dimensions, a tRNA molecule is drawn as a planar cloverleaf. Its three-dimensional structure is more complex, as is shown in Figure 10-5. Panel (a) shows a skeletal model of the yeast tRNA molecule that carries phenylalanine, and panel (b) shows an interpretive drawing.

Three regions of each tRNA molecule are used in the decoding operation. One of these regions is the **anticodon,** a sequence of three bases that can form base-pairs with a codon sequence in the mRNA. No normal tRNA molecule has an anticodon complementary to any of the stop codons UAG, UAA, or UGA, which is why these codons are recognized as stop signals. A second site is the **amino acid attachment**

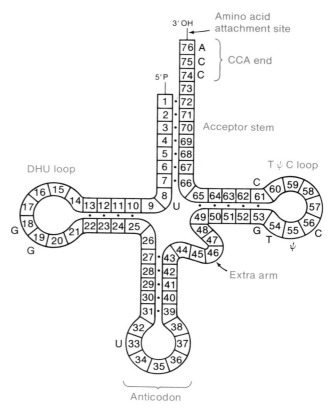

Figure 10-4
The consensus tRNA cloverleaf structure with its bases numbered. The hydrogen-bonded regions are indicated by dots between the bases. A few bases present in almost all tRNA molecules are shown. tRNA contains unusual bases; one of these, dihydrouracil (DHU), is in the DHU loop, and another, pseudouridine (ψ), is in the T ψ C loop. The extra arm varies in length in different tRNAs.

site. The amino acid corresponding to the particular mRNA codon that base-pairs with the tRNA anticodon is covalently linked to this terminus. These bound amino acids are joined together during polypeptide synthesis. A specific aminoacyl tRNA synthetase matches the amino acid with the anticodon; to do so, the enzyme must be able to distinguish one tRNA molecule from another. The necessary distinction is provided by an as yet ill-defined region encompassing many parts of the tRNA molecule called the **recognition region.**

The different tRNA molecules and synthetases are designated by stating the name of the amino acid that can be linked to a particular tRNA molecule by a specific synthetase; for example, leucyl-tRNA syn-

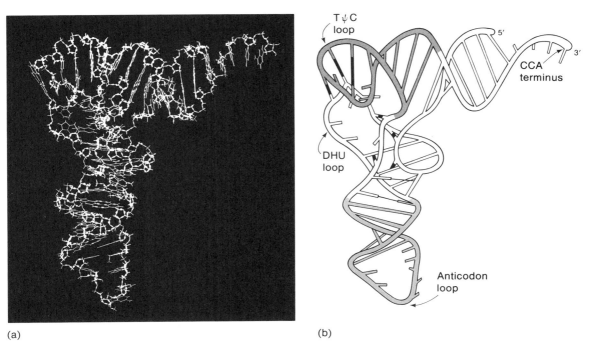

Figure 10-5
(a) Photograph of a skeletal model of yeast tRNA^{Phe}. (b) Schematic diagram of the three-dimensional structure of yeast tRNA^{Phe}. (Courtesy of Dr. Sung-Hou Kim.)

charging

thetase attaches leucine to tRNA^{Leu}. When an amino acid has become attached to a tRNA molecule, the tRNA is said to be **acylated** or **charged.** An acylated tRNA molecule is designated in several ways. For example, if the amino acid is glycine, the acylated tRNA would be written glycyl-tRNA or Gly-tRNA. The term **uncharged tRNA** refers to a tRNA molecule lacking an amino acid, and **mischarged tRNA** to one acylated with an incorrect amino acid.

At least one (and usually only one) aminoacyl synthetase exists for each amino acid. For a few of the amino acids specified by more than one codon, more than one synthetase exists.

Accurate protein synthesis (the placement of the correct amino acid at the appropriate position in a polypeptide chain) requires:

1. attachment of the correct amino acid to a tRNA molecule by the synthetase;
2. accuracy in codon-anticodon binding.

An important experiment showed that codon-anticodon binding is based only on base-pair recognition and that the identity of the amino acid attached to the tRNA molecule does not influence this recognition. In this experiment tRNA^{Cys} was charged with radioactive cysteine. The cystine was then chemically converted to radioactive alanine, yielding alanyl-tRNA^{Cys}. This mischarged tRNA molecule was then used in an *in vitro* system containing hemoglobin mRNA and capable of synthesizing hemoglobin, the complete amino acid sequence of which was known. Hemoglobin containing radioactive alanine was made, but the radioactive alanine was present at the sites normally occupied by cysteine rather than at normal alanine positions. This experiment confirmed the hypothesis that tRNA is a carrier molecule, and that the amino acid recognition region and the anticodon are distinct regions. The anticodon in the tRNA molecule is responsible for inserting the amino acid into the right position. This was demonstrated by chemical alterations of a single base of the anticodon which was sufficient to change the specificity of the tRNA. Glycine tRNA with the anticodon UCC was chemically altered to UCU, which pairs with the codeword for arginine—AGA. The altered glycine tRNA, when charged with radioactive glycine, inserted the glycine into the arginine positions in the newly-synthesized polypeptide, directed by a synthetic mRNA (AGA)n. The native glycine (unaltered) tRNA was not active with this message. Native arginine tRNA was also capable of inserting its labeled arginine into polypeptide directed by (AGA)n synthetic mRNA.

Several amino acids are structurally similar and it is to be expected that synthetases might make occasional mistakes. Valine and isoleucine constitute such a possibly ambiguous pair of amino acids and, in fact, isoleucyl-tRNA synthetase activates valine with ATP to form valyl-AMP, which remains bound to the synthetase, at a frequency of about one per 225 activation events. This would mean that 1/225 of all isoleucine positions in proteins could contain a valine. For a typical protein containing 500 amino acids of which 25 were isoleucines, about one copy in nine could be altered. Since there are at least ten known examples of misacylation, there could be an error in almost every molecule, although this does not occur. An editing mechanism, as we saw in Chapter 7 with DNA synthesis, exists.

The editing mechanism that corrects the valine-isoleucine error is a hydrolytic step in which valyl-AMP is cleaved and removed from the enzyme. The hydrolysis is carried out by the isoleucyl-tRNA synthetase itself. Interestingly, the signal that activates the hydrolytic function is the attempted binding of valine to tRNA^{Ile}. The number of times this editing system fails and valyl-tRNA^{Ile} is formed is about 1 in 800. Thus, the overall error frequency—that is, the fraction of isoleucine sites occupied by valine—is $(1/225)(1/800) = 1/180,000$. If all possible amino acid misacylations occur at this frequency, only about 0.17% of the proteins would be defective. A similar case, in which methionyl-tRNA synthetase forms threonyl-AMP and homocysteyl-AMP, has also been analyzed.

The Wobble Hypothesis

The pattern of redundancy in the code suggests that something is missing in the explanation of codon-anticodon binding; the most striking aspect of the redundancy is that with only a few exceptions, the identity of the third codon-base appears to be unimportant. That is, XYA, XYB, XYC, and XYD usually correspond to the same amino acid.

wobble hypothesis

In 1965, Francis Crick made a proposal, known as the **wobble hypothesis,** that explains how some tRNA molecules respond to several codons and also provides insight into the pattern of redundancy in the code. Up to that time it was generally assumed that no base-pair other than G · C, A · T, or A · U would be found in a nucleic acid. This is true of DNA because the regular helical structure of double-stranded DNA imposes two steric constraints:

1. Two purines cannot pair with one another because there is not enough space for a planar purine-purine pair.

2. Two pyrimidines cannot pair because they cannot reach one another.

Crick proposed that since the anticodon is located within a single-stranded RNA loop, the codon-anticodon interaction might not require formation of a structure with the usual dimensions of a double helix. By model-building he showed that the steric requirements were less stringent at the third position of the codon. By allowing a little play in the structure (this play is called wobble), Crick demonstrated that other base pairs can exist between codon and anticodon. He required, first that the first two base-pairs be of the standard type in order to maximize stability, and second, that the third base pair not produce as much distortion as a purine-purine pair might cause. He included inosine (to replace G) in his model because it was known to be in the anticodons of several tRNA molecules, and he proposed that the base-pairs listed in Table 10-2 were possible in the third position of the codon.

Table 10-2

Allowed Pairings According to the Wobble Hypothesis

Third Position Codon Base	First Position Anticodon Base*
A	U,I
G	C,U
U	G,I
C	G,I

* A is not allowed because an enzyme exists in cells which converts A to I.

The possibility of forming the four base pairs shown in the table: A · I, U · I, C · I, and G · U, explains how a single tRNA molecule can respond to several codons.

There are two major species of the alanine tRNA of yeast. One of these responds to the codons GCU, GCC, and GCA. Its anticodon is IGC, which is consistent with the entries in Table 10-2, and shows that only inosine can pair with U, C, and A. (Remember the convention for naming the codon and the anticodon—*always with the 5' end at the left.* Thus the codon 5'-GCU-3' is matched by the anticodon 5'-IGC-3'). Similarly, yeast $tRNA_{II}^{Ala}$ responds only to GCG; there are two possible anticodons (CGC and UGC) because both C and U can bond to G. If the anticodon were UGC, $tRNA_{II}^{Ala}$ would respond to both GCG and GCA, which is not the case. Thus, the anticodon cannot be UGC. The anticodon must, therefore, be CGC, as indeed it is.

The most striking achievement of the wobble hypothesis is that it explains the arrangement of all the synonyms in the code.

Polycistronic mRNA

Many prokaryotic mRNA molecules are polycistronic; they contain sequences specifying the synthesis of several proteins. A polycistronic mRNA molecule must possess a series of start and stop codons. If a mRNA molecule encodes three proteins the minimal requirement would be the sequence:

Start, protein 1, stop—start, protein 2, stop—start, protein 3, stop.

Actually, an mRNA molecule is probably never so simple—the leader sequence preceding the first start signal may be several hundred bases long, and there is usually a sequence called a **spacer** of from 5 to 20

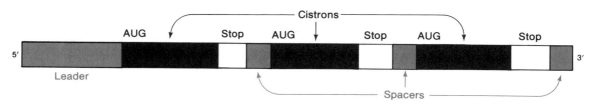

Figure 10-6
Arrangement of cistrons (black) and untranslated regions (red) in a typical polycistronic mRNA molecule.

bases between one stop codon and the next start codon. The structure of a tricistronic mRNA more typically resembles that shown in Figure 10-6.

Overlapping Genes

In all that has been said so far about coding and signal recognition, an implicit assumption has been that the mRNA molecule is scanned for start signals to establish the reading frame, and that reading then proceeds in a single direction within the frame. The idea that several reading frames might exist in a single segment was not considered for many years, primarily because a mutation in a gene that overlapped another gene would often produce a mutation in the second gene, and mutations affecting two genes had not been observed. The notion of overlapping reading frames was also rejected on the grounds that severe constraints would be placed on the amino acid sequences of two proteins translated from the same portion of mRNA. However, because the code is highly redundant, the constraints are actually not so rigid.

If multiple reading frames were present in an organism, a single DNA segment could be utilized to maximum efficiency. However, a disadvantage is that evolution might be slowed. Single-base-change mutations would be deleterious more often if multiple reading frames were present than if there were a unique reading frame. Nonetheless, some organisms—namely, small viruses and the smallest phages, have evolved with overlapping reading frames.

The *E. coli* phage φX174 contains a single strand of DNA consisting of 5386 nucleotides whose base sequence is known. If a single reading frame were used, at most 1795 amino acids could be encoded in the sequence and, if we take 110 as the molecular weight of an "average" amino acid, at most 197,000 molecular weight units of protein could be made. However, the phage makes eleven proteins and the total molecular weight of these proteins is 262,000. This paradox was resolved when it was shown that translation occurs in several reading frames from three mRNA molecules (Figure 10-7). For example, the sequence for protein B is contained within the sequence for protein A, but translated in a different reading frame. Similarly, the protein E sequence is totally within the sequence for protein D. Protein K is initiated near the end of gene *A*, includes the base sequence of B, and terminates in gene *C*; synthesis is not in phase with either gene *A* or gene *C*. Of note is protein A′ (also called A⋆), which is formed by reinitiation within gene *A* and in the same reading frame, so that it terminates at the stop codon of gene *A*. Thus, the amino acid sequence of A′ is identical to a segment of protein A. In total, five different proteins obtain some or all of their

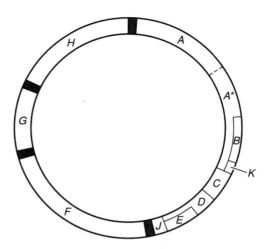

Figure 10-7
Genetic map of φX174 showing the overlapping genes. Spacers are darkened.

primary structure from shared base-sequences in φX174. This phenomenon, known as **overlapping genes,** has been observed in the related phage G4 and in the small animal virus SV40.

It should be realized that the single structural feature responsible for gene overlap is the location of each AUG initiation sequence.

Polypeptide Synthesis

Preceding sections discussed how the information in an mRNA molecule is converted to an amino acid sequence having specific start and stop points. This section examines the chemical problem of attachment of the amino acids to one another.

initiation factors
elongation factors

Polypeptide synthesis can be divided into three stages: (1) **initiation,** (2) **elongation,** and (3) **termination.** The main features of the initiation stage are the binding of mRNA to the ribosome, selection of the initiation codon, and the binding of acylated tRNA bearing the first amino acid. In the elongation stage there are two processes: joining together two amino acids by peptide bond formation, and moving the mRNA and the ribosome with respect to one another so that the codons can be translated successively. In the termination stage the completed protein is dissociated from the synthesis machinery, and the ribosomes are released to begin another cycle of synthesis.

We begin our discussion with a description of the structure of the ribosome.

RIBOSOMES

A ribosome is a multicomponent particle which contains several enzymes needed for protein synthesis. It also brings together a single mRNA molecule and charged tRNA molecules in the proper position and orientation to allow the base-sequence of the mRNA molecule to be translated into an amino acid sequence (Figure 10-8(a)). The properties of the *E. coli* ribosome are understood best, and it serves as a useful model for discussion of all ribosomes.

All ribosomes contain two subunits (Figure 10-8(b)). For historical reasons, the intact ribosome and the subunits have been given numbers that describe how fast they sediment when centrifuged. For *E. coli* (and for all prokaryotes) the intact particle is called a **70S ribosome** (S is a measure of the sedimentation rate) and the subunits, which are unequal in size and composition, are termed **30S** and **50S.** A 70S ribosome consists of one 30S subunit, and one 50S subunit.

Both the 30S and the 50S particles can be dissociated into RNA (called **rRNA,** for ribosomal RNA) and protein molecules under appropriate conditions (Figure 10-9). Each 30S subunit contains one 16S

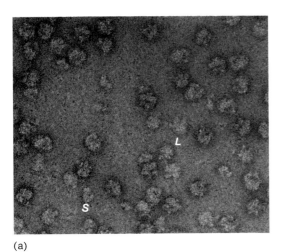

(a)

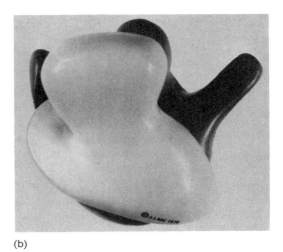

(b)

Figure 10-8
Ribosomes. (a) An electron micrograph of *E. coli* 70S ribosomes. A few ribosomal subunits are also in the field; S denotes a 30S particle and L denotes a 50S particle. (b) A three-dimensional model of the 70S ribosome. The small subunit is light, and the large subunit is dark. (Courtesy of James Lake.)

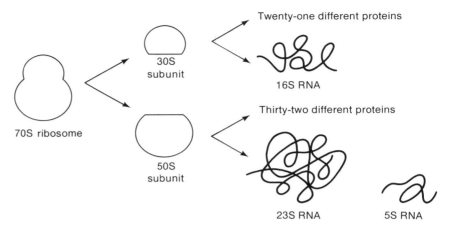

Figure 10-9
Dissociation of a prokaryotic ribosome. The configuration of two overlapping circles will be used throughout this chapter, for the sake of simplicity. The correct configuration is shown in Figure 10-8(b).

rRNA molecule and 21 different proteins; a 50S subunit contains two RNA molecules (one 5S rRNA molecule and one 23S rRNA molecule) and 32 different proteins. In each particle usually only one copy of each protein molecule is present, though a few are duplicated or modified. Like tRNA molecules, rRNA molecules are cut from large primary transcripts. In many organisms some tRNA molecules are cut also from the large transcripts containing rRNA.

The basic features of eukaryotic ribosomes are similar to those of bacterial ribosomes, but all eukaryotic ribosomes are somewhat larger than those of prokaryotes. They contain a greater number of proteins (about 80) and additional RNA molecules (four in all). The biological significance of the differences between prokaryotic and eukaryotic ribosomes is unknown. A typical eukaryotic ribosome (an **80S ribosome**) consists of two subunits, **40S** and **60S.** These sizes may vary by as much as ±10% from one organism to the next, in contrast with bacterial ribosomes, which have sizes that are nearly the same for all bacterial species examined. The best-studied eukaryotic ribosome is that of the rat liver. The 40S and 60S subunits can also be dissociated. A 40S subunit, which is analogous to the 30S subunit of prokaryotes, consists of one 18S rRNA molecule, and about 30 proteins; and the 60S subunit contains three rRNA molecules (one 5S, one 5.8S, and one 28S rRNA molecule) and about 50 proteins. The 5.8S, 18S, and 28S rRNA molecules of eukaryotes correspond functionally to the 5S, 16S, and 23S molecules of bacterial ribosomes. The bacterial counterpart to the eukaryotic 5S rRNA is very likely present as part of the 23S rRNA sequence. The eukaryotic 5S ribosomal RNA is synthesized in the nucleus while the

28S, 18S and 5.8S rRNAs are transcribed in the nucleolus as a 45S precursor and then processed to the three rRNA species.

Stages of Polypeptide Synthesis in Prokaryotes

Polypeptide synthesis in prokaryotes and eukaryotes follows the same overall mechanism, though there are differences in detail, the most important being the mechanism of initiation. The prokaryotic system is best understood and will serve as a model for our discussion.

An important feature of initiation of polypeptide synthesis is the use of a specific initiating tRNA molecule. In prokaryotes the methionine in the tRNA^{Met} is formylated to yield N-formylmethionine tRNA; the tRNA is often designated tRNA^{fMet} (Figure 10-10). Both tRNA^{fMet} and tRNA^{Met} recognize the codon AUG, but only tRNA^{fMet} is used for initiation. The tRNA^{fMet} molecule is first acylated with methionine, and an enzyme (found only in prokaryotes) adds a formyl group to the amino group of the methionine. In eukaryotes there is also a specific tRNA^{Met} for initiation. The initiating tRNA molecule is charged with methionine also, but formylation does not occur. The use of these initiator tRNA molecules means that *while being synthesized,* all prokaryotic proteins have N-formylmethionine at the amino terminus, and all eukaryotic proteins have methionine at the amino terminus. However, these amino acids are frequently removed later (this is called **processing**), and all amino acids have been observed at the amino termini of completed protein molecules isolated from cells.

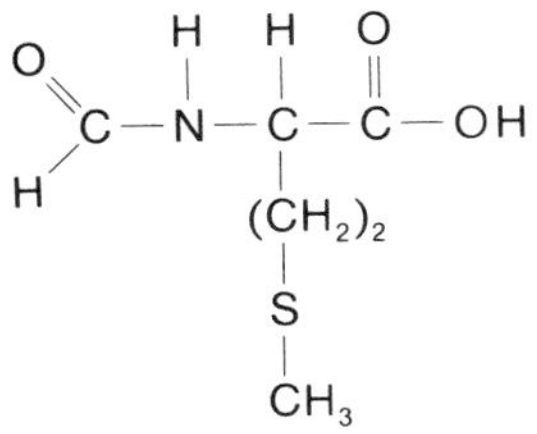

Polypeptide synthesis in bacteria begins by the association of one 30S subunit (not the entire 70S ribosome), an mRNA molecule, fMet-tRNA, three proteins known as **initiation factors,** and guanosine 5'-triphosphate (GTP). These molecules constitute the **30S preinitiation complex** (Figure 10-11). Since polypeptide synthesis begins at an AUG start codon and AUG codons are found within coding sequences (that is, methionine occurs within a polypeptide chain), some signal must be present in the base-sequence of the mRNA molecule to identify a particular AUG codon as a start signal. The means of selecting the correct AUG sequence differs in prokaryotes and eukaryotes. In prokaryotic mRNA molecules a particular base-sequence (AGGAGGU—called the **ribosome binding site** or sometimes the **Shine–Dalgarno sequence**) exists near the AUG codon used for initiation exists. It forms base-pairs with a complementary sequence near the 3' terminus of the 16S rRNA molecule of the ribosome (Figure 10-12). In eukaryotic mRNA molecules, the 5' terminus binds to the ribosome, after which the mRNA molecule slides along the ribosome until the first AUG codon in frame,

Figure 10-10
Chemical structure of N-formylmethionine. If the HC=O group at the left were an H, the molecule would be methionine.

Shine–Dalgarno sequence

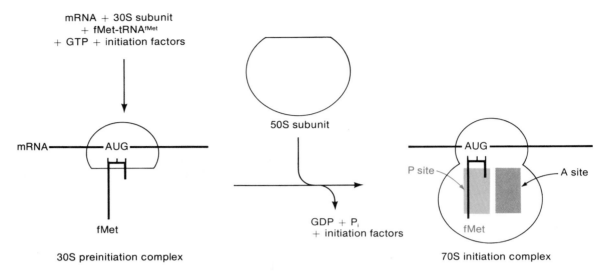

Figure 10-11
Early steps in protein synthesis in prokaryotes; formation of the 30S preinitiation complex and of the 70S initiation complex.

nearest the 5' terminus, is in contact with the ribosome; the consequence of this mechanism for initiation in eukaryotes will be explained at the end of this section.

Following formation of the 30S preinitiation complex, a 50S subunit joins with this complex to form a **70S initiation complex** (Figure 10-11).

The 50S subunit contains two tRNA binding sites. These sites are called the **A (aminoacyl) site** and the **P (peptidyl) site.** When joined with the 30S preinitiation complex, the position of the 50S subunit in the 70S initiation complex is such that the fMet–tRNA^{fMet}, which was previously bound to the 30S preinitiation complex, occupies the P site of the 50S subunit. Placement of fMet–tRNA^{fMet} in the P site fixes the position of the fMet–tRNA anticodon so that it pairs with the AUG

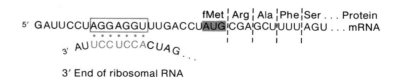

3' End of ribosomal RNA

Figure 10-12
Initiation of translation in prokaryotes. Base-pairing between the Shine-Dalgarno sequence (in box) in the mRNA and the complementary region (red) near the 3' terminus of 16S rRNA. The AUG start codon is shaded.

initiator codon in the mRNA. *Thus, the reading frame is unambiguously defined upon completion of the 70S initiation complex.*

Once the P site is filled, the A site of the 70S initiation complex becomes available to any tRNA molecule whose anticodon can pair with the codon adjacent to the initiation codon. After occupation of the A site, a peptide bond between *N*-formylmethionine and the adjacent amino acid is formed by an enzyme complex called **peptidyl transferase.** As the bond is formed, the *N*-formyl methionine is cleaved from the fMet-tRNA in the P site and transferred to form a peptide bond with the amino acid on the tRNA on the A site. Thus a dipeptide is formed on the A site.

After the peptide bond forms, an uncharged tRNA molecule occupies the P site and a dipeptidyl-tRNA occupies the A site. At this point three movements occur:

1. The $tRNA^{fMet}$ in the P site, now no longer linked to an amino acid, leaves this site.

2. The peptidyl-tRNA moves from the A site to the P site.

3. The mRNA moves a distance of three bases in order to position the next codon at the A site.

The movement of peptidyl-tRNA from the A site to the P site and the movement of mRNA in relation to the ribosome is termed **translocation** (see Figure 10-13). Step (2) requires a helper protein, elongation factor-Tu (**EF-Tu**). Step (3) requires the presence of an elongation protein **EF-G** and GTP, and it is likely that mRNA movement is a consequence of the tRNA motion. After mRNA movement has occurred, the A site is again available to accept a charged tRNA molecule having a correct anticodon.

When a chain termination codon (either UAA, UAG, or UGA) is reached, no acylated tRNA exists that can fill the A site, so chain elongation stops. However, the polypeptide chain is still attached to the tRNA occupying the P site. Release of the protein is accomplished by proteins called **release factors.** In the presence of release factors, peptidyl transferase separates the polypeptide from the tRNA; then the polypeptide chain, which has been held on the ribosome solely by the interaction with the tRNA in the P site, is released from the ribosome. The 70S ribosome then dissociates into its 30S and 50S subunits, completing the cycle.

If the mRNA molecule is polycistronic and the AUG codon initiating the second polypeptide is not too far from the stop codon of the first, the 70S ribosome will not always dissociate, but will re-form an initiation complex with the second AUG codon. The probability of such an event decreases with increasing separation of the stop codon and the next AUG codon. In some genetic systems the separation is sufficiently great that more protein molecules are always translated from the first

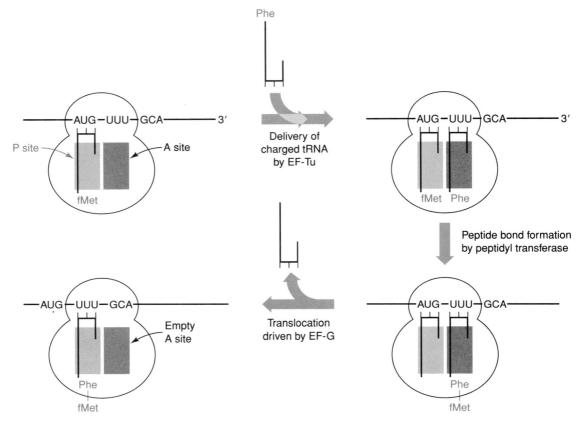

Figure 10-13
Elongation phase of protein synthesis: binding of charged tRNA, peptide bond formation, and translocation.

gene than from subsequent genes. This is a mechanism for maintaining particular ratios of gene products (Chapter 14). Mutations sometimes arise that convert a sense codon (amino-acid specifying) to a stop co-don—for example, the mRNA molecule has a UAG codon at the site of a UAC codon. Such mutations, which will be discussed further in Chapter 11, cause premature termination of a polypeptide. If the mu-tation is sufficiently far upstream from the normal stop codon of the gene containing the mutation, the distance to the next AUG codon may be so great that separation of the translating 70S ribosome and the mRNA molecule is almost inevitable. In this case, genes downstream from the mutation will rarely be translated. Such mutations are the most common type of polar mutation.

In eukaryotes reinitiation of polypeptide synthesis following an en-counter of a ribosome with a stop codon *does not occur*. Also, as pointed

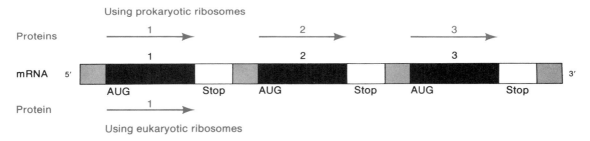

Figure 10-14
Difference in the products translated from a tricistronic mRNA molecule by the ribosomes of prokaryotes and eukaryotes. The prokaryotic ribosome translates all of the cistrons, but the eukaryotic ribosome translates only one cistron—the one nearest the 5′ terminus of the mRNA. Translated sequences are in black, stop codons are white, and the leader and spacers are shaded.

out earlier in this section, polypeptide synthesis in eukaryotes is initiated when a ribosome binds to the 5′ terminus of an mRNA molecule and slides along to the first AUG codon. There is no mechanism for initiating polypeptide synthesis at any AUG other than the first one encountered; *eukaryotic mRNA is always monocistronic* (Figure 10-14). However, a primary transcript can contain coding sequences for more than one polypeptide chain and in fact this is a frequent arrangement in animal viruses. In these cases, differential splicing generates several different mRNA molecules from one transcript (Chapter 9). For example, if the AUG nearest the 5′ terminus is excised, the second AUG will become available. Another mechanism for producing several proteins from a single transcript is protein processing. In this case, a single giant polypeptide chain, called a **polyprotein** is made and then cleaved into several component polypeptide chains, each constituting a distinct protein (Chapter 14).

Complex Translation Units

The unit of translation is almost never simply a ribosome traversing a mRNA molecule, but is a more complex structure, of which there are several forms. Some of these structures are described in this section.

After about 25 amino acids have been joined in a polypeptide chain, the AUG initiation site of the encoding mRNA molecule is completely free of the ribosome. A second initiation complex then forms. The overall configuration is two 70S ribosomes moving along the mRNA at the same speed. When the second ribosome has moved along a distance similar to that traversed by the first, a third ribosome is able to attach. This process—movement and reinitiation—continues until the mRNA

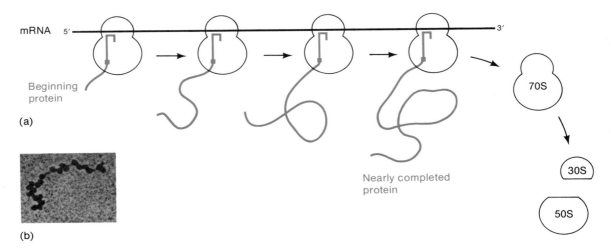

(a)

(b)

Figure 10-15
Polysomes. (a) Diagram showing relative movement of the 70S ribosome and the mRNA, and growth of the protein chain. (b) Electron micrograph of an *E. coli* polysome. (Courtesy of Barbara Hamkalo.)

polyribosome (polysome)

is covered with ribosomes at a density of about one 70S particle per 80 nucleotides. This large translation unit is called a **polyribosome** or simply a **polysome.** This is the usual form of the translation unit in all cells. An electron micrograph of a polysome and an interpretive drawing are shown in Figure 10-15.

The use of polysomes has a particular advantage to a cell; it increases the overall rate of protein synthesis compared to the rate that would occur if there were no polysomes.

An mRNA molecule being synthesized has a free 5′ terminus; since translation occurs in the 5′ → 3′ direction, each cistron contained in the mRNA is synthesized in a direction that is appropriate for immediate translation. The ribosome binding site is transcribed first, followed in order by the AUG codon, the region encoding the amino acid sequence, and finally the stop codon. Thus, in bacteria, in which no nuclear membrane separates the DNA and the ribosome, there is no obvious reason why the 70S initiation complex should not form before the mRNA is released from the DNA. With prokaryotes this does indeed occur; this process is called **coupled transcription-translation.** This coupled activity does not occur in eukaryotes, because the mRNA is synthesized and processed in the nucleus and later transported through the nuclear membrane to the cytoplasm where the ribosomes are located.

Coupled transcription-translation speeds up protein synthesis in the sense that translation does not have to await release of the mRNA from the DNA. Translation can also be started before the mRNA is degraded by nucleases. Figure 10-16(a) shows an electron micrograph of a DNA

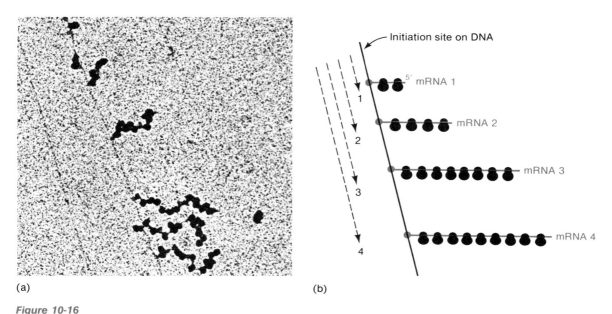

(a)   (b)

Figure 10-16
(a) Transcription of a section of the DNA of *E. coli* and translation of the nascent mRNA. Only part of the chromosome is being transcribed. The dark spots are ribosomes, which coat the mRNA. (From O. L. Miller, Barbara A. Hamkalo, and C. A. Thomas, *Science*; 1977: 169, 392.) (b) An interpretation of the electron micrograph of part (a). The mRNA is in red and is coated with black ribosomes. The large red spots are the RNA polymerase molecules; they are actually too small to be seen in the photo. The dashed arrows show the distances of each RNA polymerase from the transcription initiation site. Arrows 1, 2, and 3 have the same length as mRNA 1, 2, 3; mRNA 4 is shorter than arrow 4, presumably because its 5' end has been partially digested by an RNase.

molecule to which are attached a number of mRNA molecules, each associated with ribosomes. The micrograph is interpreted in panel (b). Note that the lengths of the polysomes increase with distance from the transcription initiation site, because the mRNA is farther from that site and hence longer.

Eukaryotic mRNA differs from prokaryotic mRNA in two major features. It is modified both by the addition of a cap at the 5' end of the message and by the addition of about 200 adenlynic residues at the *polyadenalation (poly A)* 3' end which are termed **poly A.** These modifications do not take place in mitochondrial and chloroplast mRNA. The role of the poly A in translation is not yet clear. The cap adds a 7-methylguanosine residue linked to a triphosphate at the 5' end of the mRNA. Thus this is a 5' to 5' linkage instead of the usual 5' to 3' linkage. (Figure 10-17). The cap may protect the mRNA from exonucleases, but its main role is to promote and stimulate efficient translation. The 5' untranslated region at the cap end of the eukaryotic mRNA is folded into a secondary structure held together by base-pairing. This secondary structure inhibits

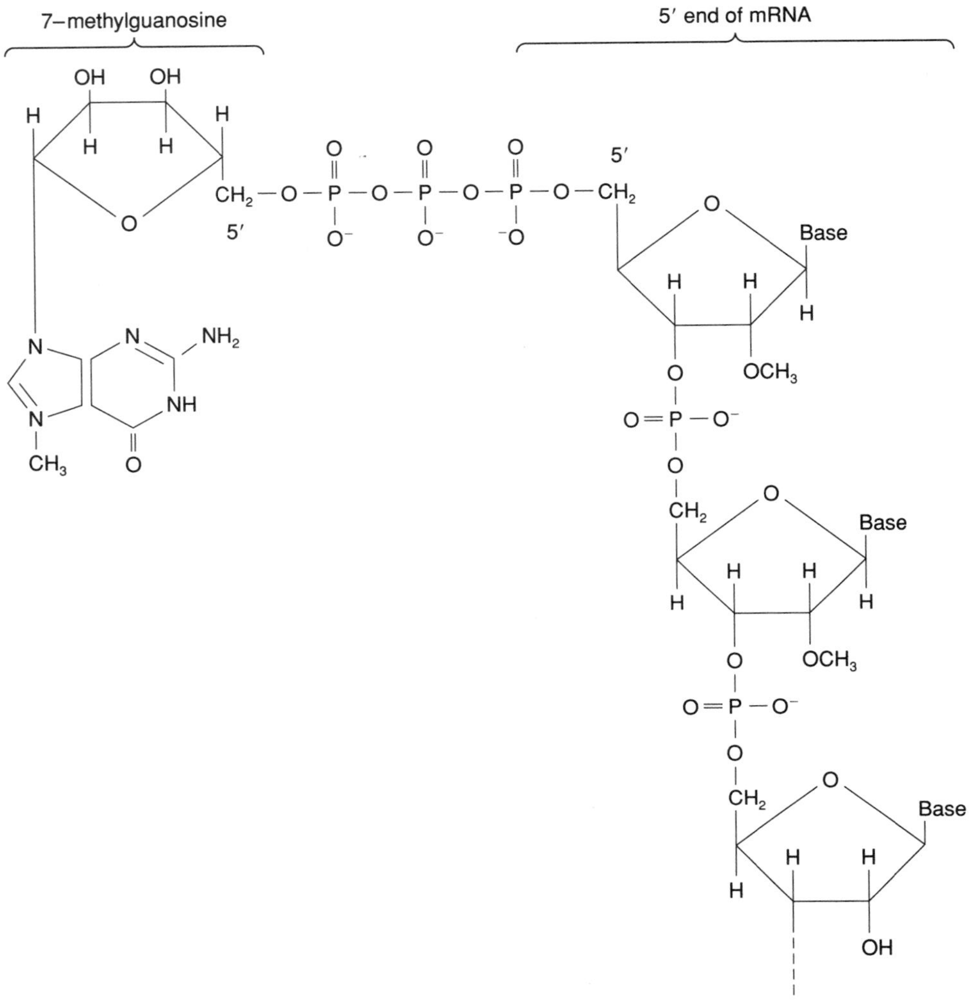

Figure 10-17
Structure of the cap at the 5' end of eukaryotic mRNA. Note the unusual 5' to 5' linkage. All caps contain 7-methylguanylate (shown in color) attached by a triphosphate linkage to the ribose of the 5' end of the mRNH. The 2' hydroxyl group of the first mRNA ribose is also methylated while the second sugar is not always methylated.

efficient translation. Specific proteins, called cap-binding proteins, bind to the cap and unwind the secondary structure of the 5' end of the mRNA. This facilitates efficient translation. These proteins are also involved in guiding the 40S eukaryotic ribosomal subunit to the cap, and thus to the 5' end of the mRNA. Chapter 9 contains further information on mRNA modification.

Antibiotics

Many antibacterial agents (**antibiotics**) have been isolated from fungi. Most of these are inhibitors of protein synthesis. For example: streptomycin and neomycin bind to a particular protein in the 30S particle and thereby prevent binding of tRNA^{fMet} to the P site; the tetracyclines inhibit binding of charged tRNA; lincomycin and chloramphenicol inhibit the peptidyl transferase; and puromycin causes premature chain termination; erythromycin binds to a free 50S particle and prevents formation of the 70S ribosome. A particular antibiotic has clinical value only if it acts on bacteria and not on animal cells; the clinically useful antibiotics usually either fail to pass through the cell membrane of animal cells, or do not bind to eukaryote ribosomes because of some unknown feature of their structure.

Some disease-causing bacteria exert their pathogenic effect because they excrete inhibitors of mammalian protein synthesis. The agent causing diphtheria is an example; it binds to a factor necessary for movement of mammalian ribosomes along the mRNA.

Summary

Translation is the production of proteins from the information encoded in DNA. That information (the genetic code) consists of the 64-three base codons standing for translational starts, stops, and amino acids. For most amino acids, more than one codon exists, differing only in the third position. Charged tRNA molecules have an amino acid attached to the 3' ends by aminoacyl tRNA synthetases, and their anticodon base pairs with mRNA codons. One tRNA reads several codons since the first base of the anticodon (wobble position) pairs with the third base of several codons. A reading frame is a series of sequential, nonoverlapping codons with one start and stop codon. Translation occurs on ribosomes and consists of three stages: (1) Initiation which brings a small ribosomal subunit, initiating tRNA, and mRNA together. The start codon is chosen by the Shine-Dalgarno sequence in prokaryotes which use formylmethionine to initiate; eukaryotes use the first AUG and methionine. The large subunit, containing the P (peptidyl) and A (aminoacyl) tRNA sites then joins; (2) Elongation, which involves the binding of charged tRNAs, peptide bond formation by peptidyl transferase, and the translocation of the mRNA by one codon. GTP and initiation and elongation factors are required; (3) Termination, which occurs when a stop codon is reached. Release factors separate the polypeptide from the tRNA in the P site. Following termination, the ribosome may dissociate, or in prokaryotes, may begin translation again at a new 3' AUG. Thus prokaryotic mRNAs can be polycistronic. Eukaryotic mRNAs are mon-ocistronic, although one mRNA can produce several proteins by dif-ferential mRNA processing or cleavage of a polyprotein. In prokaryotes, transcription and translation are coupled, while in eukaryotes, they are separate.

Drill Questions

1. Which of the following statements are true of tRNA molecules? If false explain.
   a. They are needed because amino acids can-not stick to mRNA.
   b. They are much smaller than mRNA molecules.
   c. They are synthesized without the need for intermediary mRNA.
   d. They bind amino acids without the need for any enzyme.
   e. They occasionally recognize a stop codon if the Rho factor is present.

2. Which of the following properties are part of the normal functioning of aminoacyl tRNA synthetases?
   a. Recognition of the codon
   b. Recognition of the anticodon of a tRNA molecule
   c. Recognition of the amino acid recognition region of a tRNA molecule
   d. Ability to distinguish one amino acid from another
   e. Ability to remove an incorrectly coupled amino acid from a tRNA molecule

3. What are the s values of prokaryotic and eu-karyotic ribosomes, their subunits, and their RNA molecules?

4. List three differences between prokaryotic and eukaryotic translation (ignore differences in ri-bosome structure).

5. Which of the following are steps in protein synthesis in prokaryotes? If false, explain.
   a. Binding of tRNA to a 30S particle
   b. Binding of tRNA to a 70S ribosome
   c. Coupling of an amino acid to ribosome by an aminoacyl synthetase
   d. separation of the 70S ribosome to form 30S and 50S particles.

6. What is the reading frame of a mRNA? What additional features would you expect a mRNA to have?

7. Translation has evolved in a particular polarity with respect to the mRNA molecule. What is this polarity, and what would be the disad-vantages of having the reverse polarity?

8. Which of the following is the normal cause of chain termination?
   a. The tRNA corresponding to a chain-ter-mination triplet cannot bind an amino acid.
   b. There is no tRNA with an anticodon cor-responding to a chain-termination triplet.
   c. Messenger RNA synthesis stops at a chain termination triplet.

9. Which processes in protein synthesis require hydrolysis of GTP?

Problems

1. A DNA molecule has the structure

   TACGGGAATTAGAGTCGCAGGATC
   ATGCCCTTAATCTCAGCGTCCTAG

   The upper strand is the coding strand, and is transcribed from left to right. What is the amino acid sequence of the protein encoded in this DNA molecule?

2. Which amino acids can replace arginine by a single base-pair change?

3. The amber codon UAG does not correspond to any amino acid. Some strains carry sup-pressors which are tRNA molecules mutated in the anticodon, enabling an amino acid to be

placed in at a UAG site. Assuming that the anticodon of the suppressor differs by only one base from the original anticodon, which amino acids could be inserted at a UAG site? At a UAA site?

4. There are several arginine codons. Suppose you had a protein that contained only three arginines (Arg-1, Arg-2, Arg-3). In a particular mutant Arg-1 is replaced by glycine. In another Arg-2 is replaced by methionine. In still another mutant Arg-3 is replaced by isoleucine. Suppose several hundred other mutants at various sites are isolated. Which other amino acids would you expect to find replacing Arg-1, Arg-2, and Arg-3, assuming only single-base changes?

5. Suppose that you are making use of the alternating copolymer GUGUGUGUGU ... as an mRNA in an *in vitro* protein-synthesizing system. Assuming that an AUG start codon is not needed in the *in vitro* system, what peptides are made by this mRNA?

6. The anticodon of a tRNA is 5′-IGU-3′.
   a. Which amino acid would this tRNA insert during translation?
   b. Which codons would be read by this tRNA?
   c. Assuming that only two tRNAs are needed to read this family of codons, what would the anticodon of the other tRNA probably be?

Conceptual Questions

1. How might various antibiotics be used to study the process of translation?

2. Consider the advantages and disadvantages of overlapping reading frames.

3. What do you think is the significance of the universality of the genetic code? What does this say about the biological evolution of the system and its ability to change?

Thomas Lindahl
Associate Director of Research
and Head of Clare Hall Laboratories

Birthday: **1938**

Birth Place: **Stockholm, Sweden**

Graduate Degree: **Karolinska Institute,**
MD, Stockholm, Sweden

Postdoctoral Training: **Princeton**
University and Rockefeller University,
1964–1969

Present Position: **Imperial Cancer**
Research Fund, Associate Director of
Research and Head of Clare Hall
Laboratories

Address: **London, England**

AVOIDANCE OF AN unacceptably high mutation rate is a major challenge to living cells. The various defense pathways needed to solve this problem involve at least as many proteins as are needed for the replication of DNA. We have investigated several types of spontaneous DNA damage and the repair mechanisms that counteract these potentially mutagenic lesions. Enzymes acting on DNA in unexpected ways have been discovered in this fashion. Several basic correction mechanisms appear to be present in all living cells. The extensive information obtained with *E. coli* as a model organism is now being applied to the study of cancer-prone human inherited syndromes. Some rare diseases of this type appear to be defective in DNA repair functions.

In what direction is the field of molecular biology moving?

Our increasingly detailed knowledge of key biological processes has emphasized the need for biochemical insights. Aging, cell differentiation, and more detailed topics such as acquired drug resistance in cancer cells will be investigated and clarified in the future by the techniques of nucleic acid and protein chemistry.

11

Mutagenesis, Mutations, and Mutants

In previous chapters mutants have been encountered repeatedly. During the early days of genetics, the only mutants that were available for study were those that arose spontaneously. We will learn that a wide variety of mutagenic agents are now routinely employed by molecular geneticists to produce strains of both prokaryotic and eukaryotic organisms with desired characteristics. That is, we will realize that mutants represent a powerful tool for studying molecular processes. The use of mutants can be considered as ranking alongside methods such as electrophoresis and DNA sequencing, as far as their impact on molecular biology is concerned.

Often it has been the case that metabolic pathways and mechanisms that regulate gene expression have been deduced from the study of mutants. Human intuition is sometimes insufficient for predicting which molecule or process is associated with a particular function. The study of the characteristics of an appropriate mutant organism can, however, often provide important clues.

The study of prokaryotic organisms has led the way, for these organisms provide highly convenient experimental systems for both the application of mutagenic agents and the detection of mutant phenotypes. We will, therefore, focus mostly on those simpler organisms.

In this chapter we examine the biochemical basis of being a mutant. We begin with definitions of several terms—mutant, mutation, mutagen, and mutagenesis—and a description of several types of mutations. Our ability to understand the molecular basis of mutants depends largely on our knowledge of DNA structure and the mechanism of protein synthesis. That is because mutations change the nucleotide sequence of DNA. Those changes in the informational content of DNA are ultimately expressed in terms of the production of an altered protein product.

Types of Mutations

mutant

Mutant refers to an organism or a gene that is different from the normal or wild type. His⁻ yeast, or white-eyed *Drosophila* are examples of mutants. However, when referring to biochemical properties seen in nature and the normal form is ⁻ (for example, *E. coli* isolated from nature is unable to metabolize lactose, or, Lac⁻), the ⁺ form is called a wild type, and any ⁻ form is referred to as a mutant.

mutation

Mutation refers to any change in the base sequence of DNA that gives rise to a mutant phenotype. The most common change is a substitution, addition, or deletion of one or more bases (Figure 11-1).

mutagen

A **mutagen** is a physical agent or a chemical reagent that causes mutations to occur.

Mutagenesis is the process of producing a mutation. If it occurs in nature without the addition of a mutagen, it is called **spontaneous mutagenesis** and the resulting mutants are **spontaneous mutants.** If a mutagen is used, the process is **induced mutagenesis.**

Mutations are classified in several ways. One distinction is based on the nature of the change—specifically, on the number of bases changed. Thus, we may distinguish a **point mutation** in which there is only a single changed base-pair, from a multiple mutation in which two or more base-pairs differ from the wild-type sequence. A point mutation may be a **base substitution,** a **base insertion,** or a **base deletion** (Figure 11-1), but the term most frequently refers to a base substitution.

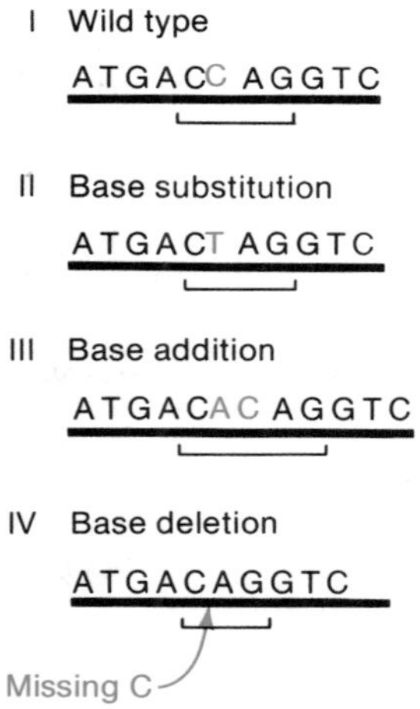

point mutation
missense mutation

I Wild type

A T G A C C A G G T C

II Base substitution

A T G A C T A G G T C

III Base addition

A T G A C A C A G G T C

IV Base deletion

A T G A C A G G T C

Missing C

A second distinction is based on the consequence of the change in terms of the amino acid sequence affected. For example, if there is an amino acid substitution, the mutation is a **missense mutation.**

If the substitution produces a protein that is active at one temperature (typically 30°C) and inactive at a higher temperature (usually 40–42°C), the mutation is called a **temperature-sensitive** or ts mutation. If the mutation generates a stop codon, protein synthesis will stop, and the mutation is called a **chain-termination mutation** or a **nonsense mutation.** Temperature-sensitive and chain-termination mutations exhibit the mutant phenotype only under certain conditions, and hence are called **conditional mutations;** they are the most versatile and useful mutations available to the molecular biologist.

With microorganisms the phenotype and genotype are written with a capital letter and lower-case italics, respectively. This convention does not apply to higher organisms. The notation for bacterial mutants is summarized in Table 11-1.

Whether produced spontaneously or by induced mutagenesis, mutants must be selected from large populations. Numerous techniques exist for selecting desired mutants; they are usually applicable to particular organisms and to specific types of mutations. A complete de-

Figure 11-1
Three types of mutations. Only the base sequence in one DNA strand is shown. Changes are shown in red. The horizontal brackets indicate the affected segment.

Table 11-1

Summary of Notation Used to Designate Bacterial Mutants and Mutations

Phenotype or Genotype	Designation*
Phenotype	
Lacking in or possessing ability to make a substance	Sub⁻, Sub⁺
Resistance or sensitivity to an antibiotic	Ant-r, Ant-s
Genotype	
Wild-type gene for making a substance	*sub⁺*
Mutant gene for making a substance	*sub⁻*†
Mutant *subA* gene; mutation number 63 in *subA* gene	*subA⁻*; *subA63*
Gene for resistance or sensitivity to a particular antibiotic	*ant*
Genotype for resistance or sensitivity to an antibiotic	*ant-r, ant-s*

* The arbitrary abbreviations sub and ant mean "substance" and "antibiotic" in this table.
† The notation sub for *sub⁻* is widespread. Because of possible ambiguity, the superscript minus is used throughout this book.

scription is beyond the scope of this book; for additional information, see texts in genetics or microbiology.

Biochemical Basis of Mutants

A mutant may be defined as an organism in which either the base-sequence of DNA or the phenotype has been changed. These definitions are the same (except for the case of a silent mutation, which will be discussed shortly) since the base-sequence of DNA determines the amino acid sequence of a protein. The chemical and physical properties of each protein are determined by its amino acid sequence, so that a single amino acid change is capable of inactivating a protein.

From the discussion of protein structure in Chapter 4 it is easy to understand how an amino acid substitution can change the structure and the biological activity of a protein. For instance, consider a hypothetical protein whose three-dimensional structure is determined entirely by an interaction between one positively charged amino acid (for example, lysine) and one negatively charged amino acid (aspartic acid). A substitution of methionine, which is uncharged, for the lysine would clearly destroy the three-dimensional structure, as would a substitution of histidine, which is positively charged, for aspartic acid. Similarly, a protein might be stabilized by a hydrophobic cluster, in which case substitution of glutamine (polar) for leucine (nonpolar) would also be disruptive.

A base substitution does not always yield a mutant phenotype. Because of the redundancy of the code, some changes do not alter the amino acid sequence, and some amino acid changes do not significantly affect the structure of a protein. A base change with these properties is said to be **silent.**

The shapes of proteins are determined by such a variety of interactions that sometimes an amino acid substitution is only partially disruptive. For instance, an isoleucine might substitute successfully for leucine and be silent, but replacement with a more bulky amino acid such as phenylalanine might cause subtle stereochemical changes, though a hydrophobic cluster is preserved. This could be manifested as a reduction rather than a loss, of activity of an enzyme. For example, a bacterium carrying such a mutation in the enzyme that synthesizes adenine might grow very slowly (but it would grow) unless adenine were provided in the growth medium. Such a mutation is called a **leaky mutation;** these mutations are not particularly useful for most genetic studies. However, several inherited human diseases are due to such leaky mutations. For example, hemolytic anemia and neonatal jaundice can be a consequence of a point mutation in the gene encoding the essential enzyme glucose-6-phosphate dehydrogenase, leading to the production of a protein with reduced catalytic efficiency.

Generally speaking, the following types of amino acid substitutions are expressed as nonleaky mutations: polar to nonpolar, nonpolar to polar, change of sign of a charge, small side chain to bulky side chain, sulfhydryl to any other side chain, hydrogen-bonding to non–hydrogen-bonding, any change to or from proline (which changes the shape of the polypeptide backbone), and any change in a substrate-binding site.

So far, only amino acid substitution mutations have been discussed. Mutations that eliminate activity of the protein totally are base deletions, which cause one or more amino acids to be absent from the protein; frameshifts, in which all amino acids starting from the mutant site are different; and chain termination mutants, in which a protein chain is prematurely terminated, will be discussed in greater detail shortly.

Mutagenesis

The production of a mutant requires that a change occur in the base-sequence. This can occur spontaneously by replication errors or chemical decay of DNA, or can be stimulated to occur in five main ways:

1. Removal of an incorrectly inserted base is prevented.

2. A base is inserted that tautomerizes and allows a substitution to occur in subsequent replication.

3. A previously inserted base is chemically altered to a base having different base-pairing specificity.

4. One or more bases are skipped during replication.

5. One or more extra bases are inserted during replication.

In the following sections we describe mutagens that act by one or more of these mechanisms.

BASE-ANALOGUE MUTAGENS

A base analogue is a substance other than a standard nucleic acid base that can be built into a DNA molecule by the normal process of polymerization. Such a substance must be able to pair with the base on the complementary strand being copied, or the $3' \rightarrow 5'$ editing function will remove it. However if it can tautomerize, or if it has two modes of hydrogen-bonding, it will be mutagenic.

The substituted base 5-bromouracil (BU) is an analogue of thymine inasmuch as the bromine has about the same van der Waals radius as the methyl group of thymine (Figure 11-2). In subsequent rounds of replication, BU functions like thymine and primarily pairs with adenine. Thymine can sometimes (but rarely) assume an enol form that is capable of pairing with guanine, and this conversion occasionally gives rise to mutants in the course of replication. The mutagenic activity of 5-bromouracil stems from a shift in the keto-enol equilibrium caused by the bromine atom; that is, the enol form exists for a greater fraction of time for BU than for thymine. If BU replaces a thymine, in subsequent rounds

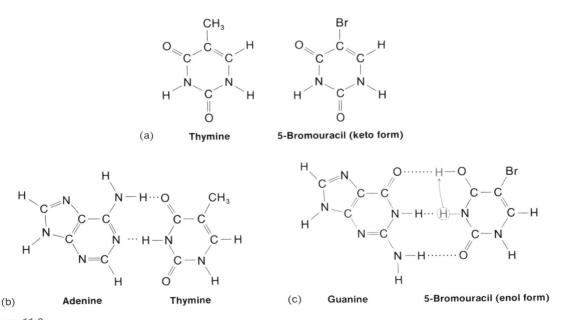

Figure 11-2
Mutagenesis by 5-bromouracil. (a) Structural formulas of thymine and 5-bromouracil. (b) A standard adenine-thymine base pair. (c) A base pair between guanine and the enol form of 5-bromouracil. The red H in the dashed circle shows the position of the H in the keto form.

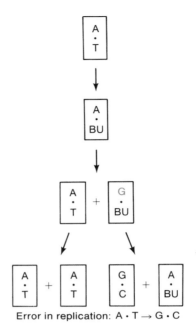

Error in replication: A · T → G · C

Figure 11-3

A mechanism of 5-bromouracil (or BU)-induced mutagenesis. (a) During replication, BU, in its usual keto form, substitutes for T and the replica of an initial A · T pair becomes an A · BU pair. In the first mutagenic round of replication the BU, in its rare enol form, pairs with G. In the next round of replication, the G pairs with a C, completing the transition from an A · T pair to a G · C pair.

of replication, it occasionally generates a guanine, which in turn specifies cytosine resulting in formation of a G · C pair (Figure 11-3).

Additional experiments suggest that 5-bromouracil is also mutagenic in another way. The concentrations of the nucleoside triphosphates in most cells is regulated by the concentration of thymidine triphosphate (TTP). This regulation results in appropriate relative amounts of the four triphosphates for DNA synthesis. One part of this complex regulatory process is the inhibition of synthesis of deoxycytidine triphosphate (dCTP) by excess TTP. The 5-bromouracil nucleoside triphosphate also inhibits production of dCTP. When 5-bromouracil is added to the growth medium, TTP continues to be synthesized by cells at the normal rate, while the synthesis of dCTP is significantly reduced. The ratio of TTP to dCTP then becomes quite high, and the frequency of misincorporation of T opposite G increases. The editing function and mismatch repair systems are capable of removing incorrectly incorporated thymine, but in the presence of 5-bromouracil the rate of misincorporation can exceed the rate of correction. An incorrectly incorporated thymine that persists will pair with adenine in the next round of DNA replication, yielding a G · C → A · T transition in one of the daughter molecules. Thus, 5-bromouracil induces transitions in both directions— A · T → G · C by tautomerization, and G · C → A · T by misincorporation. So mutations that are induced by 5-bromouracil can also be reversed by it.

Both base-pair changes induced by BU maintain the original purine (Pu)-pyrimidine (Py) orientation. The original and the altered base pairs both have the orientation Pu · Py—namely, A · T and G · C. If the original pair were T · A, the altered pair would be C · G, and the orientation would be Py · Pu in both the original and the altered pairs. A base change that does not change the Py · Pu orientation is called a **transition.** Base analogue mutations are always transitions. Later we will see changes from Pu · Py to Py · Pu, and Py · Pu to Pu · Py; when such a change of orientation occurs, the mutation is called a **transversion.**

CHEMICAL MUTAGENS

A chemical mutagen is a substance that can alter a base that is already incorporated in DNA and thereby change its hydrogen-bonding specificity. Three very powerful chemical mutagens are nitrous acid (HNO_2), hydroxylamine (HA), and ethylmethane sulfonate (EMS). Their chemical structures are shown in Figure 11-4.

Nitrous acid primarily converts amino groups to keto groups by oxidative deamination. Thus cytosine, adenine, and guanine are converted to uracil (U), hypoxanthine (H), and xanthine (X), respectively. These bases can form the base-pairs U · A, H · C, and X · C. The changes are G · C → A · T and A · T → G · C as cytosine and adenine respectively are deaminated. Since G and X both pair with C, the change should not be directly mutagenic.

Hydroxylamine reacts specifically with cytosine and converts it to a modified base that pairs only with adenine, so that a G · C pair ultimately becomes an A · T pair. The chemistry of the alteration is complex.

EMS and a related substance methylmethane sulfonate (MMS) are alkylating agents that have been used extensively in genetic research. The alkylating agents include environmental carcinogens such as nitrosamines, which require metabolic activation, and also very strong and reactive mutagens such as methylnitrosourea and N-methyl-N′-nitro-N-nitrosoguanidine (MNNG). Chemical mutagens such as nitrous acid and hydroxylamine, which are exceedingly useful in prokaryotic systems, are not particularly useful as mutagens in higher eukaryotes (because the chemical conditions necessary for reaction are not easily obtained); the alkylating agents, however, are highly effective. Many sites in DNA are alkylated by these agents. Of prime importance for the induction of mutations is the addition of an alkyl group to the hydrogen-

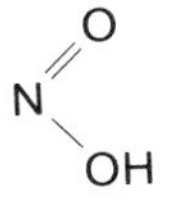

Nitrous acid

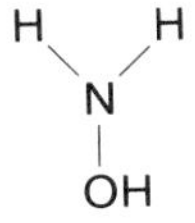

Hydroxylamine

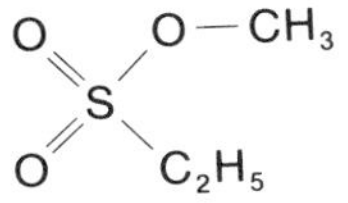

Ethyl methane sulfonate

Figure 11-4
Structures of three chemical mutagens.

bonding oxygen of guanine and thymine. These alkylations impair the normal hydrogen-bonding of the bases, and cause mispairing of G with T, leading to the transitions $A \cdot T \rightarrow G \cdot C$ and $G \cdot C \rightarrow A \cdot T$. The most important mutagenic lesion caused by simple alkylating agents is O^6-alkylguanine. The alkylation event freezes the modified guanine into the anomalous tautomeric enol form, which readily mispairs with thymine.

ULTRAVIOLET IRRADIATION

Ultraviolet light is a fairly potent mutagen. The main DNA lesions are two chemically different types of covalent dimers: cyclobutane-pyrimidine dimers, and (6-4) pyrimidine dimers. The latter only account for 20% of the total dimers but are correspondingly more mutagenic. In *E. coli* the number of mutants induced by ultraviolet light can be reduced by exposure to visible light (photoreactivation), which implicates cyclobutane pyrimidine dimers since (6-4) dimers cannot be photoreactivated. Bacterial mutants that lack the ability to carry out SOS repair (*lexA*⁻ and *recA*⁻ mutants) are not mutagenized by ultraviolet light. These and other experimental results have made it clear that mutagenesis by ultraviolet irradiation is almost exclusively a result of replication errors made during error-prone SOS repair. Error-prone repair of pyrimidine dimers leads to the production of both transitions and transversions.

MUTAGENESIS BY INTERCALATING SUBSTANCES

Acridine orange, proflavine, and acriflavine (Figure 11-5), which are substituted acridines, are planar, three-ringed molecules whose dimensions are roughly the same as those of a purine-pyrimidine pair. In aqueous solution these substances form stacked arrays, and are also able to stack with a base-pair by insertion between two base-pairs (a process called **intercalation**). Since the thickness of the acridine molecule is approximately that of a base-pair and because the two bases of a pair are normally in contact, the intercalation of one acridine molecule causes adjacent base-pairs to move apart by a distance equal to that of the

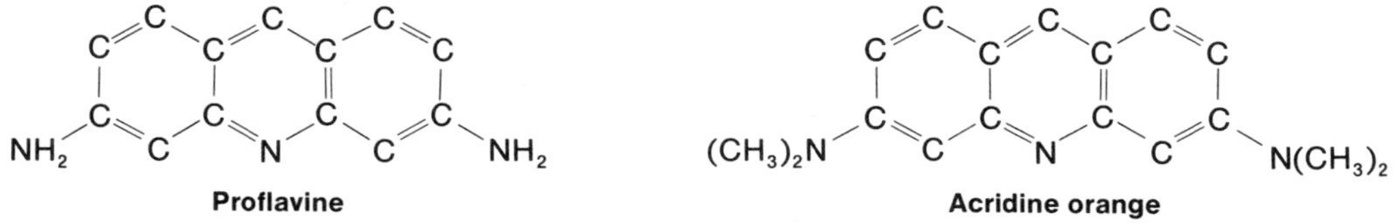

Figure 11-5
Structures of two mutagenic acridine derivatives.

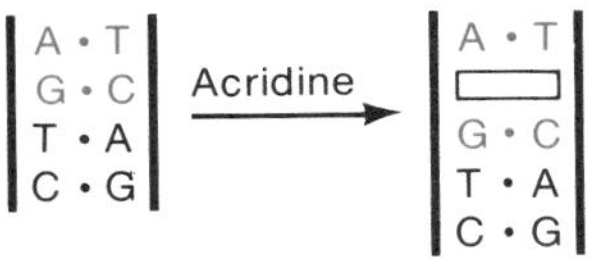

A · T
G · C
T · A
C · G

Acridine →

A · T
☐
G · C
T · A
C · G

Figure 11-6
Separation of two base pairs (shown in red) by an intercalating agent.

thickness of one base-pair (Figure 11-6). This has bizarre effects on the outcome of DNA replication, though the mechanism of action of the mutagen is not known. When DNA containing intercalated acridines is replicated, additional bases appear in the sequence (Figure 11-7). The usual addition is a single base, though occasionally two bases are added. Deletion of a single base also occurs, but this is far less common than base addition. Mutations of this sort are called **frameshift mutations.** This is because the base-sequence is read in groups of three bases when it is being translated into an amino acid sequence; the addition of a base changes the reading frame (Figure 11-7). This will be discussed in greater detail in the section on reversion.

frameshift mutation

MUTAGENESIS BY INSERTION OF LONG SEGMENTS OF DNA

(TRANSPOSABLE ELEMENTS)

E. coli, and many other organisms as well, contain mobile DNA segments that are hundreds to thousands of base-pairs long (called **transposable elements**). When a transposable element replicates, one replica remains at the original insertion site, and the other replica is inserted in another region of the chromosome in a complex process discussed in

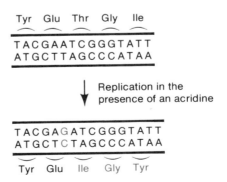

Figure 11-7
A base addition (red) resulting from replication in the presence of an acridine. The change in amino acid sequence read from the upper strand in groups of three bases is also shown in red.

Chapter 12. This process of insertion of a replica at a second site is called **transposition.** When transposition occurs, the sequence frequently inserts itself into a bacterial gene, thereby mutating that gene.

Some transposable elements contain sequences for termination of transcription. If such an element inserts between two genes that are transcribed as a polycistronic mRNA or between the promoter and the first gene, all genes downstream from the insertion site will not be transcribed. These mutations belong to a class called **polar mutations.**

MUTATOR GENES

In *E. coli* there are genes that in a mutant state cause mutations to appear very frequently in other genes throughout the genetic map. These genes are called **mutator genes.** This is a misnomer, because the function of each gene is probably to keep the mutation frequency low; only when the product of a mutator gene is itself defective, will there be widespread production of mutations.

Of the many mutators that have been observed, four types that yield strong phenotypic effects are:

1. A mutant DNA polymerase that reduces or eliminates the $3' \rightarrow 5'$ exonuclease activity of the editing function.

2. A mutant methylating enzyme (the *dam* enzyme) responsible for methylation of the sequences that the mismatch repair system uses to discriminate parental from daughter strands.

3. A mutant enzyme that cannot carry out the excision step in mismatch repair.

4. Mutations in the regulatory circuits that maintain the error-prone SOS repair system in an off state.

Mutational Hot Spots

If several hundred mutations in a single gene are mapped, they are, for the most part, distributed roughly equally over the mutated sites. However, a few sites are represented by as many as 100 times the typical number of mutations; these sites are called **hot spots.**

About 4% of the cytosines in a typical DNA molecule are in the methylated form—5-methylcytosine (MeC). The role of MeC (and other methylated bases as well) is not clearly understood in most cases, though one role is described in Chapter 8. At any rate, they are not harmful and do not change the hydrogen-bonding properties of the base; MeC pairs with guanine just as cytosine does. MeC is also subject to alteration by spontaneous deamination. When cytosine is deaminated, uracil is formed and this is removed by uracil-DNA glycosylase. However, when MeC is deaminated, the result is 5-methyluracil, which is another name for thymine; therefore the G · MeC pair becomes a G · T pair, which in subsequent replication yields an A · T pair. Note that the main mismatch repair system is certainly able to convert the G · T pair back to a correct G · C pair. However, spontaneous deamination can occur in nonreplicating DNA (e.g., in a resting cell or in a phage), and both strands will be equally methylated by the A-methylating system (Chapter 8). Thus the mismatch repair system receives no signal indicating that the G · C pair is the correct one, and could just as well convert the G · T pair to an A · T pair. The mutation frequency consequently can be very high at a MeC site. A given G · MeC → A · T transition will, of course, only produce a mutant if the change causes an amino acid substitution that affects the activity of the gene product. Such MeC sites do not occur very often, so hot spots should not be particularly frequent. Direct determination of the base-sequence of several genes and of hot spot mutants has shown that, indeed, MeC accounts for most of the hot spots for spontaneous mutagenesis.

Reversion

reversion

So far we have discussed changes from the wild type to the mutant state. The reverse process, in which the wild-type phenotype is regained, also occurs; this process is called **back mutation, reverse mutation,** or, most commonly, **reversion.** One way that the wild-type phenotype may be restored is to regain the wild-type genotype (that is, the wild-type base-sequence). However, this is not always what happens.

INTRAGENIC REVERSION

The most useful type of revertant in molecular biological studies has been a kind that does not faithfully recreate the wild-type base sequence. In these revertants the reversion does not occur at the site of the mutation, but instead entails a mutation at a second site. Such mutations are often called **second-site** or **suppressor mutations.** In some cases the reverse mutation does not even occur within the mutated gene, and we may distinguish between intragenic and intergenic reversion. We first consider the intragenic type.

Consider a hypothetical protein containing 97 amino acids whose structure is determined entirely by an ionic interaction between a positively charged (+) amino acid at position 18 and a negative one (−) at position 64 (Figure 11-8). If the (+) amino acid is replaced by a (−)

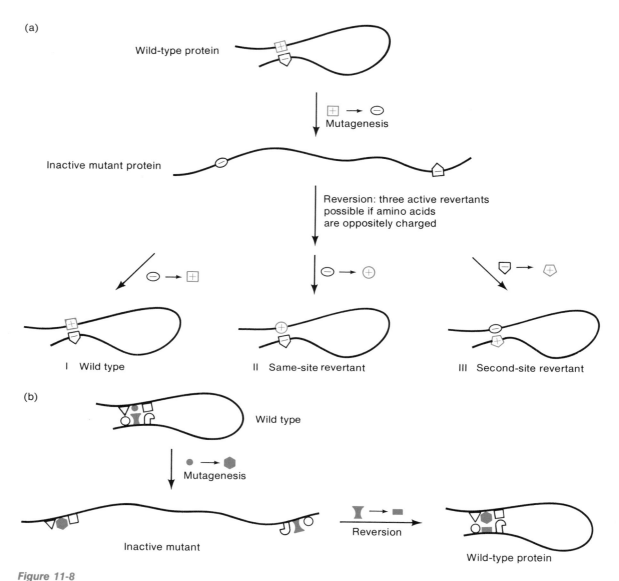

Figure 11-8
Several mechanisms of reversion. In panel (a) the charge of one amino acid is changed, and the protein loses activity. The activity is returned by (I) restoring the original amino acid, or (II) by replacing the (−) amino acid by another (+) amino acid, or (III) by reversing the charge of the original (−) amino acid. In each case the attraction of opposite charges is restored. In panel (b) the structure of the protein is determined by interactions between six hydrophobic amino acids. Activity is lost when the small circular amino acid is replaced by the bulky hexagonal one, and is restored when space is made by replacing the convex amino acid with the small rectangle.

amino acid, the protein is clearly inactive. Three kinds of reversion events would restore activity (Figure 11-8(a)):

1. The original (+) amino acid could be put back.

2. A different (+) amino acid could be put at position 18.

3. The (−) amino acid at position 64 could be replaced by a (+) amino acid; this second-site mutation would restore the activity of the protein. A possibility which would not generally work but which might work in a specific case is to insert a (+) amino acid at position 17 or 19.

Figure 11-8(b) shows another more complicated example of intragenic reversion. In this case the structure of a protein is maintained by a hydrophobic interaction. The replacement of an amino acid with a small side chain by a bulky phenylalanine changes the shape of that region of the protein. A second amino acid substitution providing space for the phenylalanine could restore the protein structure.

The analysis of second-site amino-acid substitution has been an important aid in determining the three-dimensional structure of proteins because the following rule is often obeyed: If a substitution of amino acid A by amino acid X, which creates a mutant, is compensated for by a substitution of amino acid B by amino acid Y, then A and B are either three-dimensional neighbors, or are both contained in two interacting regions.

Revertants of frameshift mutations usually occur at a second site. It is of course possible that a particular added base could be removed or a particular deleted base could be replaced by a spontaneous event, but this would not occur very often. Second-site reversion of a frameshift mutation has two requirements illustrated in Figure 11-9:

1. The reverting event must be very near the original site of mutation, so that very few amino acids are altered between the two sites.

2. The segment of the polypeptide chain in which both changes occur must be able to withstand substantial alterations.

INTERGENIC REVERSION AND SUPPRESSION

Intergenic reversion refers to a mutational change in a second gene that eliminates or suppresses the mutant phenotype. One type, which occurs when two proteins interact, is a mutation in the binding site of protein A that prevents the protein from interacting with protein B. Another mutation in the binding site of protein B alters this binding site so that the mutant B protein can bind to the mutant A protein; thus, the interaction between the two proteins is restored. The occurrence of intergenic reversion of this kind is an important indicator of the interaction

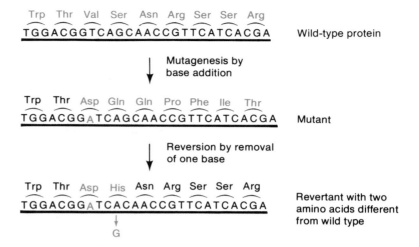

Reversion by base deletion from an acridine-induced base-addition mutant.

between two proteins and is a splendid example of a genetic result giving information about molecular structure.

A second type of extragenic reversion has the remarkable property that the second-site mutation not only eliminates the effect of the original mutation but also suppresses mutations in many other genes as well. This type of reversion, which is produced by mutations in certain tRNA molecules and aminoacyl synthetases, can be seen most clearly when reversion of chain termination mutations are examined.

Chain termination mutations are common, for they can arise in many ways. For example, a single base change in any of the codons AAG, CAG, GAG, UCG, UUG, UGG, UAC, and UAU can give rise to the chain termination codon UAG. If such a mutation occurs within a gene, a mutant protein with little or no function will result, because no tRNA molecule exists whose anticodon is complementary to UAG. Thus, only a fragment of the wild-type protein is produced, and this usually fails to function unless the mutation is very near the carboxyl terminus of the protein.

In certain bacterial strains the presence of a chain termination mutation is not sufficient to stop polypeptide synthesis. For example, a phage may have acquired a UAG codon in a gene encoding a critical protein; when the phage infects most host bacteria, no phage progeny will be produced. However, in a particular bacterial strain, the mutant phage may grow normally, indicating that the mutation is made silent by some element in the bacterium. Such a bacterium is said to be able to **suppress** the mutation and to contain a **suppressor.** A bacterium able to suppress a particular type of chain termination mutation—in this example, a UAG codon—is usually able to do so with a large number of mutations of that class, whether the mutation is in a phage or the bacterium itself. In general, other types of chain termination mutations, for example, UGA, will not be suppressed. The explanation for this phenomenon is that the bacterium (called a **suppressor mutant**) contains an altered tRNA molecule that can respond to a particular stop codon. For UAG, the altered tRNA molecule might contain the anticodon CUA, which can pair with that codon. Such a tRNA molecule is called a **suppressor tRNA,** and a mutation on which it can act is said to be suppressor-sensitive.

What has been mutated in the production of a suppressor tRNA? Clearly it must be a normal tRNA gene. Therefore, in the example just given, a $tRNA^{Lys}$ molecule whose anticodon is CUU, has been altered to have the anticodon CUA, which can hydrogen-bond to the codon UAG.

Inasmuch as a single base change is sufficient to alter the complementarity of an anticodon and a codon, there are (at most) eight tRNA molecules having a complementary anticodon that, with a single changed base, will also suppress a UAG codon. Thus, the following amino acids (whose codons are also indicated) can be put at the site of a chain termination codon: Lys (AAG), Gln (CAG), Glu (GAG), Ser (UCG), Trp (UGG), Leu (UUG), and Tyr (UAC and UAU). Note that these are the same amino acid codons that can be altered by mutation to form a UAG site. Suppressors also exist for chain-termination mutants of the UAA and UGA type. These too are mutant tRNA molecules whose anticodons are altered by a single base change.

In conventional notation, suppressors are given the genetic symbol *sup* followed by a number (or occasionally a letter) that distinguishes one suppressor from another. A cell lacking a suppressor is designated *sup*0 and *sup⁻*.

Several features of nonsense suppression should be recognized:

1. Not every UAG suppressor can restore a functional protein by suppressing each UAG chain termination mutation. Thus, a UAG codon produced by mutating the leucine UUG codon might be suppressed by a suppressor tRNA that inserts tyrosine, serine, or tryptophan; but might not be able to tolerate a substitution by the electrically charged amino acids lysine, glutamine, or glutamic acid.

2. Suppression may be incomplete in that the activity of the suppressed mutant protein may not be as great as that of the wild-type protein, and the stop codon may not always be read as a sense codon.

3. A cell can survive the presence of a suppressor only if the cell also contains two or more copies of the tRNA gene. Clearly if a $tRNA^{Ser}$ molecule that reads the UCG codon is mutated, then the UCG can no longer be read as a sense codon. This will lead to chain termination wherever UCG occurs; and a cell harboring such a mutant tRNA molecule will fail to terminate virtually every protein made by the cell.

However, as was mentioned earlier, there are multiple copies of most tRNA molecules; moreover, there are also minor tRNA molecules having the same anticodons as the major molecules. Thus, in any living cell containing a suppressor tRNA, there must always be an additional copy of a wild-type tRNA that can function in normal translation.

If a cell contains an UAG suppressor, then proteins terminated by a single UAG codon will not be terminated, and the existence of a suppressor tRNA should be lethal. There are two ways that this problem can be avoided:

1. Protein factors active in termination (see Chapter 10) respond to chain-termination codons even though a tRNA molecule that recognizes the codon is present, i.e., suppression is weak.

2. Normal chain termination often uses pairs of distinct termination codons such as the sequence UAG–UAA. Thus, the existence of a UAG suppressor would not prevent termination of a doubly-terminated protein.

Suppression of missense mutations also occurs. For example, a protein in which valine (nonpolar) has been mutated to aspartic acid (polar), resulting in loss of activity, is restored to the wild-type phenotype by a missense suppressor that substitutes alanine (nonpolar) for aspartic acid. Such a substitution can occur in three ways: a mutant tRNA molecule may recognize two codons, possibly by a change in the anticodon loop; a mutant tRNA molecule can be recognized by a noncognate aminoacyl synthetase and be misacylated; and a mutant synthetase can charge a noncognate tRNA molecule. Examples of each type of suppressor are known. Suppression of missense mutations is necessarily inefficient. If a suppressor that substitutes alanine for aspartic acid worked with, say, 20% efficiency, then in virtually every protein molecule synthesized by the cell, at least one aspartic acid would be replaced, and the cell could not possibly survive. The usual frequency of missense suppression is about 1%.

Reversion as a Means of Detecting Mutagens and Carcinogens

In view of the increased number of chemicals present as environmental contaminants and because most cancer-causing agents (carcinogens) are also mutagens, tests for the mutagenicity of these substances have become important. One simple method for screening large numbers of substances for mutagenicity is a reversion test using nutritional mutants of bacteria. In the simplest reversion test, known numbers of a mutant bacterium are plated on a growth medium containing a potential mutagen, and the number of revertant colonies that arise is counted. If the substance is a mutagen, the number of colonies will be greater than that obtained in the absence of the compound. However, such simple tests fail to demonstrate the mutagenicity of many carcinogens; these substances are not directly mutagenic (or carcinogenic), but are converted to active compounds by enzymatic reactions that occur in the livers of animals, and have no counterpart in bacteria. The normal function of these enzymes is to protect the organism from various noxious substances that occur naturally by enzymatically converting them to nontoxic substances. However, when the enzymes encounter certain manmade and natural compounds, they convert these substances (which may not be themselves directly harmful) to mutagens or carcinogens. The enzymes are contained in the microsomal fraction of liver cells. Addition of the microsomal fraction to the bacterial growth medium enables these substances to show up as mutagens and is the basis of the **Ames test** for carcinogens.

A set of histidine–requiring (His$^-$) mutants of the bacterium *Salmonella typhimurium,* which contain either a base substitution or a frameshift mutation, are used for tests of reversion to His$^+$. The frequency of spontaneous His$^+$ revertants is low in this mutant, but revertants are readily produced in one or more of these mutants by most known mutagens. Agar is prepared containing a very small amount of histidine, sufficient to initiate the growth of individual cells, but not enough for colony formation because two rounds of replication are required for a mutant to be expressed, and a known carcinogen or substance to be tested. A small amount of an extract of rat liver and about 10^8 His$^-$ cells are spread on the agar (plate A). The same number of cells is applied to another plate (B) which lacks the carcinogen. The number of colonies appearing on plate B is usually about 5 to 10 (these are the spontaneous revertants). When a known carcinogen is present, more colonies will be found on plate A. The number of colonies on plate A depends on the concentration of the substance being tested and, for a known carcinogen correlates roughly with its known effectiveness as a carcinogen.

The Ames test has now been used with thousands of substances and mixtures (industrial chemicals, food additives, pesticides, hair dyes, etc.) and numerous unsuspected substances have been found to stimulate reversion in this test. This does not mean that the substance is definitely a carcinogen, but only that it has a high probability of being so. As a result of these tests, many industries have reformulated their products to render them nonmutagenic. Ultimate proof of carcinogenicity is determined from testing for tumor formation in laboratory animals. The Ames test and several other microbiological tests are used to reduce the number of substances that have to be tested in animals, since to date only a few percent of more than 300 substances known to be carcinogens from animal experiments failed to increase the reversion frequency in the Ames test.

Summary

Mutagenesis is the process by which mutations (changes in the sequence of DNA) are created. An organism carrying a mutation is called a mutant. A base substitution is a change in a single nucleotide, and can be a missense mutation (causing the change of a single amino acid) or a nonsense mutation (causing protein-synthesis termination). Base substitution mutations may be either transitions, e.g., Py-Pu to Py-Pu or transversions, e.g., Py-Pu to Pu-Py. A frameshift mutation is an addition or loss of nucleotide causing an alteration of the reading frame. A silent mutation causes no change in the phenotype. The phenotype of conditional mutations is seen only under certain conditions. Mutations may be spontaneous (due to uncorrected replication errors), or induced by a chemical or physical mutagen. Base analogue mutagens are incorporated into DNA via normal base-pairing interactions, but in future rounds of replication they form incorrect base-pairs leading to a mutation. Chemical mutagens alter existing bases in DNA, changing their base-pairing properties or inducing error-prone replication. Ultraviolet radiation can also induce error-prone replication. Intercalating substances insert between adjacent base-pairs in DNA and cause frameshift mutations during DNA replication. Mutations may also be caused by transposable elements that insert into genes, or by mutator genes, which if mutated, lead to mutation in other genes. Mutational hot spots are highly mutable regions of DNA. Reversion or suppression is the process by which a mutant phenotype is returned to wild type, and may occur by a change at another site in the same gene (intragenic) or in a different gene (intergenic). The Ames test uses reversion to test for the possible mutagenic effects of chemical compounds.

Drill Questions

1. Distinguish the terms nonsense mutation and missense mutation.

2. In what two ways does 5-bromouracil (BU) function as a mutagen? What type of mutations would occur from its use?

3. Several hundred independent missense mutants, altered in the A protein of tryptophan synthetase, have been collected. Originally, it was hoped that at least one mutant for each of the 186 amino acid positions in the protein would be found. However, fewer than 30 of the positions were represented with one or more mutants. Suggest some possibilities to explain why this set of missense mutants was so limited.

4. *E. coli* polymerase I possesses several enzymatic activities. Two important activities are the polymerizing function and the $3' \rightarrow 5'$ exonuclease. Mutant polymerases have been found that either increase or decrease mutation rates in an organism containing the mutant enzyme. A mutant which increases the mutation rate is called a mutator; a mutant which decreases the mutation rate is called an antimutator. The mutator and antimutator activities are usually a result of changes in the ratio of the two enzymatic activities described above. How do you think the ratios change in a mutator and in an antimutator?

5. What exactly is a silent mutation? How does it differ from a leaky mutation?

6. Define the terms transition and transversion.

7. Since nonsense suppressors are mutant tRNA molecules, how does the cell survive loss of a needed tRNA by such a mutation?

8. An enzyme contains 156 amino acids. (This number has no significance in the problem.) Suppose amino acid 28, which is glutamic acid, is replaced by asparagine in a mutant, and as a result, all activity is lost. In this mutant protein, amino acid 76, which is asparagine, is replaced by glutamic acid and full activity of the enzyme is restored. What can you say about amino acids 28 and 76 in the normal protein?

Problems

1. One variety of a temperature-sensitive mutation is the cold-sensitive (Cs) mutation, which has a mutant phenotype below a particular temperature. Table 11-2 describes several mutations in an essential bacterial gene *ess*,

Table 11-2

	32°C	37°C	42°C
ess2(Ts)	+	−	−
ess5(Cs)	−	−	+

If + and − refer to colony formation or the lack thereof, respectively, what would be the phenotype of an *ess2*(Ts)*ess5*(Cs) double mutant?

2. Which of the following amino acid substitutions would probably yield a mutant phenotype? Explain your reasoning.
   a. Pro to His
   b. Lys to Arg
   c. Ile to Thr
   d. Ile to Val
   e. Ala to Gly
   f. Phe to Leu
   g. Tyr to His
   h. Arg to Ser

3. Consider a bacterial gene containing 1000 base-pairs. As a result of treatment of a bacterial culture with a mutagen, mutations in this gene are recovered at a frequency of one mutant per 10^5 cells. One of these mutants is grown, and a pure culture of this mutant is obtained. This culture is then treated with the same mutagen and revertants are found at a frequency of one per 10^5 cells. Would you expect the gene product obtained from the revertant to have the same amino acid sequence as the wild-type cell? Explain.

4. A particular mutant shows absolutely no activity. Despite an exhausting search for revertants by your colleagues using such mutagens as 5-bromouracil, nitrous acid, and UV light, none is found. What sort of mutation do you think exists in the original mutant? Why? What mutagen would you use to try to obtain a revertant?

5. What would you expect to be the difference in the observed mutation rate for an organism treated with a chemical mutagen and grown under minimal (slow), and rich (fast) growth conditions? Why?

Conceptual Questions

1. What kinds of information could be deduced from a series of intergenic suppressors (*not* in tRNA molecules or tRNA aminoacyl synthetases) of a known protein gene?

2. Propose one or more explanations for why mutations are allowed to occur in genes. That is, explain why the DNA polymerase editing function and various repair enzymes do not remove all misincorporated nucleotides.

3. Consider the effects of life on Earth of the breakdown of the ozone layer (which would allow more UV light to penetrate to the Earth's surface). What might be the eventual outcome of that situation?

Barry Polisky
Professor of Biology
Department of Biology

Birthday: **15 March 1946**

Birth Place: **Chicago, Illinois**

Undergraduate Degree: **University of Chicago**
Major: Biology 1967

Graduate Degree: **University of Colorado, Boulder, Ph.D. 1973, Molecular Biology**

Postdoctoral Training: **University of California, San Francisco, 1973–1977**

Present Position: **Indiana University, Biology Department**

Address: **Bloomington, Indiana**

MY LABORATORY STUDIES the mechanism by which a small DNA molecule regulates initiation of replication. We use the multicopy plasmid of *E. coli,* ColE1, as a model system to investigate what molecular signals are sent and sensed in making the decision to start DNA replication in a controlled manner. Because plasmids are stably inherited at well–defined copy numbers in cells, it has been assumed that molecular circuitry must exist to modulate replication. We try to understand this circuitry using a combination of genetics and biochemistry. Analysis of this system has revealed surprises and elegance at every turn, increasing our knowledge about the world and our respect for the profound complexity of life.

In what direction is the field of molecular biology moving?

Towards a full description of the structure and function of relevant macromolecules of life along with acquisition of tools sufficient to manipulate them in rational ways.

12

Plasmids and Transposons

Plasmids are extrachromosomal elements capable of autonomous replication in cells. They are found in most bacterial species. Most plasmids are circular, double-stranded DNA molecules that are isolated as supercoiled molecules; however other forms, such as linear double-stranded molecules, have also been described. Plasmids are not confined to bacteria; they have been isolated from yeast, protozoa, and plants. In humans, certain viruses including human papilloma virus and Epstein-Barr virus, can also exist as plasmids within cell nuclei, and replicate along with the nuclear chromosomes. The molecular biology of plasmids has been most intensively studied in bacteria. Table 12-1 summarizes some of the key features of several well-studied plasmids.

Certain bacterial plasmids have the ability to transfer themselves physically from one cell to another, spreading in the population. Special properties of certain plasmids even permit interaction between two bacterial cells of different species. These transfer abilities are central to the role that plasmids play in bacterial evolution. The genes carried by plasmids represent a mobile reservoir of genetic information that can endow the host bacterial cell with the ability to grow in otherwise hostile environments. An example of immediate relevance to humans is that bacterial antibiotic resistance is often conferred on bacterial cells by plasmid genes. Certain plasmids are capable of efficient transfer to other antibiotic-sensitive cells, resulting in rapid **horizontal** conversion of an entire bacterial population to drug resistance. Moreover, it is common

Table 12-1

Features of Selected Plasmids of E. coli

Plasmid	Size (Kb)	Copy Number	Conjugative	Other Phenotype
ColE1	6.6	10–20	No	Colicin production and immunity
F	95	1–2	Yes	E. coli sex factor
R100	89	1–2	Yes	Antibiotic-resistance genes
P1	90	1–2	No	Plasmid form is prophage; produces viral particles
R6K	40	10–20	Yes	Antibiotic-resistance genes

for a single plasmid to confer resistance to multiple antibiotics.

Bacterial plasmids span a wide range of sizes, from those containing one or two genes (several Kb in size), to those carrying several hundred genes (more than 500 Kb). Because certain plasmids are so large (and capable of carrying so many genes) it is possible for them to greatly alter the metabolic capabilities of their host cell. In the absence of selective pressure however, having the plasmid present is irrelevant to the host, and the plasmid is genetically dispensable. Under these circumstances the plasmid could conceivably represent a genetic burden to the cell and would therefore likely be lost from populations. However, the evolutionary mission of the plasmid is to propagate itself and insure its own stable transmission in the population. Despite being nothing more than naked DNA molecules incapable of survival outside a cell, bacterial plasmids have evolved quite sophisticated mechanisms to ensure their own survival.

An important property of bacterial plasmids is that they exist in characteristic copy numbers per cell. A central problem faced by all plasmids is the need to coordinate their replication with that of the host chromosome, so that they are not diluted by cell reproduction. Bacteria are capable of reproducing at widely different rates. In the lower intestine of a mammal, bacteria might divide once per 24 hours; in a rich medium in a culture flask, as mentioned in Chapter 1, they can divide every 20 minutes. Bacterial plasmids must be able to sense the division rates of their hosts, and alter their own replication rates accordingly. Plasmid replication must also be regulated so that it does not compromise host chromosome replication and the life of the cell.

Since plasmids are stably inherited DNA molecules, they have served as valuable research models to investigate the molecular mechanisms that control DNA replication. Because of their small size and genetic dispensability, they are easy to manipulate with the techniques of recombinant DNA technology. As will be described later, plasmids have played a key role in the development and use of recombinant DNA technology (see Chapter 13).

In this chapter, we will briefly describe some of the central features of plasmids, concentrating on bacterial plasmids because these are most widely understood at present. We will focus on the replication properties of plasmids and the important property of plasmid transfer. In addition, we will describe features of a special class of DNA sequences, often associated with plasmids, which have the ability to catalyze their own movement between different genomes. This process is called **transposition** and is mediated by transposable elements.

Plasmid-borne Genes

Bacterial plasmids play an important role in human disease. This was illustrated dramatically by the discovery during the 1960s of bacterial strains that had acquired simultaneous resistance to four antibiotics in heavy use in Japanese hospitals. The multiple drug-resistance traits were transferable to sensitive strains and were responsible for an epidemic of bacterial dysentery. The antibiotic-resistant bacteria harbored a large plasmid, called an **R factor** (for resistance), carrying genes that encoded proteins which destroyed or inactivated the antibiotics. Other R plasmids have been discovered that render bacteria resistant to heavy metals found in the environment, such as mercury and lead. The presence of these plasmids in bacteria allows for bacterial growth in heavily contaminated environments.

A bacterial cell's ability to resist the action of an antibiotic or to neutralize toxic environmental substances usually depends upon the production of large amounts of a few key enzymes. Those enzymes act by attacking the harmful substance. By virtue of encoding those key enzymes on plasmids, which function essentially as minichromosomes, the bacterium is well suited to produce large amounts of the required enzyme quickly, because the plasmid is often present in multiple copies (Table 12-2). With multiple copies of the relevant genes, transcription quickly generates sufficient enzyme to combat the antibiotic or chemical challenge the bacterium faces.

E. coli is a normal resident of the human large intestine. Certain plasmids endow *E. coli* with the ability to grow in new environments, and thereby convert an otherwise harmless bacterial resident into a pathogen. Such plasmids encode proteins that can alter the bacterial cell surface permitting adhesion of the bacteria to the lining of the small

Table 12-2

Plasmid-encoded Proteins

Protein	Function
Colicin	Secreted protein, kills bacteria lacking plasmid that encodes colicin-immunity protein.
Enterotoxin	Secreted protein, alters ion balance of eukaryotic cells. Responsible for water loss from cells.

intestine. A second gene sometimes present on a plasmid encodes a secreted protein toxin (called an **enterotoxin**) which can damage intestinal cells and is directly responsible for the symptoms of dysentery (Table 12-2).

Other plasmids, called **Col plasmids,** provide a selective growth advantage to cells carrying them because the plasmid encodes a secreted anti-bacterial protein (called a **colicin**) that kills cells lacking the plasmid. Immunity to the effects of the colicin is conferred on the host cell by a special protein also encoded by the plasmid. The mechanisms by which colicin proteins act to kill cells are varied. Certain types (such as colicin E1) produce holes in the membranes of sensitive cells, permitting ions to flow out. Others, such as colicin E3, enter cells and specifically cleave ribosomal RNA, blocking protein synthesis.

Plasmids can provide advantages to their hosts by other means, such as carrying **DNA restriction/modification genes.** These genes encode sequence-specific DNA binding proteins that recognize and degrade DNA foreign to the cell (restriction), while physically altering the host DNA so that it is not degraded (modification). Several hundred distinct restriction/modification systems have been described in bacteria. Their natural function is to protect the host bacterium from infection by an unwanted virus. In certain instances the viral DNA is chemically modified so that its ability to replicate is impaired. In other cases the DNA of the infecting virus is destroyed, thus short-circuiting the phage replication cycle (see Figure 1-3). Destruction of the virus DNA is achieved by specific nuclease enzymes. Those enzymes recognize specific nucleotide sequences and cleave DNA at those sites. These systems are the source of restriction endonucleases, which are essential molecular tools of recombinant DNA technology (see Chapter 13). Many of these restriction/modification systems are plasmid borne, while others are carried in the host genome.

Plasmid Transfer

There has been speculation that bacterial viruses evolved from plasmids. Small fragments of the bacterial cell's chromosome may have existed in the form of primitive plasmids, which eventually evolved into complete bacteriophage. The ability to transfer genes from one bacterial cell to another, a property possessed both by plasmids and viruses, no doubt contributes to the evolution of bacterial species.

From this evolutionary perspective perhaps no plasmid-borne functions are more important than those permitting plasmid transfer from cell to cell, thereby mediating the flow of genes through a bacterial population. This process of plasmid transfer is called **conjugation.** Conjugative plasmids carry a large block of transfer genes responsible for the specialized cell structures and enzymes required to physically move

conjugation

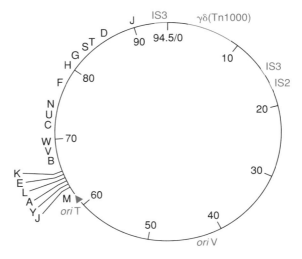

Figure 12-1
A map of the sex plasmid F. Points are given in kilobases. The single capital letters refer to the midpoints of the locations of the corresponding *tra* genes. The insertion sequences, γδ (Tn1000), IS2, and IS3 are shown in red. The red arrow (*ori*T) shows the location of the origin for transfer replication as well as the direction of transfer. *ori*V shows the location of the origin for vegetative (nontransfer) replication.

F plasmid

the plasmid genome from a donor cell to a recipient. Such plasmids are said to be **self-transmissible.** The prototype conjugative plasmid of *E. coli* is the **F** plasmid (or **sex factor**), a large circular plasmid of approximately 95 Kb present in 1–2 copies per cell (Figure 12-1). Cells carrying the F plasmid are called **males,** and are capable of mating with cells lacking the F plasmid (**females**). Cells lacking the F plasmid are often represented by the notation of F^-, while F^+ is used to represent cells having an F plasmid.

Conjugation is a complex process involving the gene products of a large F gene complex, called the *tra* (for transfer) operon, consisting of more than 20 genes (Figure 12-1). (As will be explained in more detail in Chapter 14, an operon is a set of contiguous genes coordinately expressed through a large polycistronic mRNA.) When male cells are mixed with female cells, mating pairs form, joined together by a special structure called a conjugation bridge. Certain F genes encode proteins for a specialized male surface structure called a **pilus.** The F pilus (essentially a tubular extension of the bacterial cell surface) probably interacts with specific receptors on the female outer membrane surface to establish a conjugation bridge. Following cell–cell contact, one strand of the supercoiled F DNA is nicked at a specific sequence called ***ori*T.** This is accomplished by a special DNA endonuclease encoded by the F plasmid. Other proteins then bind to the exposed 5′-terminus and guide that DNA strand into the female cell. As the DNA strand enters the female cell, it is replicated coordinately with the other parent strand

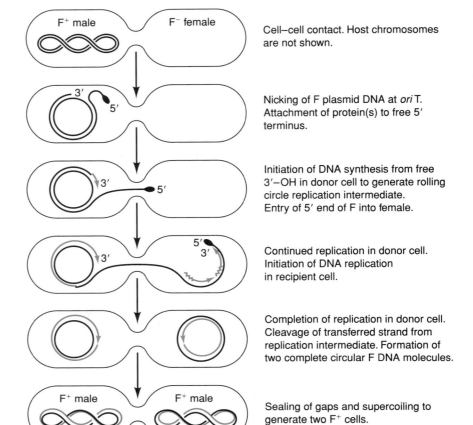

F⁺ male / F⁻ female
Cell–cell contact. Host chromosomes are not shown.

3′ / 5′
Nicking of F plasmid DNA at *ori* T. Attachment of protein(s) to free 5′ terminus.

3′ / 5′
Initiation of DNA synthesis from free 3′–OH in donor cell to generate rolling circle replication intermediate. Entry of 5′ end of F into female.

3′ / 5′ / 3′
Continued replication in donor cell. Initiation of DNA replication in recipient cell.

Completion of replication in donor cell. Cleavage of transferred strand from replication intermediate. Formation of two complete circular F DNA molecules.

F⁺ male / F⁺ male
Sealing of gaps and supercoiling to generate two F⁺ cells.

Figure 12-2
A model for transfer of F plasmid DNA from an F⁺ male (donor) cell to an F⁻ female (recipient) cell using a rolling circle replication mechanism. Black lines indicate parental DNA strands. Straight red lines indicate newly-synthesized DNA. Sawtooth red lines indicate RNA primers that are subsequently removed. The bacterial chromosomal DNA is omitted for clarity.

remaining in the donor. Following DNA replication in both donor and recipient cells, the new F plasmids are converted into circular forms. The closed circular DNA is then converted to a supercoiled form by the enzyme DNA gyrase. The process is shown diagrammatically in Figure 12-2. The result of conjugation is the semi–conservative replication of F in both cells. Initially a female, the daughter cell is converted into a male cell by this process. It then is capable of transferring the F plasmid. Thus, the F plasmid can be thought of as an infectious agent, capable of rapid spread through a population of female cells.

The ability to mediate conjugation is not confined to the F plasmid. Other plasmids, particularly the R plasmids, also carry the genetic information required for conjugation. When these plasmids are transferred, the female recipient becomes resistant to antibiotics which can be inactivated by the products of certain R plasmid genes. As with F plasmid transfer, the recipient female becomes a donor male capable of subsequent transfer. Plasmids such as ColE1 (discussed later in this chapter) are not self-transmissible by themselves. However, they have evolved the ability to take advantage of the genetic free-ride offered by a co-residing conjugative plasmid. When the conjugative plasmid is transferred, smaller plasmids such as ColE1 can also move into the recipient, and are said to be **mobilizable.**

TRANSFER OF CHROMOSOMAL GENES BY THE F PLASMID

Conjugation as described above is genetically manifested only through the conversion of the recipient cell into a donor. However, conjugation was initially discovered in a successful search for genetic processes in which chromosomal genes were transferred from one cell to another. Thus, we must account for the ability of the F plasmid to transfer genes other than its own. This ability was understood when it became clear that the F plasmid could exist in another form in certain cells *in addition to its autonomous existence as a separate plasmid. The F plasmid was found to be able to undergo recombination with the host chromosome,* and since both molecules are circular, a single recombination event results in integration of the plasmid into the chromosome. The conversion of free F into integrated F is a rare event, occurring in about one in 10^5 cells. These rare cells are called **Hfr** cells (for *high frequency of recombination*). The F plasmid can integrate into the chromosome at many different locations, so a variety of bacterial strains can be generated, each containing a single integrated F plasmid at different spots in the host genome. The molecular basis for the entry of the F plasmid into the chromosome will be described later in this chapter.

Hfr

The F plasmid genes responsible for making the cell male are expressed in Hfr cells. Thus, the Hfr cell can mate and transfer DNA (Figure 12-3). When the integrated F DNA in an Hfr strain is nicked by the endonuclease and transfer of the DNA strand into the recipient cell begins, DNA from the host chromosome can also be transferred by virtue of its physical continuity with the F DNA. This transfer of host DNA into the recipient potentially can have profound genetic consequences. DNA from the donor is transferred at a rate of about 50 Kb/min. About 100 min are required for transfer of the entire bacterial chromosome. In most cases, however, the mating pairs are not stable long enough for the entire chromosome to be transferred, and the cells break apart, leaving the recipient with a partial copy of the F plasmid and a segment of the donor genome. Thus, most often only a subset of chromosomal genes are transferred. Because the origin of transfer is within the F genome, a complete F copy is not present in the recipient unless the entire chromosome is transferred. As a result, the recipients

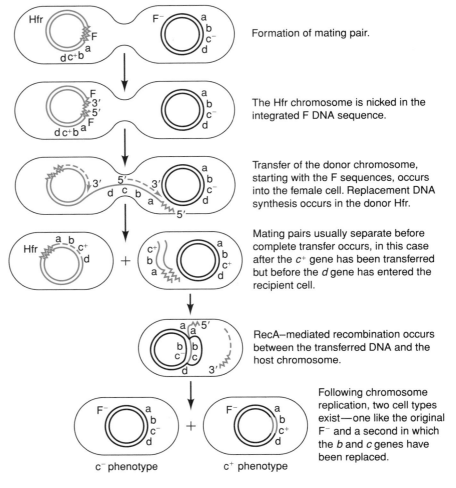

Figure 12-3
A diagram showing conjugation between an Hfr male (donor) bacterium and an F⁻ female
bacterium. The solid red and black circles represent the chromosomal DNA of the Hfr and
F⁻ cells, respectively. Dashed red lines indicate DNA synthesis. Sawtooth lines represent
the integrated F plasmid. Lower case italic letters represent genes on the chromosome.
The F⁻ cell carries a mutation in the *c* gene (*c⁻*) while the Hfr strain has a wild-type
copy of this gene. Following Hfr transfer and replication, a population of F⁻ cells has
been converted to a *c⁺* phenotype. These genes and the F plasmid DNA are not drawn
to scale relative to the chromosomal circles. Supercoiling is omitted for simplicity.

of most Hfr transfers are not capable of being donors. As with autonomous F plasmid transfer, replication (in both cells) accompanies single-
strand transfer.

    Left to its own devices, the piece of chromosomal DNA transferred
from an Hfr donor has little chance of survival and propagation in the
recipient. However, because of sequence homology with genes of the

recipient chromosome, it can undergo recombination mediated by the *rec*A system of *E. coli*. (The RecA protein is responsible for homologous recombination in *E. coli*). The recombination event can be detected by appropriate selection or screening. For example, in a mating between certain Hfr *leu*$^+$ cells and an F^-*leu*$^-$ recipient, F^-*leu*$^+$ cells can be detected. In these cells, the recombination event results in the replacement of the *leu*$^-$ gene with the incoming *leu*$^+$ gene.

The analysis of many Hfr strains revealed an important feature concerning transfer. Different Hfr strains reproducibly transfer different bacterial genes at high frequency. For example, the hypothetical strain Hfr1 might transfer genes *A B C* at a high frequency and genes *X Y Z* at a low frequency, while another strain, Hfr2, may transfer genes *X Y Z* at a high frequency and genes *A B C* at a low frequency. This behavior is due to different insertion points for the F plasmid in the Hfr donor's chromosome. Those bacterial genes oriented to one side of the **ori**T transfer point are the first bacterial genes to enter the recipient, and will be transferred at a higher frequency than genes further away from the *ori*T.

Since genes are transferred in a linear order to the recipient in an Hfr transfer, and the time of transfer of a particular gene reflects its relative position on the bacterial chromosome to the origin of transfer, Hfr strains are important tools for determining a map of the bacterial chromosome. That is, the order of the linear array of genes on the chromosome can be established. By determining the time of transfer of a particular gene relative to the transfer of other genes, the relative physical locations of the genes can be determined. The technique of *interrupted mating* was developed to exploit these characteristics. In this technique mating pairs are physically disrupted at different times after mating has initiated; under these conditions only part of the donor chromosome is transferred. Analysis of a large number of donor genetic markers revealed that characteristic times are required for their transfer (Figure 12-4). Interruption prior to this time prevented transfer. The physical location of a new gene can therefore be determined by defining its time of transfer relative to other well characterized genes in these so called "interrupted mating" experiments.

What determines the F plasmid insertion point? The F plasmid carries several special DNA sequences which are also present in several copies in the bacterial chromosome. These short sequences are called **insertion sequences** (abbreviated as **IS** sequences). They will be described in more detail later in this chapter, in connection with transposable elements. The transfer properties of most Hfr strains can be understood as a consequence of a recombination event between an IS sequence on the F plasmid, and the homologous sequence on the chromosome.

We have seen that the F plasmid can display two distinct lifestyles in a cell: either as a free plasmid, or integrated into the chromosome to form an Hfr. A third type of F plasmid has also been observed. This

insertion sequence (IS)

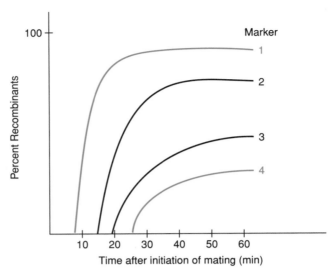

Figure 12-4
Time course of entry of different genetic markers into an F⁻ recipient. Marker 1 is transferred early; marker is transferred later. Note that the plateau level is related to the time of entry. Markers transferred early appear in a larger fraction of the recipients than markers transferred later. This difference is due to spontaneous breakage of the Hfr chromosome during transfer. The percent recombinants is expressed as a percentage of Hfr donors.

form, called **F′** (F prime) results from excision of an F from the chromosome and re-establishment in a plasmid form. The excision can also remove bacterial genes adjacent to F which are then carried in a plasmid form. The excision itself is the result of a recombination event between sequences within F and sequences outside. Many different F′ plasmids have been described; some carry very large blocks of the bacterial chromosome. Like F plasmids, F′ plasmids retain the ability to transfer themselves to female recipients.

F′ plasmid transfer results in the efficient introduction of the associated chromosomal genes into the recipient. However, the genetic consequences are different from Hfr transfer. After transfer of the F′, the F′-associated bacterial genes can be stably maintained by virtue of the ability of the F plasmid replication origin to function. Thus, recipient cells carry two copies of the genes present on the F′, and are genetically described as **merodiploid** (i.e., partially diploid), in contrast to the normal haploid state of the bacterial genome. Such merodiploids are useful for the complementation analyses described in Chapter 1. Since F′ plasmids are much smaller than the bacterial chromosome, their complete transfer usually occurs before mating pairs spontaneously break apart. Consequently, the female recipients are converted into F′ males. F′ plasmids are usually designated by the chromosomal genes they carry;

for example F'*lac pro* carries the genes for lactose utilization and proline biosynthesis.

Plasmid DNA Replication

As mentioned earlier, a fundamental property of bacterial plasmids is that they exist at characteristic copy numbers in cells. Plasmid copy numbers are usually expressed as a ratio of the number of plasmid molecules to the number of chromosomal copies per cell. Naturally occurring plasmids exist at a wide range of copy numbers. Certain plasmids, such as the sex factor F or the prophage of the *E. coli* virus P1, exist at copy numbers of 1–2, while those of the ColE1 family exist at copy numbers of 20–40. Before addressing the molecular mechanisms responsible for the establishment and maintenance of these characteristic copy numbers, a description of some general features of plasmid replication is necessary.

Plasmids are generally dependent on host enzymatic machinery for their own replication. The ColE1 plasmid, for example, carries no genetic information specifying replication-related enzymes—it relies entirely on host components. Other plasmids may encode certain enzymes involved in particular steps of replication, but in no known case is a plasmid completely independent of host replication enzymes. *E. coli* contains three different enzymes that catalyze DNA synthesis—DNA polymerases I, II and III. Most *E. coli* plasmids, as well as the *E. coli* chromosome itself, use DNA polymerase III as the central replication enzyme. (An important exception is the ColE1 plasmid family, in which DNA polymerase I plays the key replication role. Refer to Chapter 7 for more details on DNA replication.)

PLASMID COPY NUMBER

Plasmid copy number is regulated by controlling the frequency of initiation of DNA replication. Initiation of plasmid replication occurs at an *ori*. Replication forks starting at *ori* move either uni-directionally or bi-directionally around the circular plasmid (see Figure 12-5 and Figures 7-23 and 7-25). Most plasmid replication forks are bi-directional. The ColE1 plasmid, however, replicates using a uni-directional fork. Two questions of importance are:

1. What special sequence features of *ori* enable it to serve as a replication starting point? and

2. How is the frequency of initiation events controlled so as to maintain a specific copy number?

Many plasmid *ori* sequences are organized similarly to *ori*C, the bacterial replication origin. A key functional feature of origins in general is that the DNA double helix must open in that region, permitting access to the bases by DNA polymerases and other enzymatic replication machinery. The site-specific opening of the helix is accomplished in several steps. The first step is the recognition of the *ori* region by a specific DNA-binding protein. For *ori*C of the bacterial host, the key protein involved in replication initiation is encoded by the *dna*A gene. Many plasmids encode an analogous protein, which is required specifically for their replication. Called a Rep protein, it plays a key role in the initiation of plasmid replication. If the *rep* gene is missing or mutated, plasmid replication cannot occur from the plasmid origin. Moreover, plasmid copy number is related to the intracellular level of the Rep protein. The function of the Rep protein is to recognize specific short DNA sequences at the plasmid *ori*. At the *ori,* other host-encoded proteins interact with the bound Rep protein to assemble a DNA replication enzymatic machine called a **replisome.** Once bound to the *ori* the replisome causes local unwinding of the helix, followed by RNA primer formation and the replication of leading and lagging DNA strands. The specificity for assembling the replisome at *ori* comes not from sequence-specific binding of the replisome itself, but rather from an initial DNA sequence recognition by the Rep protein which directs subsequent replisome assembly at the *ori.*

Based upon that knowledge, molecular biologists used intuitive reasoning (see Chapter 1) to speculate that intracellular levels of the Rep protein would be carefully regulated. In fact, plasmids have evolved highly sophisticated molecular mechanisms to control the production of this critical protein. These control mechanisms sense the concentration of plasmid DNA in the cell, and regulate the production of the Rep protein accordingly. When a plasmid is introduced into a cell, initiation of plasmid replication occurs quickly to establish the plasmid at its steady-state copy number. This requires relatively high-level expression of the plasmid *rep* gene.

After the plasmid is established, Rep protein levels must be controlled to prevent runaway plasmid replication, which can be lethal to the cell. The amount of active Rep protein is controlled at the level of

Figure 12-5
Schematic diagram of unidirectional and bidirectional plasmid replication. Red lines indicate newly synthesized RNA (sawtooth) and DNA (straight). Black lines indicate the parental DNA strands. Black arrows show the direction of movement of replication forks. Both pathways begin with a D (displacement) loop at the origin of replication (*ori*). The key difference is that in bidirectional replication, the initial lagging strand proceeds through the origin of replication to become the leading strand for a second replication fork (fork B), while in unidirectional replication, this lagging strand is prevented from advancing through the origin of replication. Supercoiling and enzymes of the replisome are omitted for simplicity.

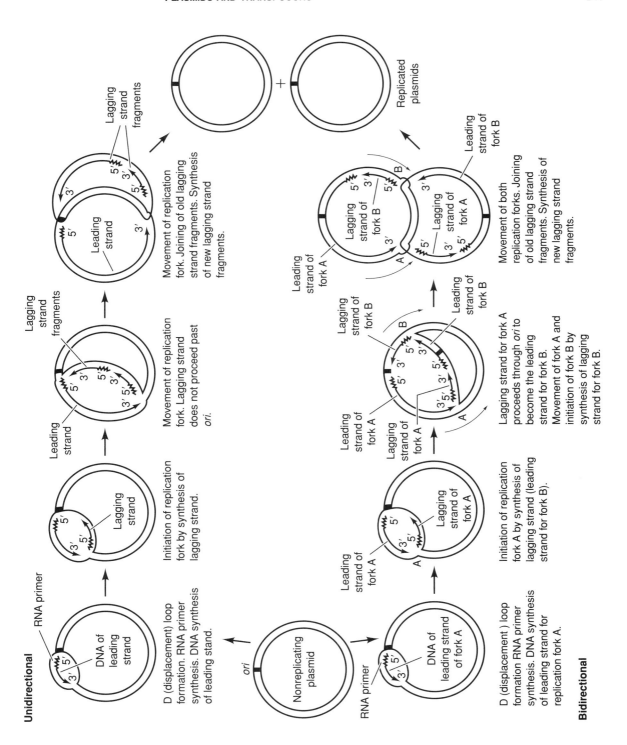

Unidirectional

RNA primer

DNA of leading strand

Lagging strand fragments

Leading strand

Leading strand

Lagging strand

Leading strand

Lagging strand fragments

Replicated plasmids

D (displacement) loop formation. RNA primer synthesis. DNA synthesis of leading strand.

Initiation of replication fork by synthesis of lagging strand.

Movement of replication fork. Lagging strand does not proceed past *ori.*

Movement of replication fork. Joining of old lagging strand fragments. Synthesis of new lagging strand fragments.

ori

Nonreplicating plasmid

RNA primer

DNA of leading strand of fork A

Leading strand of fork A

Lagging strand of fork A

Lagging strand of fork A

Lagging strand of fork B

Leading strand of fork A

Leading strand of fork B

Leading strand of fork A

Lagging strand of fork B

Lagging strand of fork A

Leading strand of fork B

D (displacement) loop formation RNA primer synthesis. DNA synthesis of leading strand for replication fork A.

Initiation of replication fork A by synthesis of lagging strand (leading strand for fork B).

Lagging strand for fork A proceeds through *ori* to become the leading strand for fork B. Movement of fork A and initiation of fork B by synthesis of lagging strand for fork B.

Movement of both replication forks. Joining of old lagging strand fragments. Synthesis of new lagging strand fragments.

Bidirectional

translation rather than by post-translational alteration, for example. The expression of *rep* genes of some plasmids, such as the *rep*A gene of the R factor, R1, is controlled both at the transcriptional and post-transcriptional levels by separate plasmid–encoded gene products that repress *rep* gene transcription. One, the *cop*B gene's protein product, blocks transcription by binding near the *rep*A gene's promoter. A second, the *cop*A gene's RNA product binds to the *rep*A gene near its initiation codon, thus blocking translation. Since the *cop*A RNA hybridizes to the *rep*A DNA, it is called "antisense" RNA.

ColE1 REPLICATION CONTROL

An alternative replication control mechanism, which exhibits several unexpected and novel features, is used to modulate copy number of the multicopy plasmid ColE1. As with the R1 system, the key regulatory molecule is a small antisense RNA produced by the plasmid. This repressor RNA is called RNA I. Like the CopA RNA of plasmid R1, ColE1 RNA I is partially complementary to another plasmid–encoded RNA that is involved in replication initiation. However, the target for RNA I is the primer RNA for synthesis of the leading DNA strand in replication, known as RNA II. RNA II is a large, structurally complex molecule. The formation of RNA II involves two unusual features. First, RNA II is able to form a stable RNA–DNA hybrid with the DNA template strand in the vicinity of *ori*. This stable RNA–DNA hybrid is unusual, because newly made RNA is normally displaced by the transcribing RNA polymerase, and the two DNA strands reform a helix. However, the newly-transcribed ColE1 primer precursor has special structural features that resist its displacement from the template strand.

The hybridized primer precursor RNA II is not used directly as a primer. Instead, it must be acted upon by a special host ribonuclease called RNase H, which specifically recognizes RNA–DNA hybrid structures. RNase H cleaves the primer precursor RNA to generate a 3'–OH terminus at a particular position. This step is an example of RNA processing. The new primer terminus is then recognized by DNA polymerase I, the host polymerase responsible for synthesizing the leading DNA strand.

The formation of the mature primer is a step that is regulated by RNA I. RNA I is complementary to the 5'-terminal region of the primer precursor RNA. RNA I acts as a negative regulator of replication by preventing the primer precursor RNA from forming a hybrid with the DNA template strand. To do this, RNA I must interact physically with the complementary region in the primer RNA during primer transcription, before the primer RNA has formed the structures necessary for hybrid formation. Because it forms an RNA–RNA duplex with the primer, RNA I physically blocks certain primer domains from folding as they normally would. This alteration in the folding of primer RNA leads to an inhibition of formation of the RNA-DNA hybrid essential for initiation of replication. Without hybrid formation, the newly made primer RNA-RNA I complex is released from the template and replication initiation is prevented (Figure 12-6).

The structure of the primer RNA is critical to its function. The same is true for RNA I, which folds into a distinctive conformation containing three stem-loop regions. It is important to realize that the RNA-RNA interactions described here occur between highly folded RNA species. This is known because mutations that alter the conformation of these RNAs can eliminate their function.

Copy number of the ColE1 plasmid is inversely related to the concentration of RNA I. RNA I is unstable in cells, decaying within a half-life of several minutes. Thus, RNA I must be synthesized constantly in an amount that is related to the number of plasmid copies in the cell. RNA I is therefore called a diffusible inhibitor—produced by a given plasmid molecule, it is also capable of inhibiting primer formation on a second ColE1 molecule.

The discovery of RNA I and its mechanism of action have provided a molecular explanation for a phenomenon called **plasmid incompatibility.** Plasmid incompatibility refers to the observation that closely related plasmids cannot be maintained together in the same cell; instead, clones of cells arise that contain one plasmid or the other, but not both (Figure 12-7). In contrast, distantly related plasmids can co-reside indefinitely (Figure 12-8). Incompatibility relationships among plasmids can be easily determined using different drug-resistance markers. Consequently, incompatibility properties have been used to classify plasmids into groups. One way to explain incompatibility is to propose that all plasmids of the same incompatibility group produce a diffusible inhibitor that can inhibit replication initiation of such plasmids. The concentration of the inhibitor sets an upper limit to the total number of plasmids of that type within a cell. Incompatibility would arise because members of the same group would be inhibited by one anothers' inhibitors, which cannot discriminate between them. In contrast, members of different incompatibility groups would be compatible (i.e., stably reside together) because their inhibitors are specific for their respective plasmids, and do not cross-react with one anothers' targets.

PLASMID PARTITION

Naturally occurring plasmids in bacteria are maintained with a high degree of stability. This is particularly dramatic for low copy-number plasmids such as F and P1. Despite the fact that these plasmids confer no apparent selective advantage to cells carrying them, and their copy numbers are about two per dividing cell, they are lost at a very low

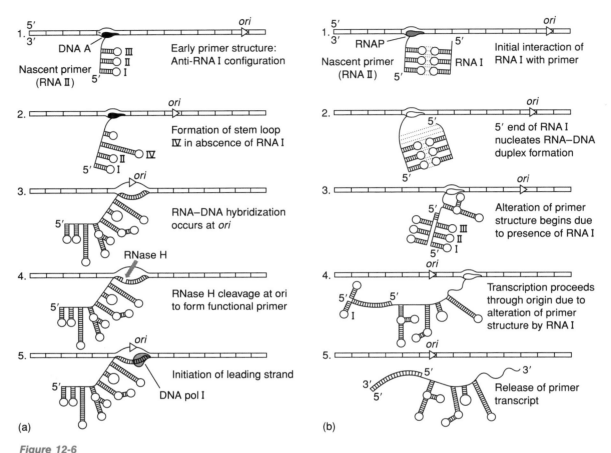

Figure 12-6

Control of ColE1 replication initiation. The fate of the primer RNA (RNA II) is shown in the absence (a) and in the presence (b) of RNA I. Much of the RNA structure shown is theoretical. 1. Transcription of primer RNA by RNA polymerase (RNAP) shown in red begins upstream of the origin of replication, *ori*. The primer RNA folds into an initial conformation which can base pair with RNA I via the loops of structures I, II, and III. 2. If no RNA I is present, new base-pairing arrangements can occur between sequences previously in the III stem-loop structure and newly transcribed sequences. These interactions are blocked if RNA I has bound to primer RNA and has begun to anneal to the 5′ end of primer RNA. Remember that RNA I is complementary to the 5′ end of primer RNA. 3. Transcription and folding of the primer RNA continues. In the absence of RNA I, the secondary structure of the primer RNA is such that a stable RNA-DNA heteroduplex is formed at *ori*. 4. The secondary structure of primer RNA in the presence of RNA I is such that transcription proceeds through *ori*. In the absence of primer RNA, the previously formed DNA-RNA heteroduplex is a substrate for RNase H which cuts the RNA strand, forming a free 3′-OH. 5. The free 3′-OH serves as a primer for DNA synthesis of the leading strand for replication by DNA polymerase I (DNA pol I) shown in grey. In the presence of RNA I, now fully annealed to the 5′ end of primer RNA, the primer RNA transcript is released from the plasmid without replication initiation occurring.

rate. This behavior suggests a precise mechanism to ensure segregation, or partition, of a single plasmid copy to each daughter cell. What molecular mechanisms might be responsible for such an efficient delivery system? The best known example for a mechanism to ensure high-fidelity

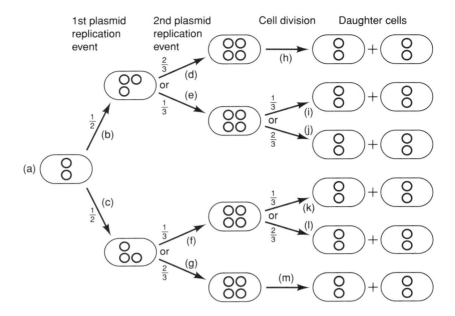

Figure 12-7
Possible segregation of two incompatible plasmids each with a copy number of two. (a) A
cell born with one each of two incompatible plasmids. Two independent replication
initiation events must occur during the cell cycle as determined by the level of copy
number control regulators. These regulators cannot distinguish between the two parental
plasmids. (b) and (c) Either of two replication events which are equally likely among the
population of cells. (d)–(g) The various possible replication events of the second plasmid
replication cycle and their probabilities. The replication initiation events in (b)–(g) are all
distributed randomly among the existing plasmid population. In other words, a plasmid
molecule that has previously replicated can undergo a second initiation event, while some
plasmids may not replicate at all during the cell cycle (e.g., pathway (d) where the red
plasmid does not replicate, and the black plasmid replicates twice). Alternatively, each
plasmid may replicate once as in pathway (e) or (f), though in either order (red first or
black first). (h)–(m) The various possible segregation events and their probabilities
(assuming random segregation) leading to daughter cells which each have two plasmids.
Note that random replication and random segregation results in some daughter cells
which differ in their plasmid contents relative to the parental cell and a loss of one or the
other of the two parental plasmids.

partition of newly-made DNA to daughter cells is mitosis in eukaryotic
cells. However, mitosis does not occur in *E. coli*, so we must look for
another mechanism. While the precise details are unknown, plasmids
like F and P1 have specific DNA sites that are required for efficient
partition. The so-called *par* sites are recognized by certain plasmid-encoded proteins, called Par proteins, that are also required for high efficiency partition.

   A plausible model for partition involves a pair of newly-replicated
plasmids which are held together as a unit by interactions of the Par

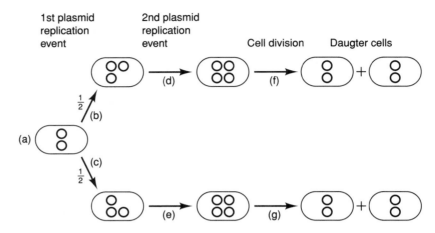

Figure 12-8
Possible segregation of two compatible plasmids, each with a copy number of one. (a) A cell born with one each of two compatible plasmids. Two replication events must occur during the ensuing cell cycle: (b) or (c) and (d) or (e). Unlike incompatible plasmids (Figure 12-7), the copy number control elements made by each parental plasmid are specific for each plasmid type (do not work on the other plasmid). Therefore, each plasmid replicates once during the cell cycle, producing identical plasmid populations after the second plasmid replication event, (d) or (e). Also unlike incompatible plasmids, segregation is not completely random. One black plasmid and one red plasmid go to each daughter cell. Thus, all daughter cells have the same plasmid make-up as the parent cell.

proteins bound to their respective *par* sites. The pair of plasmids then could interact with certain host proteins acting as receptors. These host receptors would be positioned such that they span the plane of cell division. The association of plasmid pairs with specifically located host components would insure that one plasmid copy enters each daughter cell. The nature of the host receptor is unknown.

Transposable Elements

Geneticists are able to construct genetic maps because of the stability of the linear array of genes on the chromosomes of individuals of a given species. These maps tend to give the impression that the genomes are rather static, suffering only the occasional base-pair changes due to replication errors or other mutagenic insults. However, this is not the case. A special class of opportunistic DNA sequences has evolved in all organisms. These DNA sequences have the unique ability to move as a discrete unit from one position in the genome to another. This ability

transposon

to relocate is called **transposition,** and the elements themselves are called **transposable elements** (abbreviated **Tn elements**). Transposable elements found in bacteria are simply called **transposons.**

Tn elements are found in plasmids and viruses, as well as in the genomes of prokaryotic and eukaryotic organisms. Tn elements were first detected in bacteria in the *gal* operon as mutations that eliminated the function of all the genes in the operon. These mutations were caused by the insertion of specific DNA sequences into the *gal* operon. Certain insertions were about 800 base-pairs in length; others were about 1300 base-pairs long. Moreover, in wild-type revertants derived from these mutations, the entire block of inserted DNA was missing. These inserted DNA segments are called insertion sequences, as first mentioned on page 237.

A variety of different IS sequences have been described in *E. coli* (see Table 12-3). These elements range in size from several hundred to several thousand base-pairs long. They are characterized by their ability to move from one point in the genome to another. However, they do not carry a detectable phenotype; genetically, they can be detected only by the effects of their insertion on the function of a gene in the chromosome or on a plasmid. Physically, their presence can be detected at a particular position by hybridization experiments, if the relevant region has been isolated and purified.

IS sequences contain promoters, and transcription and translation-termination signals. Depending on the position and orientation of the IS insertion, IS control signals can interfere with proper expression at either the transcription or translation level. Most often, the gene product is inactivated because the added sequences interfere with proper protein folding. If the gene into which the insertion has occurred is part of an operon, the expression of downstream genes may also be blocked. However, if the insertion point of the IS sequence is upstream of a gene or operon, the control of expression of these genes can be changed—they may be expressed at a considerably higher (or lower) rate. Alternatively (as in *gal* operon mutations) expression of all genes of the operon may

Table 12-3
Properties of *E. coli* Insertion Elements

Element	Number of Copies and Location	Size in Base-pairs
IS1	5–8, chromosome	768
IS2	5, chromosome; 1 in F	1327
IS3	5, chromosome; 2 in F	Approx. 1400
IS4	1 or 2, chromosome	Approx. 1400
IS5	Unknown	1250
$\gamma\delta$	1 or more, chromosome; 1 in F	5700

be blocked by an appropriately placed IS element. Such changes are consequences of transcriptional control signals in the IS sequence itself.

What properties of the IS sequences are responsible for their remarkable ability to transpose? Analysis of the DNA sequences of many different IS elements has revealed a striking feature: all elements contain a **terminally inverted sequence** (Figure 12-9). The size of the inverted terminal sequence may vary from one type of IS element to another; however, all members of the same type carry the same sequence. For example, IS1 has 23 base-pair terminal repeat sequences; IS2 has a 41 base-pair repeat. The two copies of the terminal sequences are perfect, or nearly perfect repeats. Moreover, the integrity of the inverted terminal repeats are essential for IS elements to transpose.

All IS elements are capable of encoding a special DNA binding protein called a **transposase.** Transposase is the key protein component responsible for carrying out the transposition process. Transposases are DNA-binding proteins that have the ability to specifically recognize the terminally inverted sequences of the IS sequence. This ability to recognize the borders of the IS element is an important feature of the mechanism, permitting the IS element to move as a discrete unit. In general, transposases are under the control of promoters within the IS element. Given the importance of the reaction they carry out, it is not surprising that the expression of transposases is carefully controlled, or that the level of transposases in cells is usually very low. We will discuss later how transposase catalyzes transposition.

direct repeats

A second important sequence feature involving IS elements concerns the site of insertion of the element. Analysis of large numbers of different insertion points for a given IS element revealed that insertion generates a short **direct repeat** flanking the element. The length of the direct-flanking repeat is characteristic of the type of IS element. The sizes range from about 3 to 13 base-pairs. The repeat sequence is not part of the IS sequence itself. Rather, it is a repeat of sequences present in the target. The fact that a **target sequence** is repeated provides an important clue to the mechanism by which the IS element transposes. It is most likely

Figure 12-9
An example of a terminal inverted repeat in a DNA molecule. The arrows indicate the inverted base-sequences. Note that the sequences AGTC and CTGA are *not* in the same strand. In a direct repeat, the sequence in the upper strand would be AGTC . . . AGTC.

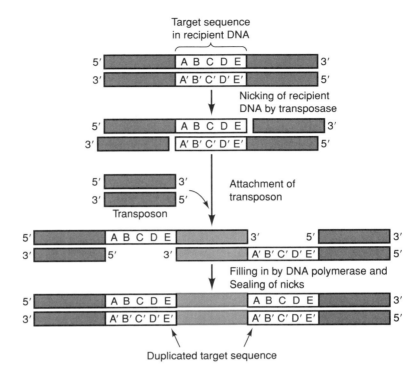

Figure 12-10
A schematic diagram indicating how target sequences might be duplicated by formation of a staggered cut at the target sequence. Capital letters represent the nucleotides of the target sequence (A′ represents the nucleotide that is the complement of A). Following attachment, the free 3′ ends on the transposon can serve as primers for the DNA synthesis which duplicates the target sequence.

that the direct repeat is a consequence of a staggered cut in the target DNA made by the transposase (see Figure 12-10).

The DNA-recognition properties of the transposase are most clearly responsible for determining where the IS element will insert. Most transposases have very broad sequence recognition, and mediate transposition into a wide variety of target sequences. For many IS elements, the sequence of the site chosen for insertion is close to random; however, transposases usually have some preference for insertion into sites with certain sequence features. Such sites are known as **hot-spots** for insertion.

COMPLEX TRANSPOSONS

IS1, IS2, IS3, etc., are examples of relatively simple transposable elements. They each encode a specific location transposase and are capable of movement from one location to another, but they carry no detectable phenotype. Two types of more **complex transposons** have been observed as natural constituents of plasmids. The first class of complex Tn elements consist of a gene or genes, usually encoding resistance to an antibiotic, sandwiched between two IS elements or IS–like elements. For example, the transposon Tn9 consists of a gene conferring resistance to chloramphenicol, flanked by IS1 elements in direct orientation. Tn10 contains a gene conferring tetracycline resistance, flanked by IS10 elements in an inverted orientation. Regardless of the orientation of the flanking IS elements, the termini of the composite element as a whole are inverted because each of the component IS termini are inverted (Figure 12-11). Both IS elements flanking the resistance gene can be capable of independent transposition. In some Tn elements, however, one of them has become mutated and can be considered vestigial, providing only a terminal repeat to the transposition process. Like their simpler relatives, the IS sequences, transposition of the complex Tn elements generates a short, direct repeat in the target genome. The length of the repeat is characteristic of the Tn element.

Thinking about how such elements arose, one could imagine that complex elements were selected under conditions of exposure to antibiotics. Prior to selection, two IS elements happened to flank the antibiotic resistance gene. Selection pressure resulted in the capture of the gene by the two IS elements. The transposon provided a means for horizontal transmission of resistance to other members of the population. Genes other than those conferring antibiotic resistance can be part of such elements. Tn1681 contains a gene encoding an enterotoxin that

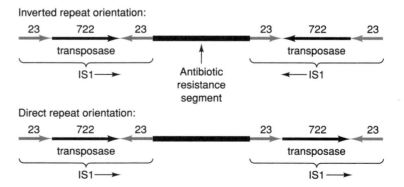

Figure 12-11
Two types of complex transposons showing an antibiotic resistance gene flanked by two IS1 insertion sequences in inverted and direct repeat orientations. (The numbers indicate the sizes of various portions of the transposon in base-pairs.) Red arrows indicate the inverted repeats of the IS1 elements. Note that the figure is not drawn to scale.

plays a role in causing human diarrhea. It is likely that in response to appropriate selective pressures, a complex transposon containing essentially any bacterial gene can be detected. The ability of this type of mechanism to respond to environmental changes provides genetic plasticity, and has profound implications for bacterial evolution.

The second type of complex transposon has a more complicated genetic structure. The best known of these elements is a 5 Kb transposon called Tn3 (see Figure 12-12 for a schematic outline). Elements of this class are not flanked by IS sequences, but instead have shorter inverted repeats; the inverted repeats of Tn3 are 38 base-pairs long. Between the inverted repeats are three genes, two of which are similar in function to those already discussed. The *bla* gene encodes an enzyme, called β-lactamase, that confers resistance to ampicillin. The second gene, called *tnp*A, encodes a transposase that specifically recognizes the inverted repeats of Tn3 during the transposition process. The third gene, called *tnp*R, encodes a bi-functional protein, called **resolvase**, that functions in the transposition process. Resolvase is a specific DNA-binding protein that recognizes a short DNA sequence called the **internal resolution site (IRS).** The consequences of resolvase binding to the IRS will be discussed later.

THE PROCESS OF TRANSPOSITION

Transposition involves the movement of a defined segment of DNA from one spot in a genome to another site either in the same genome

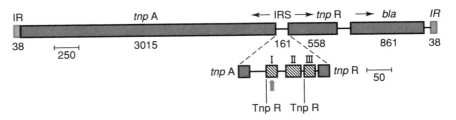

Figure 12-12
Organization of the Tn3 transposon. The solid red boxes (IR) indicate the terminal inverted repeats, the grey boxes indicate open reading frames; *tnp*A, transposase; *tnp*R, resolvase; *bla*, β-lactamase (the gene for ampicillin resistance). The black arrows show the direction of transcription. Numbers indicate the length in base-pairs. The total length of the element is 4957 base-pairs. IRS stands for internal resolution sequence (this region is also shown magnified five-fold). This region contains promoter elements for the *tnp*A and *tnp*R genes, wavy lines, as well as three sites for resolvase binding (I, II, III) indicated by red hashed boxes. The red arrow shows the site at which recombination between two Tn3 elements of a cointegrate occurs, a process mediated by the enzyme resolvase. The resolvase (TnpR) also represses synthesis of its own transcript and the *tnp*A transcript by virtue of its binding affinity for sites I, II, and III.

or in another, such as a viral or plasmid, genome. A convenient way to measure the frequency of transposition is to take advantage of the fact that movement can occur between a chromosome and a conjugative plasmid, such as the F plasmid. If this occurs, the F plasmid acquires the ability to confer an easily detectable phenotype: antibiotic resistance. The frequency at which the F plasmid is used as a transposition target can be determined by measuring the frequency at which females become antibiotic-resistant males after conjugation. Only if the transposon has moved from the chromosome to the F plasmid will the F plasmid confer antibiotic resistance on the female after conjugation. Experiments have indicated a frequency of movement of about 10^{-5} to 10^{-7} per element per generation for transposons such as Tn10. This frequency is in the same range as the spontaneous mutation rate seen for a typical bacterial gene.

Transposition occurs via two types of mechanisms depending on the element. In the simpler of the two mechanisms, transposition involves excision of the Tn element through cleavages at its inverted termini by transposase. This excision is followed by insertion of the element at a new site. Such a cut and paste mechanism is a **conservative** process, i.e., it involves no replication of the original element, except for the small amount needed to create the direct repeats flanking the element at its new location. The donor molecule (having suffered a double-strand DNA break at both ends of the element) can be lost in the process. To enter the recipient molecule, the transposase makes a staggered double-stranded cut, the length of the stagger is characteristic for each type of element. For example, the 9-base-pair direct repeats that flank Tn10 are generated by the Tn10 transposase making a 9-base-pair staggered cut in target DNA. The two ends of the element are then ligated to the cut ends of the target DNA by transposase. The ligation creates 9-base-pair single-strand gaps flanking the element—these are filled in by DNA polymerase of the host (Figure 12-10). This conservative mechanism of transposition is used by Tn10.

The existence of a second transposition mechanism was shown by observing that the transposition of certain transposons, such as Tn3, involved a special intermediate DNA structure called a **cointegrate.** The cointegrate form is a result of fusion of the donor and target replicons mediated by transposase. An important feature of the cointegrate is that it contains two copies of the transposon in a direct orientation at the junctions of the donor and target sequences. Thus in the cointegrate, the transposon has been replicated. The cointegrate is an intermediate in transposition. It can be resolved by a special recombination process into two separate DNA molecules, each containing a copy of the transposon (Figure 12-13). This step is carried out not by the transposase but by the product of the Tn3 *tnp*R gene, resolvase. The recombination step, called **resolution,** occurs at the IRS sites in the paired copies of Tn3 in the cointegrate called *res* sites (Figure 12-13). Transposition of Tn3 is thus a **replicative** process in which a second copy of the element is produced. Tn3 produces 5–base–pair direct repeats in its target sequence, implying that its transposase makes a staggered cut in the target DNA, like Tn10 transposase.

MODELS FOR TRANSPOSITION

A specific model for transposition should define the nature of the cleavages made in the donor DNA, the target DNA, and in the transposon, and explain how the target and transposon molecules are subsequently

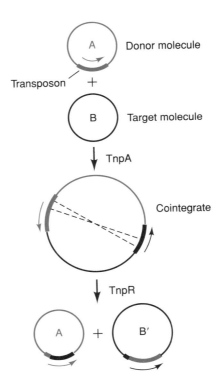

Figure 12-13

A simple two step model for replicative transposition. In the first step, transposase (TnpA) encoded by the donor molecule (A) mediates fusion with the target molecule (B) to generate a cointegrate molecule. Note that the transposon has been duplicated in the cointegrate. In the second step, called resolution, recombination between specific sites in the two transposon copies (red lines) is mediated by the TnpR protein (resolvase) to generate two separate molecules. One molecule is identical to the donor; the other (B′) is the target molecule which now contains a copy of the transposon.

covalently joined. Many models have been proposed to account for the properties of the forms detected. It might seem unlikely that a single model could account for the behavior of both classes of transposons, i.e., those which transpose conservatively, such as Tn10, as well as those which transpose replicatively, such as Tn3. However, such a model has been proposed (Figure 12-14). A key aspect of the model involves the formation of an intermediate DNA structure which can mature along two alternative paths. The intermediate results after the following steps have occurred:

1. The transposase makes a single-strand cleavage at each end of the donor transposon (remember that *each* breakage creates a 5' and 3' end).

2. The transposase makes staggered cuts in the target DNA (at this point, four single-strand breaks have been made, generating eight ends).

3. The transposase carries out two ligation steps (one at each end of the transposon) so that a 5' end of the target DNA is linked to a 3' end of the transposon. The result is a linked DNA molecule containing four free ends (two 5' and two 3').

This transposition intermediate can mature in two different ways, depending on the properties of the transposon. Note that two nicks were made in the donor DNA initially at opposite ends of the transposon in

---

Figure 12-14

A detailed model for conservative and replicative transposition utilizing a common intermediate capable of maturing in one of two ways. The donor molecule is drawn with thin lines, while the target (recipient) molecule is drawn with thick lines. The transposon is shown in red. Sawtooth regions indicate the target sequence. DNA synthesis is shown as dashed lines. (a) The two parent molecules. (b) The donor and target molecules following the nicking by transposase. Two single-strand cuts (arrows) are made which flank the transposon, and two staggered, single-strand cuts are made in the target sequence, generally about 4–10 base-pairs apart. Four free 5' ends are formed (2, 3, 6, and 7) and four free 3' ends are formed (1, 4, 5, and 8). (c) The common intermediate for conservative or replicative transposition is formed by two ligations, 4–7 and 1–6, carried out by transposase. This intermediate can then follow one of two maturation pathways. A key feature is that the 3' ends (5 and 8) can serve as primers for new DNA synthesis, as indicated by the arrow heads. (d) The first step in the conservative pathway. Two more single-strand nicks are made, separating the donor molecule from the transposon-target molecule. (The donor molecule is lost.) (e) The final product of the conservative pathway. The two free ends, 5 and 8, serve as primers for DNA synthesis to fill in the remaining, short single-strand regions of the molecule (target sequences). The result is a simple insertion of the Tn element into the target molecule, and duplication of the target sequence. (f) Formation of a cointegrate molecule by DNA synthesis from the free ends (5 and 8) which continues through the transposon (dashed lines). The newly-synthesized DNA ends are ligated to the remaining free ends (2 and 3). Note that a duplication of the target sequence and the transposon has occurred. (g) The end products of the replicative pathway formed by recombination within the cointegrate molecule in (f).

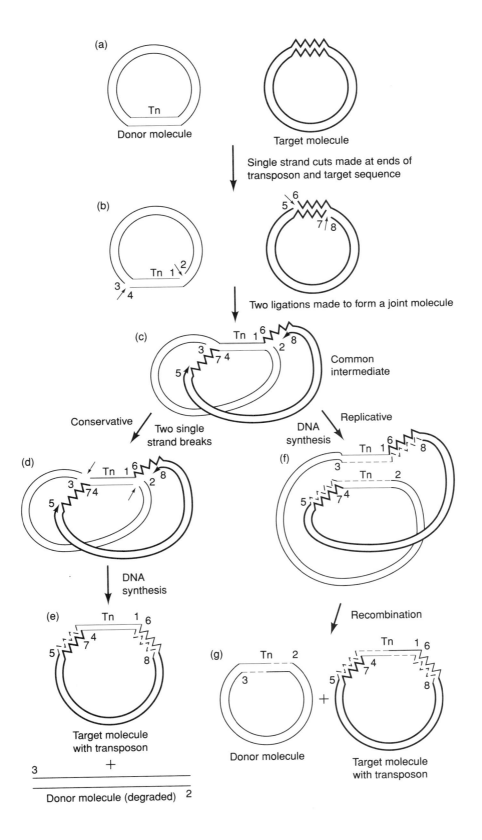

(a)

Tn

Donor molecule

Target molecule

Single strand cuts made at ends of
transposon and target sequence

(b)

Tn 1

3
4

6
5
7 8

Two ligations made to form a joint molecule

(c)

Tn 1 6
8
2
3 7 4
5

Common
intermediate

Conservative   Two single
strand breaks

Replicative
DNA
synthesis

(d)

Tn 1 6
8
3 7 4
5 2

(f)

Tn 1 6 8
3
Tn 2
5 7 4

DNA
synthesis

Recombination

(e)

Tn 1 6
4
5 7
8

Target molecule
with transposon

+

3
——————————————
Donor molecule (degraded) 2

(g)

Tn 2
3

Donor molecule

+

Tn 1 6
4
5 7
8

Target molecule
with transposon

step one above. After formation of the intermediate, two additional nicks may occur on the formerly unbroken donor strands at the transposon ends. These two nicks eliminate the linkage between donor and target DNA, and create two separate DNA molecules. One is the broken donor molecule which lacks the transposon. The other is the target molecule linked to the transposon. However, this molecule has short, single-stranded gaps flanking the transposon, which result from the initial staggered cut in target DNA made by transposase. Host DNA repair machinery can recognize the gap and fill it in, resulting in a covalently closed target containing the transposon. This is the pathway of conservative transposition—the mechanism used by Tn10.

The alternative pathway for transposition does not involve additional strand breakages, but instead utilizes the two 3′-termini present in the transposition intermediate. These termini may serve as assembly points for the DNA replication enzymes of the host. Extension of these termini by replication results in duplication of the transposon. The replication forks proceed only to the ends of the transposon, where the newly-synthesized DNA strands can be ligated to the pre-existing strands derived from the donor molecule. The result of this local replication is the creation of the cointegrate molecule—a single DNA molecule containing two directly repeated copies of the transposon. The two copies of the transposon can undergo recombination in a site-specific resolution reaction to generate the original donor molecule and a target molecule containing a single copy of the transposon. This replicative transposition pathway is used by the Tn3 family.

The IS family of transposons generally transpose via a conservative mode, generating simple insertions. Occasionally, however, an IS-mediated cointegrate form can be detected. The model described in Figure 12-14 accounts for both outcomes. However, since IS elements do not encode a resolvase, resolution of IS-mediated cointegrates probably involves recombination mediated by host components.

What is the fate of the donor molecule in conservative transposition? Two possible outcomes must be considered. The double-stranded break may lead to destruction of the donor. However, since bacteria often contain more than one copy of their chromosome, this event might not necessarily be lethal. An alternative possibility is that the double-strand gap is repaired by the host using the remaining copy of the transposon as a template.

CONTROL OF TRANSPOSITION FREQUENCY

It is clearly important to cell viability that the frequency of movement of transposable elements be tightly controlled to prevent the accumulation of additional copies of the elements in the genome. One way transposition is controlled is by limiting production of transposase. The

Tn10 transposase is produced at an extremely low level—less than one molecule per cell, per generation. This low level of production is accomplished by control both at the level of transcription of the transposase gene, and translation of the transposase mRNA. Thus, in most cells carrying Tn10, transposition is prevented simply by the absence of transposase.

A second means by which transposition frequency is controlled is that the process of transposition is largely confined to a small time window in the bacterial cell cycle. *Transposition occurs preferentially immediately after DNA replication of the element itself.* E. coli DNA normally is **methylated** at specific sites by the action of an enzyme called DNA adenine methylase. However, when a replication fork passes over a DNA region, the newly-made DNA strand is transiently unmethylated, so the recently replicated part is only partially methylated. Shortly after replication, the *dam* methylase methylates the newly-made DNA strand. This converts the daughter DNA molecules to the fully-methylated state. The transposase, however, interacts poorly with fully-methylated DNA, but interacts strongly with partially methylated DNA. During that short time interval before the DNA is fully methylated, transposition occurs. Thus, the short time available for transposase action serves to control transposition frequency.

Transposable Elements in Eukaryotes

The existence of transposable elements was first detected in Barbara McClintock's studies of maize in the 1940s. These elements are fully comparable to those in bacteria and were detected by their effects on the expression of certain maize genes. Maize contains several types of transposable elements. These elements have a sequence organization similar to their bacterial counterparts. They have inverted terminal repeats, encode a transposase, and cause short direct repeats in their targets when they transpose. They are called **autonomous,** because they have the ability to transpose. Other members of the same transposon family are often present in maize, but are unable to transpose on their own because internal deletions have destroyed their transposase gene. These copies are called **nonautonomous,** and are incapable of movement unless transposase is supplied *in trans* by an autonomous element. Nonautonomous elements retain the inverted terminal repeats at which transposase acts. The best characterized maize Tn element is a 4.5 Kb element called **Ac** (for **activator**); defective versions of Ac are called **Ds** (for **dissociation**).

As in bacteria, transposons in maize are associated with a variety of structural aberrations of chromosomes. Deletions, duplications and inversions may occur at sites where autonomous elements are present. The transposition of maize elements is regulated in part by the development of the plant; unknown signals generated at certain periods of plant development affect the frequency of movement of these elements. It is possible for an autonomous element to become reversibly inactivated, that is, incapable of transposition. Presumably such a change results from secondary modifications to the DNA sequence of the element. Cycles of methylation and de-methylation of certain DNA sequences may be involved in this process.

Transposons have been detected in the genomes of other eukaryotes, including those of yeast, *Drosophila melanogaster*, and humans. They are probably present in all genomes. Like the transposons described for bacteria, eukaryotic transposons cause unstable mutations in genes that revert to wild type when the element is deleted. In addition, these transposons cause short direct repeats at the site of their insertion. However, some of these elements have a fundamentally different genetic organization and mechanism of transposition than the bacterial elements (such as Tn10 or Tn3). These eukaryotic transposons are similar in organization to the proviral form of **retroviruses,** which are eukaryotic viruses that contain an RNA genome. Retroviruses introduce an RNA molecule into cells which can be converted into double-stranded DNA

reverse transcriptase

by a unique viral-encoded enzyme called **reverse transcriptase.** The double-stranded form of the virus can integrate randomly into the genome of the infected cell, where it resides as a **provirus.** After integration, the provirus is replicated as part of the host genome. The proviral form can express viral genes as mRNAs which are translated by the host into viral proteins that make up the capsid. Capsid proteins and the mRNA itself, which constitute the viral genome, assemble into infectious viruses that are then shed from the cell. The integration of the retroviral provirus into the host genome generates short direct repeats flanking the viral sequences (like those seen in bacterial transposition). Thus, the retroviral infectious cycle has features in common with transposons.

Imagine a situation in which a genome has acquired a number of retroviral proviruses integrated at various points. The proviral genes continue to be transcribed by the host RNA polymerase. Over time, mutations may occur which inactivate or alter the proviral DNA sequence, so that functional viral proteins can no longer be produced. The proviral element could then be regarded as functionally defective. However, as long as a functional reverse-transcriptase-like activity is produced by the element or its relatives, the mRNA can be converted into double-stranded DNA. It then will be able to integrate into new genomic locations by a process resembling bacterial transposition. The important features of this pathway are that the appearance of the proviral DNA form at new genomic locations amounts to a transposition event, and that the transposition process involves an RNA intermediate.

Summary

Plasmids are extrachromosomal DNA molecules containing an origin of replication and often one or more genes. R factors are plasmids that carry genes conferring resistance against antibiotics or other toxic substances. Col plasmids carry genes for the secreted anti-bacterial protein colicin, and an immunity protein. Some plasmids encode DNA restriction/modification systems. F plasmids carry genes for plasmid transfer or conjugation. In a primer RNA-DNA hybrid complex Hfr cells, the F plasmid can integrate into the host chromosome and cause transfer of chromosomal DNA. An F′ plasmid is an excised F plasmid containing host DNA. Plasmid replication is carried out by a replisome consisting of host replication enzymes and often a plasmid-encoded replication origin binding protein (Rep). Plasmid copy number is controlled by regulating the level of Rep protein or the formation of a primer RNA-DNA hybrid complex. Closely related plasmids are incompatible.

Transposons are mobile genetic elements. Insertion sequences or IS elements consist of a transposase gene flanked by inverted repeats. They represent simple transposons. More complex transposons consist of an antibiotic resistance gene flanked by IS elements, or genes for antibiotic resistance, transposase, and resolvase flanked by inverted repeats. Transposition may be conservative via a "cut and paste" mechanism or replicative (with a duplication of the transposon). Both types may share a single early intermediate. Transposase recognizes and cleaves the inverted repeats and makes a staggered cut in the target DNA, which leads to short flanking direct repeats of this DNA. Resolvase carries out recombination between the *res* sites of two transposons of a cointegrate molecule. The frequency of transposition is minimized by limiting the amount of transposase and occurring mainly when the host DNA is hemi-methylated following replication. Eukaryotes also contain transposons. They are similar to retroviruses.

Drill Questions

1. What would be the minimum required sequence elements for a stably inherited plasmid?

2. The following questions concern R factor R1 replication.
   a. Where does replication begin?
   b. What plasmid-encoded factor is required, and what is its function?
   c. What host-encoded polymerase is required?
   d. What is the replication assembly called?
   e. In what two ways is the level of the Rep protein regulated?

3. In plasmid transfer, what are the roles of DNA synthesis in the donor and the recipient?

4. The following questions concern F factor transfer.
   a. What genes endow F-plasmid-containing bacteria with the ability for conjugation?
   b. What bacterial structure is induced to allow transfer?
   c. What mode of DNA replication is used?
   d. What early enzymatic step is needed in transfer replication, but not in normal replication?

5. The following questions are concerned with the recombination functions of the F plasmid.
   a. What is a bacterial strain called that contains an F plasmid within the chromosome?
   b. What feature of the F plasmid allows insertion?
   c. When an integrated F initiates transfer, why is the entire *E. coli* chromosome not transferred?
   d. What is an F′ plasmid?

6. What is meant by the terms direct repeat and inverted repeat? Use the sequence A B C D as an example.

7. What is the difference between conservative and replicative transposition? What base-sequence is duplicated in both types?

8. What is an insertion element? What is a simple transposon? Describe the differences between the two types of complex transposons.

9. What determines the selection of a site for transposon insertion?

Problems

1. An F′(ts)*lac*⁺ plasmid has a temperature-sensitive 40° mutation in its replication system.
   a. What is the phenotype of an F′(Ts)*lac*⁺/*lac*⁻ cell at 42°C?
   b. An F′(Ts)*lac*⁺/*lac*⁻ *gal*⁺ strain is grown for many generations at 37°C and then plated at 42°C. Some *Lac*⁺ colonies form at 42°C. How have these formed?
   c. Some of the *Lac*⁺ colonies in (b) are *Gal*⁻. How have these formed?

2. Plasmid A contains an *amp*⁻r gene, and plasmid B contains a *tet*⁻r gene. Both have a copy number of two. An *E. coli* cell containing both plasmids is used to start a culture which is grown under selecting conditions (both antibiotics). The culture is then allowed to grow for one generation in non–selective medium (no antibiotics). Cells are plated on non-selecting medium, and 100 colonies are checked for their antibiotic resistance phenotype by plating on Amp, Tet, and Amp–Tet medium. Of the 100 colonies, 22 were Amp⁻r Tet⁻s, 23 were Amp⁻s Tet⁻r, and 55 were Amp⁻r Tet⁻r.
   a. Which plasmids do cells of each antibiotic resistance phenotype contain?
   b. What is the most likely explanation for these results?
   c. What would be the explanation if all the colonies were Amp⁻r Tet⁻r?

3. To what conclusions might you come from the following observations about ColE1 replication?
   a. ColE1 can replicate in a host cell carrying a mutation in the RNase *H* gene.
   b. Mutations in the RNA I-primer RNA overlap region can cause a very high copy number.
   c. Certain mutations in the ColE1 primer RNA which cause a decreased plasmid copy number can be suppressed by second-site changes elsewhere in the primer RNA.

4. A hypothetical transposon has inserted at the target sequence 5′-TTAGCA-3′. Its left inverted repeat sequence is 5′-GCAATGGCA-3′. This now serves as the donor molecule for a transposition event to the new target sequence 5′-GATCCA-3, in a recipient molecule. Show the structure of the DNA (both strands) of the following molecules after all steps of transposition. (Assume that the inverted repeats on the transposon are perfect).
   a. The recipient molecule.
   b. The donor molecule if transposition occurred by the conservative pathway.
   c. The donor molecule if transposition occurred by the replicative pathway.

5. The transposition of a wild type Tn3 and several mutant Tn3 elements is being followed by observing their transfer to a plasmid and screening for ampicillin resistance (don't worry about the details). Briefly describe the effect on the frequency of transposition, and the types of products produced by transposition for the following mutations relative to wild type. Assume that each mutation completely abolishes at least one particular function, but keep in mind that some proteins are bifunctional. A RecA⁻ bacterial strain is used to eliminate homologous recombination.
   a. A resolvase point mutation.
   b. A resolvase promoter mutation.
   c. A transposase point mutation.
   d. A transposase promoter mutation.

e. A point mutation in the *bla* gene.

f. A mutation in the *res* site.

6. Sometimes, in a bacterial strain carrying two different plasmids (one of which contains a transposon), a single plasmid arises that has the genetic information of both plasmids. If the cell is RecA⁻ (cannot carry out homologous recombination), this fusion can occur in two possible ways—nonhomologous recombination, or transposon-mediated formation of a cointegrate. What information would be necessary to distinguish these mechanisms unambiguously?

Conceptual Questions

1. How do you suppose transposons and plasmids evolved? What advantages do they confer on their host to allow their continued existence?

2. How have plasmid-borne genes for antibiotic resistance and toxins affected medical treatment and pharmaceutical development?

3. How might transposable elements affect the evolution of their host species?

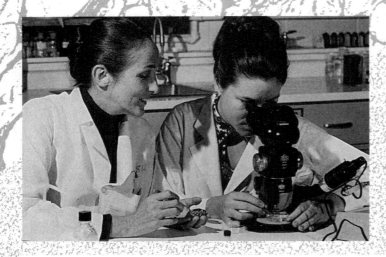

Dorothy M. Skinner (*left*)
Senior Scientist
Biology Division

Birthday: **22 May 1930**

Birth Place: **Newton, Massachusetts**

Undergraduate Degree: **Jackson College (Tufts University)**
Majors: Biology and Chemistry 1952

Graduate Degree: **Harvard University, Ph.D. Biology, 1958**

Postdoctoral Training: **Yale University and Brandeis University, 1958–1962, Biochemistry**

Present Position: **Oak Ridge National Laboratory, Biology Division**

Address: **Oak Ridge, Tennessee**

Dorothy M. Skinner

Very highly repeated (VHR) DNAs are present in large amounts in eukaryotic genomes. Blocks of VHR DNAs arranged in tandem form so-called satellite DNAs (stDNAs) under different physical conditions. The intriguing question of the function of most VHR DNAs remains unanswered. The Bermuda land crab Gecarcinus lateralis is an ideal model system for studying the molecular biology of VHR DNAs. The animal has stDNAs of the simplest repeat (poly dA–dT) as well as the most complex (a family based on a ~2.1 kb repeat). In clones of domains of the complex stDNA, we (1) analyze unusual structures (hairpins, Z-DNA, etc.), (2) characterize sequence-dependent hot spots for mutations in the divergent domains, and (3) recover from DNA libraries conserved regions that are transcribed at specific times in the growth cycle. The latter research led us to investigating growth and the mechanisms of its control in Crustacea, including cyclic muscle atrophy and resynthesis and cyclic exoskeletal degradation and resynthesis. We are recovering DNA sequences that code for cyclically functional proteins in order to define the control of their expression.

Christie Anna Holland

Certain retroviruses induce leukemia in certain strains of mice. The focus of my lab is to define the genetic mechanisms of retroviral infections that result in transformation and cell death and lead to cancer or AIDS. In addition, we are using our knowledge of these controls to exploit retroviruses as tools for gene therapy.

Christie A. Holland (*right*)
Associate Professor
Pediatrics and Biochemistry Department

Birthday: **25 August 1950**

Birth Place: **Newport News, Virginia**

Undergraduate Degree: **University of Richmond**
Major: Biology 1972

Graduate Degree: **University of Tennessee-Oak Ridge Graduate School of Biomedical Sciences, Ph.D., 1977**

Postdoctoral Training: **Worcester Foundation for Experimental Biology and MIT, 1977–1984**

Present Position: **Director, Center for Virology, Immunology, and Infectious Disease/Research, Children's National Medical Center; Pediatrics and Biochemistry Department, George Washington University**

Address: **Washington, D.C.**

What advice can you offer to undergraduates who want to become scholars of molecular biology?

In addition to biology, take all the genetics, math, chemistry, and physics you can. Spend a semester or summer at a research lab taking a course or doing research. Alternatively, take a research course early on. These activities accomplish two purposes: you see if you like research; and your mentor forms an opinion of you that can support your entering graduate or medical school. Participate in a seminar; you learn to think on your feet and to present your thoughts clearly. Learn to write: sentences, paragraphs, essays, reports. Read widely to encompass the thought processes of others. Learn a foreign language and become comfortable with computers. Develop a strongly logical approach to science; cultivate a thoughtful approach in the analysis of data. Live with it; love it.

13

Recombinant DNA and Genetic Engineering: Molecular Tailoring of Genes

Technical developments often lead to quantum jumps forward in science. In the past two decades, a recently developed technology, **recombinant DNA technology,** or **genetic engineering,** has revolutionized genetics, in both basic research and its practical aspects. The discovery of two naturally occurring and quite remarkable classes of biological molecules made possible the development of methods to isolate and manipulate specific DNA fragments. These two classes of molecules are **plasmid DNA** and **restriction enzymes.** With recombinant DNA methodology, a DNA fragment—even an entire gene and its controlling elements—can be isolated, the fragment coupled with a plasmid or phage, and the hybrid inserted into a bacterium. And since bacteria can be replicated in vast quantities, this process can be used for large-scale production of the foreign DNA insert itself; if the bacterial host is able to express or synthesize the protein product of the foreign DNA, the hybrid (or chimeric) plasmid can direct the production of quantities of otherwise scarce or expensive proteins, or even of proteins containing specific mutations created in the laboratory.

In this chapter we will learn how segments of a cell's DNA can be cut, modified, joined together, and replicated. The relative ease with which those manipulations can be carried out is remarkable. Recombinant DNA technology has provided spectacular discoveries in virtually all areas of modern biology. We will review some recent advances in medicine, agriculture, and commercial processes.

Plasmids Act as Nature's Interlopers

plasmid DNA
restriction enzymes

In the early 1950s, it was discovered that bacteria transfer genes to other bacteria by a process called **conjugation.** One bacterium attaches small projections (**pili**) on its surface to those on the surface of an adjacent bacterium. DNA from the donor bacterium is passed to the recipient through the pili. The ability to form pili and to donate genes to neighbors is a genetically controlled trait. The genes controlling this trait are not located on the bacterial chromosome but rather on separate genetic elements called **plasmids** which were described in Chapter 12. Plasmids are ideal **vectors** to accept and carry pieces of foreign DNA (including entire genes) into a bacterial host, where the foreign DNA can be replicated along with the plasmid. As vectors, plasmids are therefore one of the major tools of genetic engineering.

Restriction Enzymes Function as Nature's Pinking Shears

Restriction enzymes, many of which have been found in bacteria, recognize and degrade DNA from foreign organisms. Bacteria have confronted the invasion of foreign DNAs for millions of years, and have evolved these enzymes as protective mechanisms that preserve their own DNA while destroying the invading DNA. Each restriction enzyme recognizes only one short sequence, usually (but not always) 4 to 6 basepairs long. The sites of cutting are called **palindromes** (they read the same backwards as forwards). For example, EcoR I, one of the first restriction enzymes to be isolated (from *Escherichia coli*) cuts DNA only at the sequence:

$$\overset{\downarrow}{\text{G}}\text{-A-A-T-T-C}$$

$$\underset{\uparrow}{\text{C-T-T-A-A-G}}$$

If the sequence occurs once in a circular invading plasmid, the enzyme would open the circle by cutting it once. If the sequence occurs at several sites, the DNA would be cut into several pieces. In general, the names of restriction enzymes are derived from the first letter of the genus followed by the first two letters of the species name of their bacterial source. (For example, EcoR I isolated from ***Escherichia coli,*** Hind III from ***Hemophilus influenzae,*** Taq I from ***Thermus aquaticus***—several hundred of these enzymes have been isolated from hundreds of species of microorganisms). For the molecular biologist, these enzymes allow plasmids to be opened up *in vitro* so that foreign DNA can be inserted. Restriction enzymes also offer a way of isolating predictable fragments of any DNA molecule. With the discovery of plasmids and restriction enzymes, it became technically simple to *clone* DNA, and obtain a large quantity of exact copies of any chosen DNA fragment. This was done by using restriction enzymes to isolate it, inserting it into a plasmid, and inserting the hybrid plasmid into a host bacterium which then reproduced many copies as the bacteria proliferated. This rearrangement of DNA in a living organism is the equivalent of genetic recombination, hence the name: recombinant DNA technology.

The process of generating recombinant DNA can be divided into four steps: first, the generation of desired DNA fragments; second, the insertion of these fragments into a suitable vector such as a plasmid or phage; third, the introduction of the vector into a particular host, usually a strain of *E. coli;* and finally, the identification, selection, and characterization of recombinant clones. Some of the methods used to accomplish these steps are described in this chapter. Though not exhaustive, the description provides a basic outline of some of the major methods used in recombinant DNA technology.

ISOLATION AND CHARACTERIZATION OF DNA FRAGMENTS

Owing to their specificity, restriction endonucleases were the first nucleases found to cleave DNA at specific sequences, and have proven useful in genetic analyses from the oligonucleotide to the chromosomal levels. At the oligonucleotide level, they are used to provide specific fragments for cloning, sequencing, and custom-made mutations. Segments of DNA thought to contain regions that control gene expression can, if desired, be cut away selectively and the effect of their loss on gene expression monitored. At the chromosomal level, they permit analyses of entire chromosomes by "walking" or "jumping" (described in Figure 13-3).

Most restriction enzymes make two single-strand breaks, one in each strand (close to, but not opposite each other), thus generating two 3'-OH and two 5'-P groups at each restriction enzyme site. A technically useful property of restriction enzymes was detected by electron microscopy: because of the complementary 3 to 5 base-pair overhanging termini (called cohesive or "sticky" ends) produced by many restriction enzymes, fragments circularized spontaneously. These circles could be relinearized by heating, but if after circularization they were also treated with *E. coli* DNA ligase (which covalently joins 3'-OH and 5'-P groups), circularization became permanent. This observation was the first evidence for two important characteristics of restriction enzymes:

1. Restriction enzymes make breaks in symmetric sequences.
2. The breaks are usually not directly opposite one another; therefore sticky (or cohesive, or complementary) ends are produced.

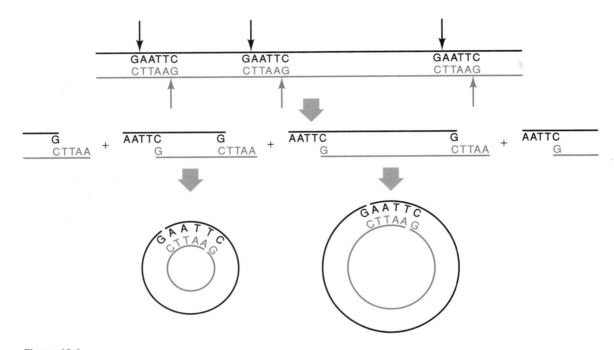

Figure 13-1
Restriction fragments that have overhanging cohesive ends following digestion with some restriction enzymes (here, at the 5′ ends) can circularize. Small arrows indicate cleavage sites of EcoR I; large arrows indicate the next step in the process.

These properties are illustrated in Figure 13-1.

The sequences recognized by restriction enzymes are **palindromes** as shown here:

A B C	C′ B′ A′		A B	B′ A′		A B X	B′ A′
A′ B′ C′	C B A	or	A′ B′	B A	or	A′ B′ X′	B A

Capital letters represent bases (A, B, and C may be the same or different), the ′ indicates the complementary base, and the vertical line is the axis of symmetry.

Examination of a very large number of restriction enzymes showed that the breaks are usually in one of two distinct arrangements: staggered, but symmetric around the line of symmetry, forming two types of cohesive ends—a single-stranded extension with a 5′-P terminus and a 3′-OH extension; or both cuts at the center of symmetry, forming blunt ends. Fragments with blunt ends cannot circularize spontaneously. These arrangements and their consequences are shown in Table 13-1 which

Table 13-1

Some Restriction Endonucleases and Their Cleavage Sites

Name of Enzyme	Microorganism	Target Sequence and Cleavage Sites
Generate flush ends Bal I	*Brevibacterium albidum*	↓ TGG|CCA ACC|GGT ↑
Generate cohesive ends EcoR I	*Escherichia coli*	↓ GAA|TTC CTT |AAG ↑
BamH I	*Bacillus amyloliquefaciens H.*	↓ GGA|TCC CCT |AGG ↑
Hind III	*Haemophilus influenzae*	↓ AAG|CTT TTC |GAA ↑
Pac I	*Pseudomonas alcaligenes*	↓ TTAA|TTAA AATT|AATT ↑

Note: The vertical dashed line indicates the axis of dyad symmetry in each sequence. Arrows indicate the sites of cutting.

lists the recognition sequences and cleavage sites for several restriction enzymes. Given their sequence specificity, each restriction enzyme generates a unique set of fragments for a particular DNA molecule.

Fragments obtained from a DNA molecule from one organism will have the same cohesive ends as the fragments produced by the same enzyme acting on DNA molecules from another organism. This unsurprising fact is one of the cornerstones of recombinant DNA technology. Furthermore, since most restriction enzymes recognize a unique sequence, the number of cuts made in the DNA from an organism by a particular enzyme is limited. The DNA in a typical bacterial chromosome, which contains roughly 2×10^6 base-pairs, is cut into several hundred to several thousand fragments, and nuclear DNA of mammals ($^\sim 3 \times 10^9$ base-pairs) is cut into more than a million. These numbers are large, but still relatively small compared to the total number of bases in the DNA of an organism. Of great usefulness in cloning are the less complex DNA molecules, such as bacteriophage λ or plasmids, which have only 1–10 sites of cutting (or even none) for particular restriction enzymes. Plasmids having a single site for each of a number of restriction enzymes are especially valuable in cloning, as we will see shortly.

If restriction sites common to two cloning partners (a plasmid and a foreign DNA) are not available, an alternative strategy can be employed. Each of the two DNAs can be cleaved with an enzyme that generates protruding termini identical to those generated by the other restriction enzyme. These **cohesive, compatible ends** can then be ligated (see Figures 13-1, 13-5, and 13-6).

A map showing the unique sites of cutting of the DNA of a particular organism by one or more restriction enzymes is called a **restriction map** (Figure 13-2(b)). The family of fragments generated by a single enzyme can be detected easily by gel electrophoresis of enzyme-treated DNA (Figure 13-2(a)).

In order to map a large segment or the entire genome, it is essential to localize experimentally-manageable mapped segments in relation to each other. To achieve this overwhelming task, chromosome walks are taken on large (~10 to 20 Kb) DNA fragments (Figure 13-3). These may be obtained by digesting genomic DNA with restriction enzymes whose cleavage sites are rarely found in DNA, or by doing partial digests so

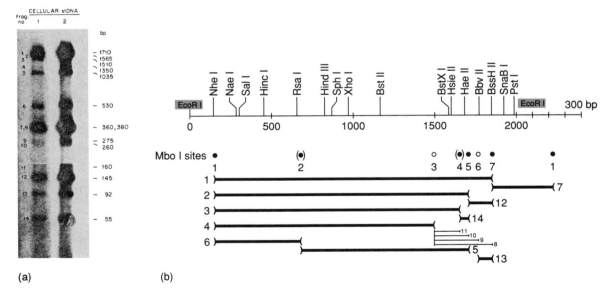

(a)   (b)

Figure 13-2
Restriction mapping: an example using a complex, G + C-rich, cellular satellite DNA of the Bermuda land crab. Satellite DNA is present in very similar, but not identical multiple copies. (a) Autoradiogram of [^{32}P]end-labeled Mbo I digest. Lane 1, normal exposure; lane 2, overexposed to visualize minor fragments. Sizes of the 14 fragments are indicated. (b) Restriction map of one repeat unit beginning at an EcoR I site and extending 2100 base-pairs to the next EcoR I site, which signals the beginning of the next repeat unit. The map indicates positions of sites of restriction enzymes that cut this repeat unit only once. However, Mbo I cuts the repeat unit at seven sites indicated by the circles numbered from 1 to 7. Numbered solid bars (1–14) represent Mbo I fragments in (a); thick bars, major fragments; thin bars, minor fragments. ● = sites present in almost all repeat units of the satellite; cuts at these sites yield the major (dark) bands seen on the gel in part (a). (●) = sites in some repeats; thus cuts between sites 1 and 2 and 1 and 4 yield bands of intermediate density of 530 and 1510 base-pairs, etc. o = sites in very few repeats, yielding correspondingly very few fragments (8–11) seen clearly only in overexposed lane 2 in (a).

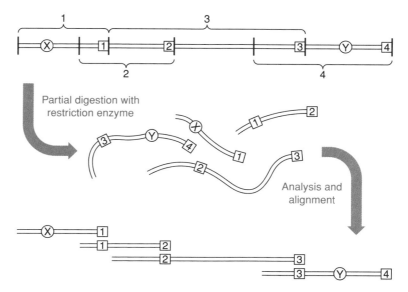

Figure 13-3 Chromosome walking and jumping
Chromosome walking using partial DNA digests with a single restriction enzyme.
Top: Original (unknown) DNA sequence, showing sequences X and Y, with dark vertical bars representing restriction sites (shown only in the top stick figure). The problem is to localize Y with respect to X. *Middle:* Restriction fragments. The DNA has been cut with the restriction enzyme so that each piece of the original DNA is cut at only some of the sites (i.e. a partial digest). In the example diagrammed, four overlapping fragments are generated, each ending in a sequence indicated by a box with the corresponding fragment number. *Bottom:* Analysis and alignment. The fragment containing X is found to end in sequence 1. A probe hybridizing to sequence 1 also hybridizes to fragment 2, which ends in sequence 2. Similarly, a probe hybridizing to sequence 2 also hybridizes to fragment 3 which ends in sequence 3, and a probe hybridizing to sequence 3 hybridizes to fragment 4 which contains Y. Knowing the orientation from restriction maps of each fragment, the sequences can be aligned and the relation of Y to X determined.

It will be apparent that fragments other than the four diagrammed will be generated in the restriction digestion. For clarity, these have been omitted. They will either be silent, i.e., will not hybridize with any of the probes, or they will be sub- or superfragments that can be identified by analyses and alignment similar to that depicted.

Instead of cutting with a single restriction enzyme, two different restriction enzymes can be used provided that overlapping fragments are generated.

that the DNA is cut at only a few of the cutting sites. The restriction fragments are then cloned into suitable hybrid plasmids as described in Figure 13-5, grown up (amplified) in bacteria, and recovered. The collection of these fragments, containing pieces of DNA from the entire genome of the organism, is called a **genomic library.**

The fragment of interest at the beginning of a chromosome walk (containing the known sequence or gene) is cut with a number of other restriction enzymes to map it, and its ends are sequenced. A radioactive probe (commonly about 50 base-pairs long) that will hybridize to the end sequence is used to identify other fragments with overlapping se-

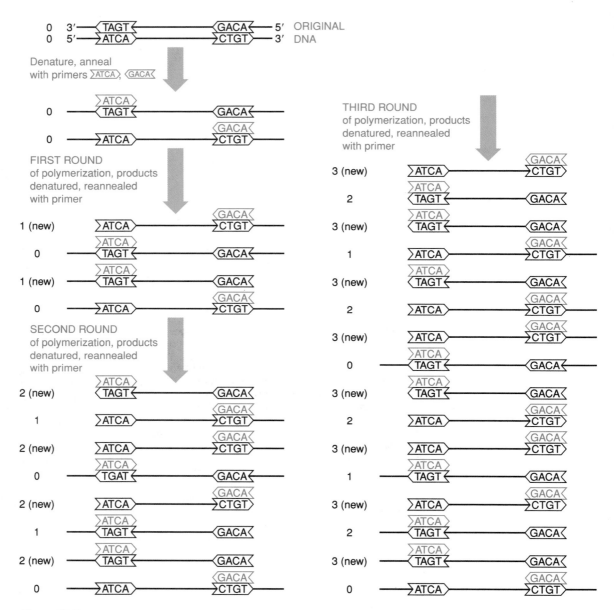

THIRD ROUND
of polymerization, products
denatured, reannealed
with primer

FIRST ROUND
of polymerization, products
denatured, reannealed
with primer

SECOND ROUND
of polymerization, products
denatured, reannealed
with primer

Figure 13-4
Polymerase chain reaction. Top diagram represents the original double-stranded DNA. The sequences at the ends of
the segment to be amplified are shown in arrow-shaped boxes pointing in the 5′ → 3′ direction, i.e., the direction of
polymerization. In practice these sequences are 15–30 base-pairs in length, although only four bases are shown in the
diagram. The DNA is denatured by heating, the primers are annealed to the end sequences of the target DNA, and the
first round of polymerization is carried out. The numbers in the left hand column indicate the polymerization round in
which a given strand is synthesized, with 0 representing the original strands, and NEW representing newly-synthesized
strands. At the end of the first round, there are 4 strands, the 2 original strands with extensions beyond both ends of
the target sequence, and the 2 copied strands with extensions beyond only the 3′ ends of the target sequence.
Annealing and polymerization are repeated, yielding 8 strands at the end of the second round, of which 6 have

quences that are adjacent to the first probe, and these are similarly mapped and sequenced at their ends. The orientation of the adjacent probes can be established by comparing overlapping restriction maps. The probe adjacent to the first can then be analyzed in a similar way to find the next fragment in the walk. The process is experimentally slow, but has been effectively used to align as many as 10^6 adjacent base-pairs. Obviously, repeated DNAs scattered throughout the genome cannot be used for identifying overlapping sequences.

Chromosome jumping is just that: instead of the shorter steps of ~10 or 20 Kb taken on walks along chromosomes, jumps of 100 Kb or more are taken. The very large fragments needed for the jumps can be recovered by electrophoresis of partial digests of restriction enzymes. Other than the size of the fragments analyzed (for which cosmids, or large plasmids, are utilized), the technique is basically the same as chromosome walking. The ends of the fragments are restriction mapped or sequenced, and matching overlaps on other DNA fragments are identified by computer analyses of the assembled sequences or restriction maps. If the chromosomal location of one of the fragments or a gene contained within the original fragment is known, then the chromosomal location of the adjacent DNA sequence or gene can also be deduced. By the use of such methods, a number of genes have been localized to specific chromosomes.

In 1983 a new technique, the **polymerase chain reaction (PCR),** was devised. It has made the detection and cloning of rare DNA sequences possible. The method is based on the amplification of target DNA sequences in genomic DNA. Two specific oligodeoxynucleotides are synthesized, one complementary to the 3′ end on one DNA strand, and one to the 3′ end on the opposite strand of the target DNA to be amplified. The oligodeoxynucleotides are hybridized to the denatured, single-stranded target DNA containing the segment of interest (Figure 13-4). The hybrids of the oligonucleotides with the genomic DNA strands, starting at the 5′ end of the primers, serve as initiation points for replication primers of the original strands by *Taq* polymerase (a heat stable DNA-dependent DNA polymerase isolated from a thermophilic bacterium, *Thermus aquaticus*). After replication of the original strands, the products are denatured by melting, the reactants cooled to annealing

---

extensions at one or both ends, and 2 encompass the target sequence only. At the end of the third round there are 16 strands, of which 8 have extensions and 8 the target sequence only. Because the original strands are always present, they are replicated with 3′ extensions on every round, and after n rounds there are $2n + 2$ strands with extensions ($2n$ is the number replicated with 3′ extensions, 2 are the original with both 5′ and 3′ extensions). The total number of strands increases exponentially, after n rounds being 2^{n+1}. Thus after 10 rounds there are 2048 strands, of which only 22 have extensions; and after 20 rounds there are more than 2 million strands of which 42 have extensions. In other words, the contribution of strands with extensions beyond the target sequence rapidly becomes negligible as the target sequence is amplified. Up to 60 rounds can be carried out successfully. That provides enough DNA to identify a target sequence from a single cell by gel electrophoresis.

temperatures to allow the primers to re-anneal to the synthesized products as well as to the original strands, and another round of replication is initiated. This process, which takes less than 10 minutes, is repeated as many times as desired. After the first few rounds, the vast majority of the products will be the amplified fragment between the 3' sites of the original DNA, as shown in Figure 13-4. The extensions of the original DNA beyond these sites will be diluted to a negligible proportion. The *Taq* polymerase is used because it is not inactivated by the heating steps used for strand separation after each round. PCR is currently carried out with commercially available machines programmed to perform the temperature cycling automatically

The products of the amplification can then be used to detect or clone the gene of interest. This technique, and many elegant variations on it, have been used to identify rare genes, such as viral DNA sequences, and to recover genes in very small amounts of DNA, even from a single cell. If an altered base (mutant) is deliberately introduced into the primer, the mutated primer will be replicated and amplified after the first round. This modification of PCR has been successfully used to generate large numbers of copies of engineered genes.

Genetic Interlopers: Vectors Function as Vehicles for Transferring Genes

Several types of vectors and some of the procedures for cloning DNA molecules are described in this section.

To be useful, a vector must have three properties:

1. It must be able to enter a bacterial host.
2. It must have an origin of DNA replication that allows it to replicate in the host.
3. Cells thus transformed must be selectable, preferably by growth of host cells on a solid medium containing the selecting agent.

Two types of vectors are commonly used in cloning. One type is derived from bacterial viruses (bacteriophages or phages). Phages act like hypodermic needles; they inject DNA into the bacterial host where it replicates. The phages most commonly used for cloning are λ and M13. Lambda phage have the unique feature of being able to accept large fragments (up to 40 Kb) of inserted foreign DNA (Figure 13-5). Vectors developed from M13 (which can accept fragments up to 10 Kb) have unique promoters that facilitate the sequencing of the inserted foreign DNA.

The second type of vector is derived from bacterial plasmids (often pBR322) which are relatively small (Figure 13-6). Some of their characteristics have been described above. Antibiotic resistance is one of the most important ones. Many plasmids have been genetically engineered

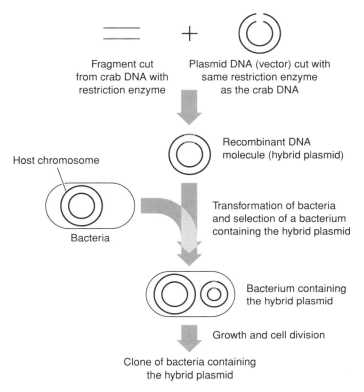

Figure 13-5 Cloning foreign DNA
A restriction fragment of crab DNA and a plasmid vector linearized by digestion with the same restriction enzyme used·
to cut the crab DNA are ligated. The hybrid (or chimeric) plasmid transforms a bacterium. Henceforth, when the
plasmid replicates so does the crab DNA.

to serve as specialized vectors. Polylinkers (also called multiple cloning
sites, are synthetic DNA oligomers that provide one or more specific
restriction sites, yield cohesive ends and/or a bacterial origin of replication) may have been inserted into them. Vectors that have received a
eukaryotic transcriptional promoter can be expressed in both bacteria
and eukaryotes. Since they can shuttle between bacterial and eukaryotic
cells, they are called **shuttle vectors.** Still other vectors have unique
promoters useful for sequencing DNA or synthesizing RNA *in vitro*.
Because of their small size, and the ease of handling them, plasmids are
the most commonly used cloning vectors.

The strategy used to ligate a fragment of foreign DNA to a vector
depends on the nature of the termini of the DNA. As described above,
restriction enzymes produce three basic types of termini (3'-OH, 5'-P
overhangs, and blunt). A strategy for cloning DNA fragments with
these termini and some of the problems or advantages associated with
their cloning are shown in Table 13-2.

Given the large variety of vectors available, the optimal cloning
strategy is to select a vector with two restriction sites that are compatible

Table 13-2

Parameters of the Ligation Reaction

Ends of DNA Fragments	Cloning Requirements	Features

Blunt ends (ex: Bal I)
 	
↓
...TGG | CCA...
...ACC | GGT...
↑
 	
| Separation of
↓ fragments
 	
...TGG ³'   ⁵' CCA...
    +
...ACC   GGC...
     ⁵'   ³'
	DNA and ligase present in high concentrations	Restriction sites at junctions of DNAs may be lost; number of wild type plasmids without inserts may be high; multiple tandem copies of the foreign DNA may be inserted

Identical overhanging cohesive sticky ends (ex: EcoR I)
 	
↓
...GAA | TTC...
...CTT | AAG...
↑
 	
| Separation of
↓ fragments
 	
  ³'   ⁵' AATTC...
...G   +
...CTTAA   G...
     ⁵'   ³'
	Phosphatase treatment of linearized plasmid DNA reduces religation of wild type plasmids	Restriction sites at junctions of ligated ends are constituted; foreign DNA inserted bidirectionally; multiple tandem copies of the foreign DNA may be inserted

Different cohesive ends (directional cloning; see Figure 13-6)	Purification of linearized plasmid increases cloning efficiency	Restriction sites at junctions of ligated DNA are usually reconstituted; recircularization of wild type plasmid is low; insertion of foreign DNA is unidirectional

with the restriction sites on the 5′ and 3′ termini of the fragment of foreign DNA. The fragment of foreign DNA can then be inserted into the vector by **directional cloning** (Figure 13–6). For example, the vector pUC19 (UC for University of California, where these plasmids modified from pBR322 were constructed) can be cleaved with EcoR I and Hind III, restriction sites introduced into the plasmid by a polylinker, after which the large fragment of the vector can be purified by gel electrophoresis from the small fragment containing the polylinker. The vector (the large fragment without the polylinker) can then be ligated to a segment of foreign DNA that has been digested by the same enzymes and therefore contains cohesive termini that are compatible with those of the vector. Since two restriction enzymes are used in the construct, the two termini of the fragments are different, and the foreign DNA can thus be inserted in only one orientation, hence the name **directional**

cloning. The resulting circular recombinant can then transform *E. coli* to antibiotic resistance by inserting the gene which is on the vector. Most of the bacterial cells resistant to ampicillin will contain plasmids that carry a foreign DNA insert between the EcoR I and Hind III sites.

Foreign DNA fragments carrying identical termini are usually cloned into a plasmid vector that has been linearized by digestion with the same enzyme(s), and the foreign fragment and the vector are covalently linked. The concentrations of the vector and the foreign DNA in the ligation reaction are adjusted to optimize the number of ligated products comprised of both the vector and the foreign DNA. Since T4 DNA ligase joins a 3'-OH and a 5'-P, recircularization of linearized plasmid DNA can be minimized by removing the 5'-phosphates from both termini of the plasmid by digestion with alkaline phosphatase. Then, a foreign DNA segment with 5'-terminal phosphates can be hybridized by complementarity, and ligated efficiently to the dephosphorylated plasmid DNA. Since circular DNA transforms much more efficiently than linear plasmid DNA, most of the transformants will contain recombinant plasmids. Blunt-ended DNA fragments are ligated much less efficiently than fragments with protruding complementary termini. Therefore, ligation reactions involving them require higher concentrations of ligase and of the foreign and plasmid DNAs.

After a fragment of DNA is cloned into a plasmid, inserted into a bacterium, and grown to large amounts in a selecting (antibiotic containing) medium, a method must be available to recover it from the hybrid plasmid. If the restriction sites in the plasmid vector are the same as those on the fragment of foreign DNA, the foreign DNA can be recovered by lysing the bacteria to release the plasmid, and digesting the plasmid with the restriction enzymes used in cloning. If, however, appropriate restriction enzymes to cut the DNA had not been identified, polylinkers could be added to the end of each of the partners during cloning. Then, the cloned fragments can be released from the recombinant plasmids by digestion with restriction enzymes that cleave in the flanking polylinker sequences.

INSERTION INTO A VECTOR OF A DNA MOLECULE THAT IS COMPLEMENTARY TO AN mRNA

Often the genes which are candidates for cloning are organized in such a way that cloning is difficult. Some genes do not produce selectable products that can serve as an indicator of their presence. Many genes are rare, or are difficult to isolate. In those cases, it is sometimes preferable to enrich or partially purify the fragment containing the gene to be cloned prior to ligating it to the vector. If the restriction map of the DNA of interest is known, then it is possible to predict the size of the DNA fragment that will be generated when that DNA is digested with a particular restriction enzyme. Fragments of the predicted size can be isolated from an agarose gel after electrophoresis and joined to an ap-

Directional Cloning

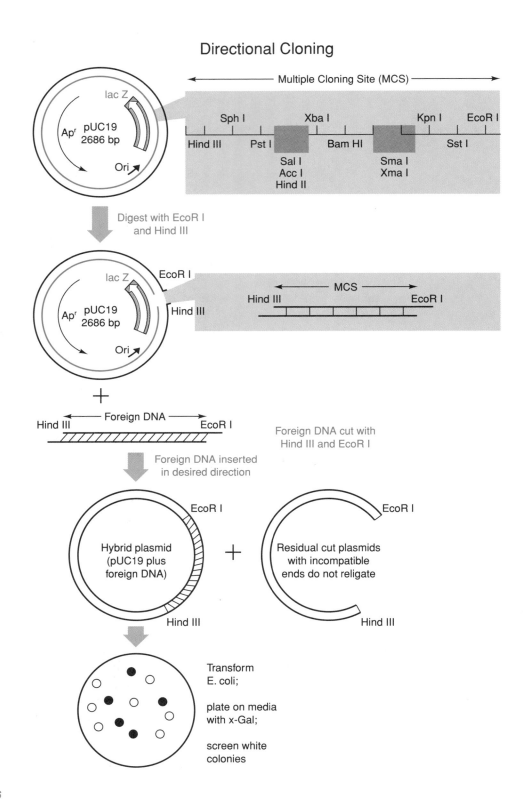

propriate vector. However, eukaryotic cells contain about 10^6 cleavage sites for a typical restriction enzyme. Direct isolation of a eukaryotic gene from a mixture of fragments separated by electrophoresis can enrich the frequency of occurrence of the gene, but is possible only if the size of the restriction fragment containing the gene is known in advance. In this section another procedure for cloning a particular segment of DNA that encodes an mRNA is described. By this technique, any coding sequence whose mRNA can be isolated in almost pure form can be cloned.

Let us assume that the chicken gene for ovalbumin (the protein in egg white) is to be cloned into a bacterial plasmid. A recombinant plasmid could be formed by treating both chicken cellular DNA and the *E. coli* plasmid DNA with the same restriction enzyme, mixing the fragments, annealing, ligating, and transforming *E. coli* with the DNA mixture containing the recombinant plasmid. However, identification of a bacterial colony containing this plasmid could not be carried out by a test for the presence of ovalbumin, because a bacterium containing the intact ovalbumin gene will not synthesize ovalbumin. Introns, sequences that do not code for proteins, will be interspersed in the gene, and bacteria lack the enzymes needed to remove these noncoding sequences (see Chapter 9).

Some specialized cells of higher organisms, such as those producing ovalbumin in chickens, make large amounts of a single protein. In some plants for example, the CO_2-fixing enzyme ribulosebisphosphate carboxylase/oxygenase, can comprise ~50% of the total protein. In such cells the specific mRNA molecules (whose introns will have been removed by processing enzymes when the mRNA is isolated from the cytoplasm) constitute a large fraction of the total mRNA synthesized in the cell. Consequently, mRNA samples can usually be obtained which consist predominantly of a single mRNA species—in the example of the chicken cells, ovalbumin mRNA. If genes of this class, that is, those whose gene products are the major cellular proteins, are to be cloned, the purified mRNA can serve as a starting point for creating a collection

---

Figure 13-6
Directional cloning into a plasmid vector (pUC19). Ap^r indicates the position in the vector of the ampicillin resistance gene; *lacZ*, that of the *lacZ* gene; Ori indicates the origin of replication. PUC19 can be cut with any of the 13 restriction enzymes with sites in the multiple cloning site (MCS, shown expanded on the upper right of the diagram). To achieve directional cloning, both the vector and the foreign DNA to be inserted into the vector are cut by the same two enzymes, here Hind III and EcoR I. The excised segment of the MCS is shown. The foreign DNA can then be inserted in only one direction determined by the compatible cohesive ends. When bacteria transformed with the hybrid plasmid are grown in X-Gal medium, if there are untransformed bacteria (viz., those lacking an insert) present, their *lacZ* gene is expressed, and these colonies are blue. An insertion of foreign DNA into the *lacZ* gene interrupts the gene which is then not expressed; these colonies are white. Only white colonies need be screened for the insertion of foreign DNA. Another advantage to directional cloning is that the linearized plasmids that have not ligated to a fragment of foreign DNA cannot circularize. The number of false positives is thereby markedly decreased.

of recombinant plasmids, many of which should contain only the protein-coding sequences of the gene of interest. In cloning such sequences, the enzyme **reverse transcriptase** is used.

Many RNA-containing animal tumor viruses (**retroviruses**) contain reverse transcriptase. This enzyme can use a single-stranded RNA molecule (such as mRNA) as a template to synthesize a double-stranded DNA copy (called **complementary DNA** or **cDNA**). If the template RNA molecule is an mRNA molecule with the introns removed from the primary transcript, the corresponding full-length cDNA will contain an uninterrupted coding sequence. This sequence will not be identical to that of the original eukaryotic gene. If, however, the purpose of constructing the recombinant DNA molecule is to synthesize a eukaryotic gene product in a bacterial cell, and the processed mRNA can be isolated, then cDNA copied from it is the material of choice for insertion into the vector. The joining of cDNA to a vector can be accomplished by any of the procedures for joining blunt-ended molecules.

cDNA clone

DETECTION OF RECOMBINANT MOLECULES

When a vector is cleaved by a restriction enzyme and allowed to ligate to fragments of the DNA from a particular organism cut with the same restriction enzyme, several types of molecules result. These include:

1. A re-ligated vector that has not acquired any fragments of foreign DNA (methods for avoiding this have already been described).

2. A vector with one or more foreign DNA fragments.

3. A molecule without a vector, consisting only of joined fragments of the organism's DNA. Molecules in this class do not contain an origin of replication, and therefore cannot be replicated in bacteria.

To facilitate the isolation of a bacterial colony containing a vector with a particular gene, a method is needed to insure, first, that the vector possesses an inserted foreign DNA fragment, and second, that it is the DNA fragment of interest. In this section several useful procedures for detecting recombinant vectors will be described.

In the cloning procedure as described so far, a foreign DNA fragment obtained by digestion with a restriction enzyme is ligated to a cleaved vector molecule, yielding a large number of hybrid vectors containing different fragments of foreign DNA. If a particular DNA segment or gene is to be cloned, the vector possessing that segment must be isolated from the set of all vectors possessing foreign DNA. For many genes, simple selection techniques are adequate for recovery of a vector containing that gene. For example, if the gene to be cloned were a bacterial leucine gene, a Leu$^-$ bacterial host (incapable of synthesizing its own leucine) would be used, and a Leu$^+$ colony selected and grown in a medium lacking leucine. This procedure is useful in cloning bacterial genes that can be selected, or genes that encode products of intermediary metabolism. Often, however, the gene to be cloned is a eukaryotic gene whose product does not change the phenotype of the bacterial host. Four methods are used to identify bacterial colonies that contain recombinant plasmids that cannot be directly selected:

1. Screening by colony hybridization
2. Insertional inactivation of a plasmid gene (often for antibiotic resistance)
3. Analysis of plasmid DNA by restriction enzymes
4. α-complementation

Each of these methods will be discussed in this section.

Recombinant clones are most commonly identified by determining if the DNA in the colonies **hybridizes** with a radioactive probe to the foreign DNA. Bacterial colonies are replica-plated onto a solid support, usually nitrocellulose or nylon filters. The bacteria are lysed with alkali which also denatures their DNA, neutralized, and allowed to hybridize to the ^{32}P-labeled probe. The location of the ^{32}P-hybridized probe is then determined by exposing x-ray film to the filters. Bacterial colonies releasing DNA which hybridizes to the DNA of the probe can then be identified by aligning the x-ray film with the original bacterial plates. In this way it is possible to screen for many hundreds of colonies simultaneously, and recognize colonies that carry recombinant plasmids (see Figure 13-7).

Insertional inactivation is used with vectors such as pBR322 that have two or more antibiotic resistance genes. This method is illustrated in Figure 13-8. The foreign DNA and the plasmid are digested with restriction enzymes that recognize sites located in one of the two antibiotic resistance genes (for example Pst I, in the ampicillin-resistance gene, Ap^r, but not in the tetracycline-resistance gene, Tc^r). After the two DNAs are ligated and the chimeric plasmids are used to transform *E. coli,* the transformants can be selected on plates that contain the

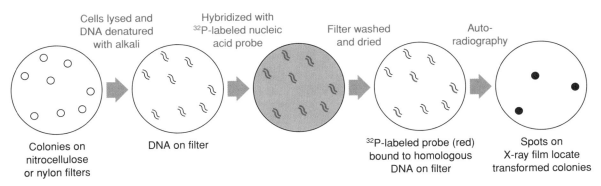

Cells lysed and DNA denatured with alkali	Hybridized with ^{32}P-labeled nucleic acid probe	Filter washed and dried	Auto-radiography

Colonies on nitrocellulose or nylon filters

DNA on filter

^{32}P-labeled probe (red) bound to homologous DNA on filter

Spots on X-ray film locate transformed colonies

Figure 13-7
Detection of transformed cells by colony hybridization. Only 8 of thousands of colonies on a reference plate (not shown) are included.

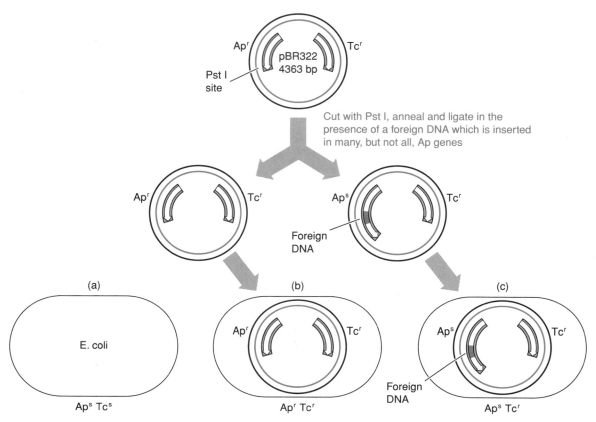

Figure 13-8
Detection of transformed cells by insertional inactivation. Arrows inside the plasmid (here, pBR322, a very common cloning vector) indicate the direction of transcription of the ampicillin (Ap^r) and tetracycline (Tc^r) resistance genes, clockwise for the Tc^r gene and counterclockwise for the Ap^r gene. If the plasmid has a fragment of foreign DNA inserted into one of its genes for antibiotic resistance (here, in the Pst I site of the Ap^r gene), it renders the gene inactive, and it no longer conveys to its bacterial host (here, a strain of *E. coli*) resistance to that antibiotic. The superscripts ^{r} and ^{s} indicate resistance and sensitivity, respectively. (a) Untransformed cells will not grow on either antibiotic. (b) Bacterial cells transformed with plasmids that have recircularized without ligating to a piece of the foreign DNA will grow on ampicillin and/or tetracycline since they are resistant to both. (c) Cells transformed with a modified (i.e., a hybrid) plasmid will grow on the antibiotic (here, tetracycline) whose gene has not been inactivated by insertion of a foreign DNA but not on the other, here ampicillin.

antibiotic whose gene was not interrupted by the insertion of the foreign DNA (the Tc^r gene in this case). To discriminate between transformants that contain foreign DNA and those that contain plasmid DNA which has recircularized without foreign DNA, separate plates containing ampicillin and tetracycline can be inoculated. The colonies which survive and grow in the presence of tetracycline contain plasmids with active Tc^r genes and may or may not contain active Ap^r genes. The colonies that do not grow in the presence of ampicillin, but grow in the presence of tetracycline should contain plasmids with inactive Ap^r genes, and probably carry foreign DNA sequences inserted into the Pst I site on that gene.

To verify the identities of positive clones, a number of putatively positive bacterial colonies are picked and grown individually in small-scale (1–2 ml) cultures (called **minipreps**). Plasmid DNA isolated from each culture is digested with restriction enzymes and electrophoresed on gels. This method is useful if there is a high probability of identifying the recombinant by its restriction map in a small number of randomly chosen transformed colonies. This technique is often used in conjunction with insertional inactivation or α-complementation to confirm the identity of the recombinant molecule.

The fourth method commonly used to identify bacterial colonies with chimeric plasmids exploits the biological properties of the *E. coli* β-galactosidase gene (*lacZ*). Many of the vectors in current use (e.g., pUC19) carry a segment of the regulatory sequences and the coding information for the first 146 amino acids of the *lacZ* gene. Within the coding region of the gene is a polylinker that does not disrupt the protein reading frame of the gene. These vectors are used with host *E. coli* cells that code for the carboxy-terminal portion of β-galactosidase. Neither the host-encoded nor the plasmid-encoded fragment is active, but together they associate to form an active enzyme. This type of complementation in which deletion mutants are complemented by host genes is called **α-complementation.** Lac$^+$ bacteria are identified because they form blue colonies in the presence of the chromogenic substrate 5-bromo-4-chloro-3-indolyl-β-D-galactopyranoside (X-Gal). Insertion of a fragment of foreign DNA into the polylinker of the plasmid almost invariably results in a transformant that is not capable of α-complementation. Such transformants usually form white colonies. With this simplified color test, many thousands of colonies can be screened visually, and those that carry recombinant plasmids are easily recognized.

When *E. coli* phage λ is the vector, one procedure used for selecting for the presence of inserted genes is based on the presence in λ of several centrally located genes that are not needed for lytic growth and on the fact that the length of DNA that can be packaged in a λ phage head must be at least 77% of the wild type length. A λ variant (one of several that have been constructed by recombinant DNA techniques) has restriction sites for EcoR I near each end of the nonessential region. The linear vector is cleaved with EcoR I, and the three fragments are separated by gel electrophoresis. The outer two fragments (called "arms") are isolated, and the central fragment (called a "stuffer," since it provides a DNA segment sufficiently long to permit packaging into phage heads, but is missing any essential functions) is discarded. The terminal fragments can be joined via the cohesive termini produced by EcoR I, but the resulting DNA molecule, with a length that is 72% that of wild-type λ DNA, will be noninfective because it is too small. However, if foreign DNA whose size is 5-35% of the size of wild-type λ DNA is inserted, the resulting recombinant DNA molecule will be packageable, and hence able to produce progeny phage. Thus, if an *E. coli* culture is transformed with DNA obtained by annealing a mixture of the two terminal λ fragments and foreign DNA, any phage that are produced will contain foreign DNA.

Applications of Genetic Engineering

Unquestionably, recombinant DNA technology has revolutionized biology. At present, its main uses are:

1. Facilitating the production of useful proteins.
2. Creating bacteria capable of synthesizing economically important molecules.
3. Supplying DNA and RNA sequences as research tools.
4. Altering the genotype of organisms (both animals and plants).
5. Potentially correcting genetic defects in animals (gene therapy).

It is used in research of all types: basic, medical, agricultural, commercial, and industrial; and can be expected to contribute to filling some of the most fundamental needs of humankind. At the same time, the technology arouses concerns about potential effects on the environment, and risks to health and society. In this section the current and future potential applications (and a few of the issues which these applications have raised) will be discussed.

USES IN RESEARCH: SITE-SPECIFIC MUTAGENESIS

If the sequence of a DNA (and the protein it encodes) is known, it is possible to change one or more nucleotides at will, thus changing the coding sequence, the sequence of a promoter, another regulatory element, or a restriction site. This engineered DNA, in proper orientation with suitable promoters, can then be inserted into a vector and the protein expressed with a mutation in the pre-selected location. Specific deletions and insertions can also be constructed. This **site-specific mutagenesis** (also sometimes called site-directed mutagenesis) is the basis of genetic engineering, and has been a powerful tool in the analysis of protein function *in vivo* and *in vitro*.

In principle the procedure can be straightforward. With commercially available DNA synthesizers, the desired sequence containing the mutated base or coding trinucleotide(s) is synthesized as one of a set of single-stranded oligodeoxynucleotides encompassing the entire gene and the desired regulatory sequences, and the fragments are then phosphorylated on their 5′ termini. Similarly the complementary sequence is also synthesized as a set of 5′-phosphorylated oligodeoxynucleotides with ends offset by 5–9 nucleotides from the first set. When each oligomer is annealed to its complement, there is single-stranded extension beyond the double-stranded region that will hybridize with the next appropriate oligomer until the entire DNA is constructed and the fragments are then ligated. The construct is designed so that at each end are restriction sequences corresponding to those in the expression vector, and these sequences are used to insert the engineered gene into the vector. The vector is then used to transform the host organism. Transformants are screened by hybridization of the DNA they synthesize with a radioactive oligonucleotide containing the mutated sequence. It is important finally to sequence the entire engineered DNA and to perform an amino acid analysis of the mutated protein (after purification) synthesized by the host cells to ascertain that the desired products have been achieved.

There are many tricks and variations on this basic scheme appropriate to the specific result that is sought. For example, with larger genes it is sometimes useful to nick the gene in one strand near the bases to be altered, remove a segment of the nicked strand with exonucleases, anneal the mutated DNA fragment to the resulting single-stranded portion, and fill in and ligate the segment to recover the double-stranded DNA that can then be inserted into the vector. When this DNA is replicated in the host, native DNA will be synthesized from the unmodified strand, but the mutant DNA can again be found by screening with the chemically mutated fragment. Also, techniques have been developed using only single-stranded fragments inserted into a single-stranded bacteriophage (M13) for a vector. PCR, as a means of propagating selected mutants to be used as primers, has already been discussed.

Three examples of how this approach has proven useful are briefly given here. The first relates to the study of protein-protein interactions. Epidermal growth factor (EGF) is a small (molecular weight = ~6000) polypeptide that binds noncovalently to receptive cells and influences many growth-associated properties. Human EGF is 53 amino acids long, and is folded into 3 loops with a tail, held together by disulfide bonds. By site-directed mutagenesis, amino acids at many positions in the molecule have been altered to other specific amino acids, and substantial quantities of the altered product have been isolated from bacteria. The properties of the mutant molecules in binding to the native EGF receptor can then be analyzed. Such studies enable investigators to determine which amino acids participate in the binding of the growth factor to its receptor, and to determine the relative roles played in the interactions by ionic charge, size, or hydrophobicity of the individual amino acids.

A second example is the EGF receptor itself. The receptor is a large protein (molecular weight = 170,000) in the cell membrane. Part of the molecule, including the EGF binding site, faces the external medium. A short segment of the molecule traverses the cell membrane. The major component of the receptor is internal, facing the cytoplasm. The internal domain has tyrosine kinase activity, which acts enzymatically (when stimulated by EGF) to transfer the terminal phosphate of ATP to tyrosine residues on itself and other cellular proteins. The internal domain thus has ATP binding sites, and also has serine that is phosphorylated by other protein kinases. When receptor-bearing cells are stimulated by EGF, many things happen; other membrane-bound and cytosolic enzymes are stimulated, internal Ca^{2+} is elevated, several membrane transport systems are activated (either immediately or following protein synthesis), and other genes are induced to synthesize a battery of growth-associated enzymes. After prolonged EGF stimulation, DNA synthesis is also induced, and finally, the cells divide. In the meantime, the EGF and its receptor are clustered together on the cell surface and internalized; most of the EGF is delivered to lysosomes and degraded, and many of the EGF receptors are eventually recycled back to the cell surface. This cascade is either regulated or at least initiated within the cell by the activities of the activated receptor. What parts of this large and complex molecule are responsible for the various components of the response? By site-directed mutagenesis, it has been possible to engineer the receptor cDNA, introduce it and allow it to be expressed in cells that have no endogenous receptor. These means have shown that replacing all of the phosphorylable tyrosines with other amino acids blocks the entire cascade: autophosphorylation is thus essential. On the other hand, large deletions and other changes do not prevent the internalization. By fine analysis of the engineered changes and their consequences, it will become possible to determine what parts of the receptor molecule are responsible for its enzymatic activity, and what parts are associated with responses of the entire cell.

A final example comes from studies on the CO_2-fixing enzyme ribulose bisphosphate carboxylase/oxygenase isolated from the photosynthetic bacterium, *Rhodospirillum rubrum*. In this organism the enzyme is a homodimer (its functional form is made up of two identical subunits). Each subunit contains two components of shared active sites (A and B), where A is a part of the site on one subunit, and B is a part on the other. When the subunits are associated in the dimeric form, component A from each subunit interacts with B from the other to form two active AB sites per dimer. Now we can construct two mutants, A★ and B★, by altering a base in the appropriate locations in two cDNA fragments so as to cause a nonfunctional amino acid to be incorporated into each mutant. If we express the A★ mutant under conditions where normal dimerization takes place, both sites in the dimer are A★B and the enzyme is inactive. The reciprocal pattern, AB★, also an inactive enzyme, is found if we express the B★ mutant. But if both types of cDNA are co-expressed in the same bacterial host so that the mutant subunits may randomly associate, four interactions at the sites would be expected: A★B, AB★, and A★B★ (which are all inactive), and AB which is native. Thus starting with cDNA that codes for two different inactive mutants, co-expression would be expected to yield a partially active (25%) product; and this is what has been found experimentally. Site-directed mutagenesis has thus been able to confirm deductions about the active site structure made on independent chemical grounds.

These examples have been selected from hundreds that are under investigation, exploring the structure and functions of hundreds of proteins *in vivo* and *in vitro*. Genetic engineering has opened whole new vistas for molecular biochemistry and physiology.

USES IN RESEARCH: MAMMALS MADE TO ORDER

The analysis of bacterial mutants has been an extraordinarily powerful means of untangling cellular functions including metabolic pathways. A major stumbling block to extending this approach to higher organisms has been the diploid nature of the eukaryotic genome; it is difficult to isolate an experimentally–induced change in both copies of a given gene. One current approach to overcoming this difficulty is the use of targeted mutagenesis via homologous recombination, first in yeast and more recently in mammals. This technique results in the replacement of an endogenous allele with a specifically engineered gene in the appropriate genetic environment under normal genetic controls, including those operative during development and differentiation.

In mammals, an essential preliminary step was the demonstration that embryonic stem cells (ES cells) from mouse preimplantation embryos could be cultured and cloned on feeder layers. The cells of the feeder layer are prevented from proliferation by treatment with mitomycin C or irradiation before the ES cells are seeded onto them. The feeder layer not only supports the growth of the ES cells, but also secretes substances that prevent the differentiation of the totipotent ES cells.

A typical protocol for the construction of a mouse mutant by homologous recombination can be briefly described (Figure 13-9). The example given here is the achievement of germ–line transmission of a disrupted allele of the C^{abl} gene. C^{abl} is the cellular homolog of an oncogene from the Abelson murine leukemia virus. Of unknown function, it is transcribed in most normal mouse tissues, although tumors do not ordinarily appear.

The first step is the construction of a plasmid containing the gene or portion of the gene of interest. This portion is engineered *in vitro* so as to alter the gene as desired, either by base-substitution, deletion, or insertion. In the C^{abl} example, a promoterless neomycin resistance gene (Neo^r) was inserted near the 3′ end of the coding region of the C^{abl} gene, but drug resistance markers with controllable promoters have also been used effectively. The 3′ end of the gene, together with the Neo^r marker, was inserted into the plasmid.

The ES cells used are homozygous for distinctive phenotypic markers, such as coat color or rare electrophoretic variants of tissue enzymes, which can be readily identified in the eventual progeny. The C^{abl}/Neo^r plasmid was introduced into ES cells by electroporation (reversible permeabilization of cells by electric shock in the presence of the plasmid) and the cells were plated onto the feeder layer. After 1–2 days (to allow

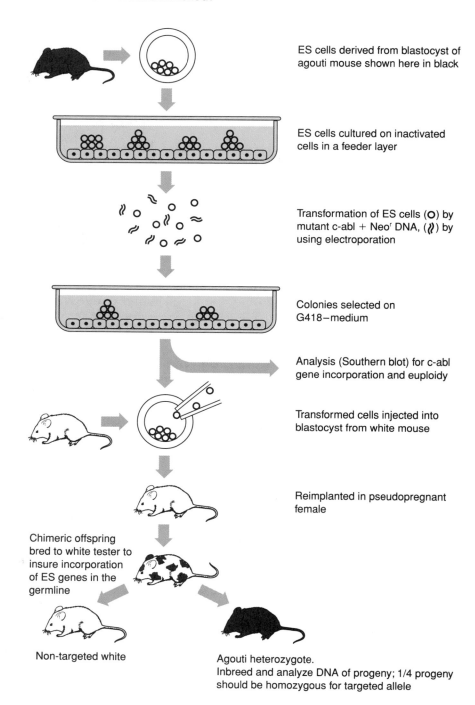

ES cells derived from blastocyst of agouti mouse shown here in black

ES cells cultured on inactivated cells in a feeder layer

Transformation of ES cells (O) by mutant c-abl + Neo^r DNA, ($\approx$) by using electroporation

Colonies selected on G418−medium

Analysis (Southern blot) for c-abl gene incorporation and euploidy

Transformed cells injected into blastocyst from white mouse

Reimplanted in pseudopregnant female

Chimeric offspring bred to white tester to insure incorporation of ES genes in the germline

Non-targeted white

Agouti heterozygote.
Inbreed and analyze DNA of progeny; 1/4 progeny should be homozygous for targeted allele

integration of the plasmid into host DNA) the culture medium was replaced with fresh medium containing G418, a drug that kills cells not expressing the Neo^r gene, and colonies were grown in G418-containing medium. Since the Neo^r gene did not include its promoter, Neo was expressed only in cells where the gene was next to a cellular promoter, including of course, cells in which the engineered construct had replaced one of the endogenous C^{abl} alleles by homologous recombination. Drug resistant ES colonies were isolated and their DNA was analyzed by Southern blots, in which DNA/DNA hybrids were formed between restriction digests of the cellular DNA and a probe derived from the C^{abl} gene, outside the region of homologous recombination. Since the parent line of the ES cells contains its own C^{abl}, one criterion for homologous integration was two bands on the Southern blots, one band in the same position as in the parent, and the other at a position 0.6 Kb larger. The difference was due to the 0.6 Kb Neo^r insert in the integrated gene. Only the parental band is observed in nonhomologous recombinants; these are recovered from the drug-selection step in far greater abundance than homologous recombinants.

There are other variations in the technology up to this point, each appropriate to the corresponding details of constructing the original plasmid. When clones with the desired characteristics have been identified, karyotype analysis of the selected cells is performed to insure that they are euploid. Cells (10–15) from selected clones are injected into blastocysts of embryos that are homozygous for dissimilar markers, e.g., if the ES cells are from black or agouti (grizzled brown) mice, they are injected into embryos of white or nonagouti stock. The injected blastocysts are then reintroduced into pseudopregnant females (prepared by mating 2–3 days previously with vasectomized males). In successful experiments, the totipotent ES cells become incorporated into the inner cell mass of the recipient blastocyst and colonize various tissues including the germ cells of the developing embryo. The resulting chimeric animals can be preliminarily identified by chimeric coat color patterns.

Figure 13-9 Mammals made to order
Homologous recombination in mammals. In this example, totipotent embryonic stem cells (ES cells) are derived from the blastocyst of an agouti donor and proliferated in culture by growth on a feeder layer of metabolizing cells that have been treated (radiation, mitomycin, or other irreversible DNA synthesis inhibitor) to prevent growth of the feeder cells. Engineered gene(s), including an antibiotic resistance gene, are introduced into the ES cells by electroporation, and the cells are cultured further in the presence of the antibiotic which kills the nontransformed cells. Clones are picked and assessed for euploidy and the presence of the targeted allele of the desired gene. 10–15 cells meeting the criteria are injected into the blastocyst from a white mouse, where they become incorporated into the inner cell mass, and the blastocyst is reimplanted into a pseudopregnant female. Offspring in successful cases are chimeric for coat color (white from the maternal embryo and agouti from the established ES cells). Chimeric animals are mated to white tester stock; if the engineered ES cells have entered the germ line, the traits from the ES cells (for example, coat color) will also be present in the offspring. DNA from such offspring is analyzed by Southern blotting for presence of the targeted allele. Heterozygous animals are mated, with one quarter of their offspring expected to be homozygous for the targeted mutation; homozygosity can be identified by further analysis.

Male chimeras are then allowed to develop to sexual maturity and crossed with tester females to determine whether the engineered gene has been incorporated into the germ line. The progeny are again examined for properties, such as hair pigmentation or electrophoretic variants of enzymes derived from the ES cells. Finally, these mice are tested, by Southern blot analysis, for the property that was of interest in the first place, in this example, the C^{abl} gene with the Neo^r insert. By continued inbreeding of mice heterozygous for the targeted gene, homozygous mice for the engineered gene can be derived.

Methods closely similar to this have been used to derive a strain of mice in which the β_2-microglobulin gene has been disrupted. Normally, β_2-microglobulin is required for bringing certain histocompatibility antigens to the cell surface, and also for the surface expression of antibody-binding receptors in intestinal epithelia. Mice homozygous for the nonfunctional gene product have remarkable characteristics. First of all, they breed and develop normally, showing that β_2-microglobulin is not required for development *in utero*. The histocompatibility antigens are not to be found, and the antibody-binding receptors are absent from intestinal epithelia. This latter observation is of importance because these receptors are required for the uptake of intact immunoglobulins from the mother's milk, yet the offspring survive apparently without maternally derived antibodies. The β_2-microglobulin-deficient animals are also deficient in immunologically important killer T cells. Nevertheless, in a pathogen-free (but not germ-free) environment, these animals survive very well.

The β_2-microglobulin example shows the power of the technology. In principle, a wide spectrum of genes could be engineered *in vitro* and introduced into recipient animals, and the physiology of the gene products studied in affected offspring. Such an approach will be of immense benefit, allowing analysis of integrated functions in intact organisms, not just cellular functions in cell isolates in culture.

There are still major concerns and unsolved problems remaining in the technology. First and most obvious, if the engineered gene is essential to embryonic development, homozygous offspring may die *in utero*. Second, it is apparent that the recipient (or host) organism plays an important role not yet understood, in the successful incorporation of the donor cells into the developing embryo and subsequent germ-line transmission. Third, in the current methodology nonhomologous recombination is far more frequent than homologous, and rigorous selection and careful analysis of the cloned, transformed ES cells are necessary before they can be used as donors. The latter point is particularly significant if homologous recombination is eventually to be used in human gene therapy. Having an otherwise normal insulin gene in the wrong location could lead to its inappropriate expression or even no expression at all, which could be more harmful than helpful. Another example, related to the protocol described above, is that the expression of the human C^{abl} gene is associated with at least two human cancers, chronic myelogenous leukemia and acute lymphocytic leukemia. In these diseases the gene is usually activated by chromosomal translocation. The potential danger of recombination of an introduced gene in the wrong location is evident.

One way around these problems, particularly when the host is to be a differentiated organism or individual rather than a preimplantation embryo, is the genetic manipulation of cells that have first been removed from the body, selection of the desired genotype *in vitro,* and the reintroduction of the cells into the original donor or a suitable recipient. One can envisage that defective bone marrow stem cells could be removed from a patient and their genetic problems corrected *in vitro,* including selection of the desired genotype, and the cells returned to the patient. Similarly, sheets of cultured skin cells in which genes responsible for immune recognition have been inactivated could possibly become "universal donor" cells that are invisible to the recipient's immune system. Cultures of such cells could be maintained in the clinic for ready use in severe burn cases.

The long range goal, however, remains the development of a method that results in only homologous recombination with high efficiency in virtually any cell type. Such an achievement would offer the utopian potential of using normal genes to correct genetic disorders that may not be recognized until after the birth of a fully differentiated individual.

USES IN MEDICINE

The goal of medicine is to alleviate and ultimately eliminate human disease. More than one hundred inheritable human diseases have been identified and characterized. Many of these are the result of a defect in a single gene; some involve a mutation in a gene that encodes for an enzyme or a structural protein. Here again, because of the development of recombinant DNA methodologies, the ability to identify genetic defects linked to specific diseases has increased.

One of the most useful applications of recombinant DNA technology is to provide novel and precise methods for detecting preclinical cases of genetic disease. The first such use was in the prenatal diagnosis of α-thalassemia, a disease in which there is a defect in the α-globin gene. Methods to detect deletions, mutations, and genetic rearrangements have been used to identify many of the hemoglobinopathies. Here, a deletion or point mutation which alters a restriction enzyme site can be detected directly on a Southern blot with an appropriate probe, or indirectly through linkage to another genetic marker associated with the genetic defect. Linkages between genes can be demonstrated by using the separate genetic markers as probes and determining whether they hybridize to the same restriction fragment on a Southern blot. For linkage to be used as a diagnostic tool, it must first be demonstrated that DNA from patients with the genetic disorder shows a characteristic restriction pattern that is not observed in healthy persons.

Genomes, including human, are polymorphic. Between any two individuals there is a difference of about 1 base per thousand. Some of these may represent point mutation differences, but many are phenotypically "silent" because they occur in the variable third coding base, and do not alter the amino acid being coded, or (probably more commonly) the differences lie in sequences between genes. This polymorphism has well-recognized implications for sequencing the human genome. When these differences occur in restriction sites, they generate restriction fragments of different sizes. This **Restriction Fragment Length Polymorphism (RFLP)** is readily observed by gel electrophoresis of DNA that has been cut by restriction enzymes and probed with an appropriate marker. Scattered throughout the genome of any individual eukaryote there are also different numbers of copies of repeated DNA (**minisatellites**). Digestion of genomic DNA with restriction enzymes followed by Southern blotting, and probing with repeated sequences detects these kinds of RFLPs, which are sometimes called **Variation in the Number of short Tandem Repeats, or VNTRs.**

RFLP can be used to design indirect tests that indicate the presence of defective genes. Linkages identified by suitable probes for RFLP on Southern blots have already been found for such disorders as phenylketonuria, congenital adrenal hyperplasia, retinoblastomas, Huntington's chorea, cystic fibrosis, and for X-linked disorders such as Duchenne muscular dystrophy, Becker muscular dystrophy, Lesch–Nyhan syndrome, hemophilia B, retinitis pigmentosa, and the fragile X-mental retardation syndrome. The international initiative to map and sequence the human genome makes it likely that either direct probes, or indirect gene linkages will be identified for many human disorders.

Like a gene, the restriction pattern of an individual is inherited in strictly Mendelian fashion. Thus RFLP, especially as analyzed with a number of different restriction enzymes, can constitute a "DNA fingerprint" that provides strong evidence for the identity of the parents of a child.

Several therapeutic approaches are available for the treatment of genetic diseases. One of these is **drug or dietary treatment** to circumvent a metabolic block. In this case, the missing hormone or metabolite is simply provided. Therapeutically important products are being synthesized by recombinant DNA technology. In brief, a gene that codes for a particular peptide, along with an appropriate promoter and a sequence that instructs the cells to secrete the gene product, is inserted into an expression vector (i.e., λ gt11); the gene is oriented in such a way that its coding strand is linked to the strand containing the promoter in the vector. Bacterial cells can be prepared that contain several hundred copies of the gene, and this can result in synthesis of a gene product that reaches a concentration of about 1 to 5% of the cellular protein. The gene product, which is secreted by the bacteria containing the recombinant

DNA, can be harvested from the medium in which the bacteria have grown. Somatostatin, insulin, growth hormone, interferon, opioid peptides (enkephalins and endorphins), interleukins, thymosin, tissue plasminogen activator, urokinase and hemophiliac factor VIII, have all been synthesized in this way. The list continues rapidly to lengthen. Methods to deliver these protein products to the patient slowly or in sequential doses have also improved, so that in most cases it is now possible to replace the defective gene product with a functional protein at concentrations that mimic the physiological dose of these proteins.

However, if the gene is from a eukaryote, special problems must be considered:

1. Eukaryotic promoters are not usually recognized by bacterial RNA polymerases; hence the eukaryotic gene must be linked to a bacterial promoter.

2. The mRNA transcribed from eukaryotic genes may not be translatable on bacterial ribosomes.

3. Introns may be present, and bacteria are unable to excise eukaryotic introns. In such cases cDNA, or a synthetic DNA with a sequence deduced from the amino acid sequence of the desired protein product, is used rather than the eukaryotic gene.

4. The protein itself often must be processed (for example, insulin), and bacteria cannot recognize processing signals from eukaryotes.

5. Eukaryotic proteins may be recognized as foreign material by bacterial proteinases and degraded.

There are several approaches to solving these problems in addition to those mentioned. One procedure uses a yeast artificial chromosome (YAC) as a cloning vehicle for eukaryotic genes. Recent studies suggest that when a recombinant YAC containing a eukaryotic gene is put into yeast, also a eukaryote, some of the problems described no longer exist.

A second approach to treating genetic disorders is **supportive or replacement therapy.** In this case, malfunctioning cells are replaced with normal cells. A notable instance is the transplantation of bone marrow cells. Here, for example, a patient with bone marrow cancer can be treated with radiation or chemotherapy to destroy all endogenous bone marrow cells. The bone marrow can then be reconstituted with normal cells pre-selected from the patient, or with cells from a matched donor.

Although these specific treatments may benefit the patient, they only alleviate some symptoms of the disease rather than cure it. Furthermore, neither drug nor replacement therapy is available for many diseases. In addition, some genetic diseases are thought to involve mechanisms that control the expression of a number of genes (i.e. cancer) rather than the loss or malfunction of a single protein. The current treatments (chemotherapy, radiation, immunological intervention) are ineffective for some of these diseases (e.g., malignant melanoma). One promising approach to curing such diseases is *gene therapy,* or restitution of the normal gene *in vivo.*

Several recent advances have enhanced the potential to eliminate human disease through gene therapy. One is the ability to identify mutant genes that cause disease and to clone them without having to isolate the protein they encode, for example, Duchenne muscular dystrophy. This powerful technique has been called **reverse genetics.** It is likely that there will be a dramatic increase in the number of genetic diseases identified as being caused by specific defects in single known and cloned genes, especially given the rate at which genes are being assigned to chromosomes, which materially assists in their identification. (At the first Human Gene Mapping Workshop in 1973, about 75 genes had been localized to 19 autosomes and the X chromosome; in the 1990 report of the tenth meeting, ~5,100 genes had been described, and more than 2000 had been mapped. The maps covered all of the chromosomes!)

Another advance is the development of techniques for the efficient transfer of DNA into human cells. This achievement is the result of an increased understanding of the biology of cells as well as of a variety of viruses. Still in early experimental stages is the introduction of cloned genes directly into tissues; expression of several genes so introduced has been detected. For example, plasmids containing the bacterial gene coding for chloramphenicol acetyltransferase have been injected directly into rat muscle. It has subsequently been shown that a number of the muscle fibers expressed the bacterial gene. A major problem here is that, as noted above, such genes are most frequently not incorporated into the recipient genome by homologous recombination, and thus are not under normal control. Such genes are also frequently unstable in the recipient tissue, and their expression may be lost over time.

Nevertheless, all of these advances increase the probability that genetic treatment can eventually lead to cures for a number of devastating inherited human diseases. The development of efficient gene therapy is a compelling goal, because although some genetic disorders occur in only a small percentage of the population (adenosine deaminase deficiency, for example) others, such as sickle-cell anemia and the thalassemias, affect millions of people throughout the world.

USES IN AGRICULTURE

For decades, genetics has played an important role in plant breeding—that is, in the development of new strains of plants with desirable characteristics. For the most part, plant breeders have had to rely on the selection of mutants, and hybridization between different varieties of the same species. Although these approaches have been highly successful, organisms specifically constructed by recombinant DNA technologies can have much more precisely tailored genes and therefore technology is likely to become the preferred approach. Creating new varieties of plants by direct alteration of the genotype is becoming an increasingly more important application of genetic engineering. Genes for fixing nitrogen, improving photosynthesis, and providing resistance to pests, pathogens, and herbicides, have been transferred and successfully expressed in the hosts. A bacterium has been developed that increases the tolerance of crops to frost because the bacterium can seed ice crystals on the surface of the plant rather than allowing the damaging crystals to form within plant cells. Resistance to drought and increased salinity are other traits that are being engineered.

One of the novel developments in plant genetic engineering involves crown gall, a plant tumor associated with infection by the bacterium *Agrobacterium tumefaciens*. When a segment of a plasmid carried by the bacterium is inserted into host DNA, tumor formation is induced. The plasmid, which is therefore called a **tumor inducing (Ti)** plasmid, has been exploited as a vector to transfer foreign genes into host plants. Unfortunately, *A. tumefaciens* infects only broad-leaved flowering plants (dicots). Alternative, but similar, strategies are being developed for other plants. The supplement entitled "The Molecular Biology of Plants" offers additional insights into applications of recombinant DNA technology to plants (see page 419).

Recombinant DNA is being applied to animal husbandry for the production of vaccines, and the introduction of genes for commercially-important traits like milk yield and body composition. The successful creation of these transgenic organisms is heralded on the front pages of the newspapers daily.

OTHER COMMERCIAL AND INDUSTRIAL APPLICATIONS

For many years genetic variance of plants, animals, and microorganisms has been exploited for a variety of commercial applications—such as developing tetraploid corn, breeding fast race horses, and selecting suitable yeasts for use in brewing and baking—usually by identifying and isolating naturally occurring, or artificially produced mutants.

Apart from the more obvious applications in the pharmaceutical industry, there are a number of other industrial uses of recombinant DNA technology. For example, attempts are being made to broaden the range of substrates of biological fermentation to include cellulose or even plastic waste. It may be possible one day to engineer organisms that can ferment at high temperatures, so that fermentation and part of the distillation process may take place simultaneously. A DNA-dependent polymerase isolated from a thermophilic bacterium, *Thermus aquaticus,* is already being used successfully in PCR at high temperatures. Furthermore, several genes from oil-metabolizing bacteria have been inserted into a plasmid which was then inserted into a marine bacterium to yield an organism capable of metabolizing petroleum at sea. These organisms, naturally occurring oil-eaters, were used to help clean up the Alaskan shoreline after an 11-million-gallon oil spill in 1989, as well as a 1990 open-ocean spill off the coast of Texas. These bacteria metabolically convert the oil to an emulsion of fatty acids which can be consumed by marine organisms. Although there is certainly a need for similar organisms to degrade other toxic products in the environment, the problems of the approach are numerous. For example, the concentration of pollutants often is too low to maintain the growth of engineered microorganisms. The growing concern for cleaning our planet may provide sufficiently strong incentives to persuade industry to develop the technology required to overcome these obstacles.

In the food industry, there is the likelihood of using engineered organisms to convert waste into usable food for animals or humans. There is also the possibility of using recombinant DNA technology to redesign the tests currently used to detect harmful or unwanted substances in food and other products. In the future, genetic engineering will affect many industrial and commercial ventures, and offer applications as yet unimagined.

Problems and Future of Recombinant DNA

Although the advent of recombinant DNA technology has revolutionized biology, its widespread application in some areas has raised a number of issues. These include fears of possible biohazards, legal issues concerned with priorities and patents, and ethical issues about permanently modifying the genomes of both eukaryotes and prokaryotes that are to be released into the environment. Although the possible biohazards of the technology appear to have been initially exaggerated, caution is appropriate and necessary when dealing with potential pathogens, especially in industry where production is on a different and much larger scale than that of the research laboratory. Legal problems have arisen over priorities and patents, particularly the patenting of genetically modified organisms. A different sort of legal problem has arisen over the admissability in some courts of RFLP evidence in establishing disputed paternity or identification of rapists. New and very profound ethical questions raised by recombinant DNA technology include the possibility of one day altering the germ-line by treating human embryos, and deciding which genes are suitable or acceptable candidates for replacement. Introducing engineered microorganisms into ecosystems raises both legal and ethical issues. These are formidable problems, but future developments hold promise of far-reaching applications. Exciting future applications for the technology lie in defining the control of gene action in specific tissues and at particular stages of development, and in understanding the molecular basis of human disease and, possibly even human behavior.

Summary

Recombinant DNA technology (or genetic engineering), which has revolutionized biological research, has been made possible by the use of restriction enzymes and cloning vectors. Restriction enzymes allow the cutting of DNA molecules at specific sites. The location of restriction sites along a DNA molecule is a restriction map. The resulting pattern of fragments is a restriction map. Some genetic defects can be identified by restriction fragment length polymorphisms. Cloning vectors, such as plasmids, phage, and cosmids, allow DNA to be amplified, modified, or expressed. The polymerase chain reaction may first be used *in* vitro to amplify rare DNA sequences before insertion into the cloning vector. DNA fragments inserted into vectors can transform a suitable host. Transformants are often selected by antibiotic resistance conferred by the vector. The correct insert is screened by a combination of colony hybridization, insertional inactivation of a vector gene, analysis by restriction digest, or α-complementation of a vector-encoded enzyme activity. Genomic libraries are made from restriction digests of total genomic DNA, while cDNA libraries are made from mRNA which has been transcribed to DNA by the enzyme reverse transcriptase. The order of sequences in a genomic library may be determined by chromosome walking or jumping. Once cloned, genes may be altered by site-specific mutagenesis. The mutant gene may then be expressed and studied. Some cloned genes may be reintroduced into mammals and expressed, with the hope of someday curing human genetic defects by replacing mutant genes (gene therapy). Cloned genes may also be expressed by bacteria in large amounts and the products used pharmaceutically. Reintroduction of altered or novel genes into plants or animals may provide better strains for agricultural purposes.

Drill Questions

1. These questions concern restriction endonucleases.
   a. What is a restriction enzyme?
   b. What is the biological function of a restriction enzyme?
   c. For what main purpose do scientists use restriction enzymes?
   d. What common feature is present in every base-sequence recognized by a restriction enzyme?

2. Two types of cuts are made by restriction enzymes. What are they? These two types of cuts produce three possible types of DNA termini. Describe them.

3. What property must two different restriction enzymes possess if they yield identical patterns of breakage?

4. List some properties of an ideal cloning vector.

5. Explain insertional inactivation and α-complementation. How are they used in cloning?

6. Once a series of recombinant plasmids or phages has been produced, describe several ways in which plasmids or phage containing a particular insert may be identified.

7. A Kan-r Amp-r plasmid is treated with the Bgl II enzyme which cleaves the *amp* gene.

The DNA is annealed with a Bgl II digest of *Drosophila* DNA and then used to transform *E. coli*.

   a. What antibiotic would you put in the agar to insure that a colony has a plasmid?
   b. What antibiotic-resistant phenotypes will be found on the plate?
   c. Which phenotype will have the *Drosophila* DNA? How would you screen for this phenotype?

8. When cloning into plasmid vectors, re-ligation of vectors without inserts is always a problem. Describe two ways in which this problem may be lessened.

9. What is the difference between a genomic library and a cDNA library? What are the major differences in the structure of a gene cloned into either type of library? Give an advantage of each type of clone.

10. What is site-directed mutagenesis? What advantages and disadvantages does this technique have compared to the conventional form of mutagenesis described in Chapter 11? What types of information can be gained from experiments utilizing these techniques?

Problems

1. The restriction enzyme, EcoR I, was used to cut (1) a linear DNA molecule containing ten EcoR I sites and (2) a circular DNA molecule also containing ten EcoR I sites.

   a. How many fragments are generated from the linear molecule and from the circular molecule?
   b. Can all the fragments of the linear molecule and the circular molecule circularize? If not, how many can? Which ones cannot and why?

2. A plasmid contains a single BamH I site within an antibiotic gene. The recognition sequence for BamH I is G$\downarrow$GATCC, where the sequence is written 5′ to 3′ and the arrow indicates the point of cleavage.

   a. A fragment of DNA containing a gene of interest is to be cloned into the BamH I site of the plasmid described above. Unfortunately, the gene sequence contains a

BamH I site. The gene is flanked by Bgl II sites (A$\downarrow$GATCT), however, as follows:

$$5'----AGATCT----\overrightarrow{gene}----AGATCT----3'$$

Diagram the cuts made by BamH I on the plasmid DNA and the cuts made on the gene-containing fragment by Bgl II. Show both strands and the cohesive ends produced. Are the BamH I and Bgl II cohesive ends compatible (i.e. can the Bgl II fragment be cloned into the BamH I site)? If so, show the resulting construct. Are the BamH I and/or Bgl II sites regenerated?

   b. Repeat the above analysis for the BamH I digested plasmid and the following insert sequence:

$$5'-----CGATCC-----AGATCG----3',$$
   cut with Sau3A ($\downarrow$GATC).

   c. Repeat the above analysis for the BamHI digested plasmid and the DNA sequence in (b) cut with DpnI (GA$\downarrow$TC).

3. Restriction enzymes A and B are used to cut the DNA of phages T5 and T7, respectively. A particular T5 fragment of size 2.0 Kb is mixed with a particular T7 fragment of size 1.5 Kb and treated with low concentrations of DNA ligase. Three major circular forms are produced, and all three can be cut with both restriction enzymes A and B, producing one, one, and two fragments, respectively. Diagram the most likely structures of these circles. What can you conclude about restriction enzymes A and B?

4. Plasmid pBR607 DNA is circular, double-stranded, and has a molecular weight of 2.6×10^6. This plasmid carries two genes whose protein products confer resistance to tetracycline (Tc^r) and ampicillin (Ap^r) in host bacteria. The DNA has a single site for each of the following restriction enzymes: EcoR I, BamH I, Hind III, Pst I, and Sal I. Cloning DNA into the EcoR I site does not affect resistance to either drug. Cloning DNA into the BamH I, Hind III, and Sal I sites abolishes tetracycline resistance. Cloning into the Pst I site abolishes ampicillin resistance. Digestion with the following mixtures of restriction enzymes yields fragments with the sizes listed in

the table below. Position the Pst I, BamH I, Hind III, and Sal I cleavage sites on a restriction map, relative to the EcoR I cleavage site. Show also the approximate locations of the *amp* and *tet* genes, and the nucleotide distances between restriction sites.

Enzymes in Mixture	Molecular Weights of Fragments (millions)
EcoR I, Pst I	0.46, 2.14
EcoR I, BamH I	0.2, 2.4
EcoR I, Hind III	0.05, 2.55
EcoR I, Sal I	0.55, 2.05
EcoR I, BamH I Pst I	0.2, 0.46, 1.94

5. For this problem, refer to the C^{abl} gene replacement experiment described in the chapter.
   a. Diagram the wild type C^{abl} gene and what the insert of the plasmid construct with the inserted Neo^r gene might look like relative to the wild-type gene. Also indicate a region of the wild-type gene which could be used as a hybridization probe in the Southern blot screening procedure.
   b. Show possible homologous and nonhomologous recombination events which would lead to G418-resistant cells.
   c. Assuming that the restriction sites used for the Southern blot analysis flank the C^{abl} gene (are in the chromosomal DNA upstream and downstream of the gene), show the chromosomal organization of transformed strains (G418-resistant) for both homologous and nonhomologous recombination events. Indicate the approximate positions of the restriction sites and the hybridizing and nonhybridizing fragments. Remember that this is a diploid organism.

6. You have the cDNA clones for two human genes, A and B, which you believe may be closely linked on a human chromosome, since these two genes are known to be close together in rats. Briefly describe the tools and procedures you would use to answer this question.

Conceptual Questions

1. Consider the issue of gene therapy in the treatment of human diseases and the potential of creating a world in which diseases such as muscular dystrophy or hemophilia would be curable, and our elderly need not become senile. Given the time and difficulty involved in identifying mutant genes which cause disease and replacing the mutant gene with a wild-type one in homologous fashion, is it worth the time and effort spent for the good of only a small group of people in many cases? Who takes responsibility if a gene is integrated incorrectly during treatment?

2. One concern about recombinant DNA technology is the danger of creating recombinant organisms hazardous to the environment, or to life on earth. Consider such dangers for yourself, versus the current and potential good for society. Which side should prevail, and how great a threat do you think recombinant DNA technology is to life as we know it? Imagine doing scientific research today without recombinant DNA technology. What types of findings would be impossible to obtain? Consider also the legal application of recombinant DNA technology such as using PCR or RFLP mapping to confirm relatedness, or identify criminals from a few cells left at the scene of a crime. What are the potential benefits and dangers of using recombinant DNA technology in this manner?

3. The entire human genome is being sequenced. What types of information can be gained? With the potential for identifying a great number of genes, how will their functions be determined? In your opinion, would it be better to use the total sequencing approach, or more conventional gene isolation techniques to unravel the mysteries of the human genome?

Sankar L. Adyha
Head, Developmental Genetics Section
Laboratory of Molecular Biology

Birthday: **4 October 1937**

Birth Place: **Calcutta, India**

Undergraduate Degree: **University of Calcutta**
Major: Chemistry 1958

Graduate Degree: **University of Calcutta, Ph.D. 1963; University of Wisconsin, Ph.D. 1966**

Postgraduate Training: **University of Rochester, 1966–1968; Stanford University, 1968–1969**

Present Position: **National Cancer Institute, National Institutes of Health**

Address: **Bethesda, Maryland**

MY LABORATORY STUDIES the control of gene expression during growth and development, with emphasis on structure-function of regulatory proteins and DNA elements in transducing information from environmental signal to gene transcription.

In what direction is the field of molecular biology moving?

Many biologists are curious about evolution—how we came about. I wonder where our present and constantly exploding knowledge in molecular biology will take us, say, even 50 years from now. How will it influence our thinking? How will it be applied to our living? I even feel envy of the biologists of the future when I think about their research, even though I haven't the slightest idea about what they will be working on.

14

Regulation of Gene Activity in Prokaryotes

In earlier chapters, we have learned that chromosomal DNA is organized into genes, and how the encoded one-dimensional information is first transcribed into RNA and then translated into protein—as three-dimensional information. This three-dimensional information is responsible for all biological processes—essential or auxiliary. However, the behavior of an organism depends not only on the nature of the proteins that are expressed, but also on the extent to which they are expressed and the environmental conditions under which this expression occurs. Of course, some housekeeping proteins—proteins which build cell architecture, for example—must be made all the time. If all proteins were made all the time, there would not be any variation. Remember that organisms consume large amounts of energy in the process of transcription and translation. Thus, the ability to synthesize materials only as needed would make sense for economy and adaptation. There are devices to insure that proteins are synthesized in exactly the amounts they are needed and only when they are needed. In general, the synthesis of particular gene products is controlled by mechanisms that are collectively called gene regulation. Regulation of gene expression not only exists, but makes cellular adaptation, variation, differentiation, and development possible.

The regulatory systems of prokaryotes and eukaryotes are somewhat different. Prokaryotes are generally free-living unicellular organisms that grow and divide indefinitely as long as environmental conditions are suitable and the supply of nutrients is adequate. Thus, their regulatory systems are geared to provide the maximum growth rate in a particular environment, except when such growth would be detrimental. This strategy also seems to apply to the free-living unicells such as yeast, algae, and protozoa.

The requirements of tissue-forming eukaryotes are different from those of prokaryotes. In a developing organism—for example, in an embryo—a cell must not only grow and produce many progeny cells, but also must undergo considerable change in morphology and biochemistry, and then maintain that changed state. Furthermore, during the growth and cell division phases of the organism these cells are challenged less by the environment than are bacteria, since the composition of their growth media does not change drastically with time. Some examples of such media are blood, lymph, other body fluids, or, in the case of marine animals, sea water. Finally, in an adult organism, growth and cell division in most cell types has stopped, and each cell needs only to maintain itself and its properties. Many other examples could be given; the main point is that because a typical eukaryotic cell faces different contingencies than a bacterium, the regulatory mechanisms of eukaryotes and prokaryotes are not the same. In this chapter we consider regulation in prokaryotes. Eukaryotes are examined in Chapter 16.

Principles of Regulation

Since the best understood regulatory mechanisms are those used by bacteria and bacteriophages, these are the basis of all important present-day concepts of such regulation. Obviously, one can regulate gene expression by controlling synthesis of mRNA or protein. Mechanisms for both exist in prokaryotes. In *E. coli*, the maximum number of copies of a protein present depends upon its nature and need. For example, a cell may contain only 20 copies of *lac* repressor, a protein described later, but possesses over 100,000 copies of Ef-Tu, a protein factor needed for polypeptide synthesis. Regulatory signals exist to determine the maximum amount of each protein by controlling the rate of transcription or translation. *E. coli* cells have enzymes that give them the capability of using compounds other than glucose for food, protecting themselves from toxic environments, synthesizing useful compounds only when needed, etc. However, the enzymes that participate in these adaptive processes are normally absent; they are made only when the need arises. The on-off regulatory activities are obtained by allowing synthesis (or translation) of a particular mRNA when the gene product is needed, and prohibiting synthesis when the product is not needed.

In bacteria, few examples are known in which a system is completely switched off. In the off state, a basal level of gene expression almost always remains, often consisting of only one or two transcriptional events per cell generation; hence, very little synthesis of the gene product occurs. For convenience, when discussing transcription, the term "off" will be used; but it should be kept in mind that usually levels are very low rather than absent. In only one known case in bacteria (namely, in spores) is expression of most genes totally turned off. In eukaryotes complete turning off of a gene is quite prevalent.

In bacterial systems, when several enzymes act in sequence in a single metabolic pathway, usually either all or none of these enzymes are produced. This phenomenon, **coordinate regulation,** results from control of the synthesis of a single polycistronic mRNA molecule that encodes all of the gene products. This type of regulation does not occur in eukaryotes, because eukaryotic mRNA is usually monocistronic, as discussed in Chapter 9.

Transcriptional Regulation

Several mechanisms for regulation of transcription are common; the particular one used often depends on whether the enzymes being regulated act in degradative or synthetic metabolic pathways. For example, in a multistep degradative system, the availability of the molecule to be degraded frequently determines whether or not the enzymes involved in the pathway will be synthesized. In contrast, in a biosynthetic pathway the final product is often the regulatory molecule. The molecular mechanisms for each of the two regulatory patterns vary widely, but usually fall into one of two major categories—**negative or positive regulation** (Figure 14-1).

repressor protein

inducer

In negatively-regulated systems, a specific protein (called a **repressor protein**) that inhibits transcription of a specific gene (or genes) may be present in the cell. In some cases the repressor alone acts to prevent transcription, and a molecule (called an **inducer**) which is an antagonist of the repressor is needed to allow transcription. In other instances of negative regulation, the repressor on its own does not inhibit transcription—it does so only when complexed with a specific signal molecule. In a positively regulated system, a protein called an **activator** works to increase the frequency of transcription of an operon. To be effective inducers, some activator molecules must first associate with a signal molecule.

Repressors and activators are not mutually exclusive in their action and nature. Some systems are both positively and negatively regulated; such a system can respond with a high degree of sensitivity to different conditions in the cell. In addition, systems exist in which the same regulatory protein acts as a repressor under one set of circumstances, and as an activator under another.

A degradative system may be regulated either positively or negatively. In a biosynthetic pathway, the final product usually negatively

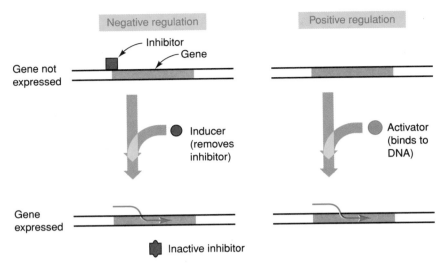

Figure 14-1
The distinction between negative and positive regulation. In negative regulation an inhibitor, bound to the DNA, must be removed before transcription can occur. In positive regulation an activator molecule must bind to the DNA. A system may also be regulated both positively and negatively; then, the system is "on" when the positive regulator is bound to the DNA and the negative regulator is not bound to the DNA.

regulates its own synthesis; in the simplest type of negative regulation, the absence of the product results in an increase in its synthesis, and the presence of the product decreases its synthesis.

In both prokaryotic and eukaryotic systems enzyme activity rather than enzyme synthesis may also be regulated. That is, the enzyme may be present, but its activity is turned on or off. When this occurs, the product of the enzymatic reaction (or in the case of a biosynthetic pathway, the final product of a sequence of reactions) usually inhibits the enzyme activity. This mode of regulation is called **feedback inhibition.** Small molecules other than reaction products are also frequently used either to activate or inhibit a particular enzyme. Such a molecule is termed an allosteric effector molecule.

THE *E. COLI* LACTOSE SYSTEM AND THE OPERON MODEL

In *E. coli* two proteins are necessary for the metabolism of lactose—the enzyme β-galactosidase (which cleaves lactose to yield galactose and glucose), and a carrier called **lactose permease** (required for the entry of lactose into a cell). The existence of two different proteins in the lactose-utilization system was first shown by a combination of genetic experiments and biochemical analysis.

First, hundreds of mutants unable to use lactose as a carbon source (Lac⁻ mutants) were isolated. Some of the mutations were in the *E. coli* chromosome, and others were in F'*lac,* a plasmid carrying the genes for lactose utilization. By performing F' × F⁻ matings, partial diploids having the genotypes F'*lac*⁻/*lac*⁺ or F'*lac*⁺/*lac*⁻ were constructed, as described in Chapter 12 (Figure 12-3). (The genotype of the plasmid is to the left of the diagonal line, and the chromosome to the right.) It was observed that these diploids always produced a Lac⁺ phenotype (that is, they made both permease and β-galactosidase). Other partial diploids were then constructed in which both the F'*lac* plasmid and the chromosome carried *lac*⁻ genes. When these partial diploids were tested for the Lac⁺ phenotype, it was found that all of the mutants initially isolated could be placed into two groups; *lacZ* mutants and *lacY* mutants. The partial diploids F'*lacY*⁻ *lacZ*⁺/*lacY*⁺ *lacZ*⁻, and F'*lacY*⁺ *lacZ*⁻/*lacY*⁻ *lacZ*⁺ had a Lac⁺ phenotype (producing both β-galactosidase and permease), but the genotypes F'*lacY*⁻ *lacZ*⁺/*lacY*⁻ *lacZ*⁺, and F'*lacY*⁺ *lacZ*⁻/*lacY*⁺ *lacZ*⁻ had the Lac⁻ phenotype. The existence of two groups was good evidence that the *lac* system consisted of at least two genes (*lacY* and *lacZ*), and that a cell must have a functional copy of each in order to metabolize lactose.

Further experimentation was needed to establish the precise function of each gene. Experiments in which cells were placed in a medium containing ¹⁴C-labelled lactose showed that no lactose would enter a *lacY*⁻ cell, whereas it readily entered a *lacZ*⁻ mutant. Treatment of a *lacY*⁻ cell with lysozyme (an enzyme that destroys part of the cell wall and makes it permeable) enabled labelled lactose to enter a *lacY*⁻ cell, indicating that the product of the *lacY* gene is involved with the transport of lactose into the cell through the cell membrane (i.e. lactose permease). Enzymatic assays showed that β-galactosidase is present in *lacZ*⁺ cells, but not *lacZ*⁻ cells; this provided the initial evidence that the *lacZ* gene is the structural gene for β-galactosidase. Genetic mapping showed that the *lacY* and *lacZ* genes are adjacent. Another gene called *lacA* is located next to the *lacY* gene. It codes for a galactoside transacetylase enzyme whose physiological role is unknown.

These observations show a very sensitive on-off mechanism that allows the production of enzymes involved in lactose metabolism only when the cell needs them; they also provide a classic example of inducible enzyme synthesis.

1. When a culture of Lac⁺ *E. coli* was grown in a medium that contained no lactose or any other β-galactoside, the intracellular concentrations of β-galactosidase and permease were extremely low—possibly one or two molecules per bacterium. However, when lactose was present in the growth medium, the number of these molecules was increased about 10^5-fold.

2. When lactose was added to a Lac⁺ culture growing in a lactose-free medium (that also lacks glucose for reasons which will be discussed later), both β-galactosidase and permease began to be synthesized nearly simultaneously, as shown in Figure 14-2. Analysis of the mRNA present in the cells before and after the addition of lactose showed that no *lac*

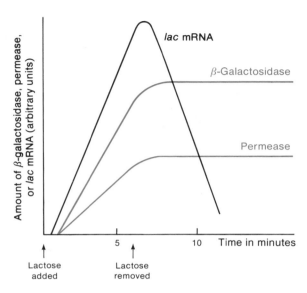

Figure 14-2

The "on-off" nature of the *lac* system. *Lac* mRNA appears very soon after lactose is added; β-galactosidase and permease appear subsequently because of the time required for sequential translation. When lactose is removed, no more *lac* mRNA is made, and the amount of *lac* mRNA decreases owing to the usual degradation of mRNA. Both β-galactosidase and permease are stable. Their concentration remains constant even though no more can be synthesized. A third protein of the *lac* system, β-galactoside transacetylase, is synthesized coordinately with β-galactosidase and permease. This protein, the product of the **a** gene, is not involved in lactose metabolism.

mRNA (the mRNA which codes for β-galactosidase and permease) was present prior to the addition of lactose, and that it was in fact the addition of lactose that triggered the synthesis of *lac* mRNA.

This kind of analysis is done by growing cells in a radioactive medium in which all newly-synthesized mRNA is radioactive, isolating the mRNA, and allowing it to renature with previously obtained *lac* DNA. Since the only RNA that will anneal to the *lac* DNA is *lac* mRNA, the amount of radioactive DNA–RNA hybrid molecules is also a measure of the amount of *lac* mRNA. These two observations led to the view that the lactose system is inducible, regulated at the level of transcription, and that lactose acts as an inducer.

Lactose itself is rarely used in experiments to study the induction phenomenon for a variety of reasons—one is that β-galactosidase catalyzes the cleavage of lactose, resulting in a continual decrease in lactose concentration. This complicates the analysis of many types of experiments (kinetic experiments, for example). Instead of lactose itself, a sulfur-containing analogue called **isopropylthiogalactoside (IPTG)** (Figure 14-3) is used; this analogue is an inducer, but not a substrate, of β-galactosidase.

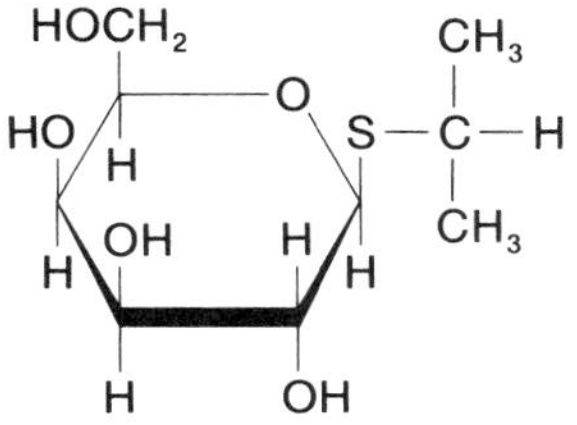

HOCH₂

HO

OH

H

H

OH

O

H

H

S—C—H

CH₃

CH₃

Isopropylthiogalactoside (IPTG)

Figure 14-3
Structure of Isopropylthiogalactoside (IPTG).

Mutants have been isolated in which *lac* mRNA is made in both the presence and absence of an inducer. When regulation of the expression of *lac* genes had been eliminated, these mutants provided the key to understanding induction. In these mutants, *lac* genes were said to be **constitutively expressed**—that is, they were "on" all the time. Mutations that lead to constitutive expression are called **constitutive mutations.** Complementation tests were performed using partial diploids carrying two constitutive mutations, one on the chromosome and one on a plasmid. These tests showed that the mutants fall into two groups, termed *lacI⁻* and *lacOᶜ*. The characteristics of the mutants are shown in Table 14-1. As shown in that table (entry 1), in the absence of an inducer (IPTG in this case) the *lacI⁺* cell fails to make *lac* mRNA, whereas this mRNA is made by a *lacI⁻* mutant (entry 2) even in the absence of an inducer. Thus, the *lacI* gene is apparently a regulatory gene whose product is an inhibitor that keeps the system turned off unless an inducer is present. Entries 3 and 4 show that the presence of a functional *lacI* gene anywhere in the cell (on the chromosome or on the plasmid) is sufficient to regulate (i.e., make responsive to an inducer) any *lacI⁻* system in the cell, whether it is on the same DNA molecule or not. The *lacI*-gene product must therefore be a diffusable entity. Wild–type copies of the *lacI* gene are present in a *lacI⁺/lacI⁻* partial diploid, so the system is inhibited (but it would not be if the *lacI⁺* gene were not present). The *lacI* gene product, a protein molecule, is called the Lac repressor. Genetic mapping experiments place the *lacI* gene adjacent to the *lacZ* gene, establishing the gene order *lacI, lacZ, lacY*. How the repressor prevents synthesis of lac mRNA will be explained shortly.

The *Oᶜ* mutations have properties different from the *lacI⁻* mutations. Mapping experiments show them to be located in a very small segment between the *lacI* and *lacZ* genes. A significant characteristic of *lacOᶜ* mutations that was found by examining the partial diploids shown in entries 6 and 7, is that β-galactosidase is synthesized constitutively in an *OᶜlacZ⁺/O⁺lacZ⁻* cell, but is made in an *OᶜlacZ⁻/O⁺lacZ⁺* cell only

Table 14-1

Characteristics of Partial Diploids Having Combinations of *lacI* and *lacO* Genes

Genotype	*lacZ* Expression		Phenotype
	without IPTG	with IPTG	
1. *I⁺ O⁺ lacZ⁺*	−	+	Inducible
2. *I⁻ O⁺ lacZ⁺*	+	+	Constitutive
3. *F' I⁺ O⁺ lacZ⁺/I⁻ O⁺ lacZ⁺* (or −)	−	+	Inducible
4. *F' I⁺ O⁺ lacZ⁻/I⁻ O⁺ lacZ⁺*	−	+	Inducible
5. *I⁺ Oᶜ lacZ⁺*	+	+	Constitutive
6. *F' I⁺ Oᶜ lacZ⁺/I⁺ O⁺ lacZ⁻*	+	+	Constitutive
7. *F' I⁺ Oᶜ lacZ⁻/I⁺ O⁺ lacZ⁺*	−	+	Inducible
8. *F' I⁺ Oᶜ lacZ⁻/I⁻ O⁺ lacZ⁺*	−	+	Inducible
9. *F' I⁺ Oᶜ lacZ⁺/I⁻ O⁺ lacZ⁻*	+	+	Constitutive

when an inducer is present. It should be noted that the O^c mutation causes constitutive expression of *lac* genes only when the mutation and the genes are on the same DNA molecule. As only the genes *cis* (adjacent) to the O^c mutation are affected by it (i.e., expressed constitutively), the O^c mutation is said to be *cis*-dominant. Since the presence of an O^+ gene cannot alter the constitutive behavior of an O^c lac system on another DNA molecule in a partial diploid, the O^+ region does not code for a diffusable product. Actually, the O region is a non-coding region of DNA, and is involved in regulation. This region is called an **operator;** the repressor protein binds to the O^+ operator and prevents expression of the *lac* structural genes.

Another class of Lac$^-$ mutants (called P$^-$) have been isolated in which no *lac* mRNA is made. These mutations have been mapped to a small region adjacent to the operator, and in experiments with partial diploids, have been shown to have a *cis*-dominant effect, in that they only prevent transcription of structural genes on the same DNA molecule as the mutant. Because of this, the P region, like the O region, must define a non-coding DNA site, and has been determined to be a promoter.

The regulatory mechanisms of the *lac* system were first explained by the operon model, which has the following features (Figure 14-4):

1. The lactose-utilization system consists of two kinds of components: structural genes that code for products needed to transport and metabolize lactose, and regulatory elements (the *lacI* gene, the operator (O), and the promoter (P)). Together, the structural genes and regulatory elements comprise the **lac operon.**

2. The products of the *lacZ* and *lacY* genes are encoded in a single polycistronic mRNA molecule. The transacetylase enzyme mentioned earlier is also encoded in this molecule.

3. The *lac* operon promoter is immediately adjacent to the operator region. RNA polymerase binds to the promoter and initiates transcription of the *lacZ, lacY,* and *lacA* genes into the mRNA molecule.

4. The *lacI* gene product (the repressor) is always produced, (at about 20 copies per cell) and binds to the operator.

5. When the repressor is bound to the operator, initiation of transcription of *lac* mRNA by RNA polymerase is prevented.

6. Inducers stimulate mRNA synthesis by binding to and deactivating the repressor, a process called **derepression.** Thus, in the presence of an inducer the operator is unoccupied, and the promoter is available for initiation of mRNA synthesis.

Note that regulation of the operon requires that the operator be adjacent to the structural genes of the operon (*lacZ, lacY,* and *lacA*); however,

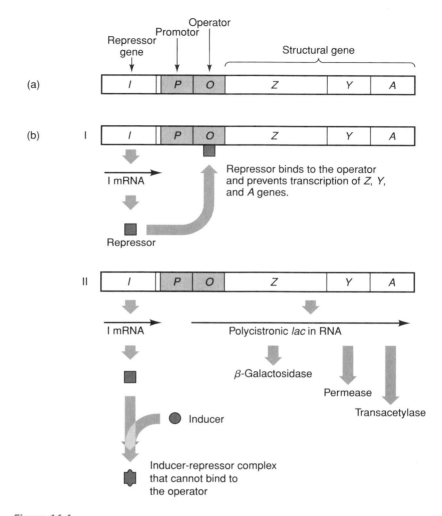

Figure 14-4
(a) Genetic map of the *lac* operon (not drawn to scale) the *P* and *O* sites are actually much smaller than the genes. (b) Diagram of the *lac* operon in (I) repressed and (II) induced states. The inducer alters the shape of the repressor, so the repressor can no longer bind to the operator.

proximity of the *lacI* gene is not necessary, because the repressor is a soluble protein and is therefore diffusable throughout the cell.

The operon model is supported by a wealth of experimental data and explains many of the features of the *lac* system, as well as numerous other negatively regulated genetic systems. One aspect of the regulation of the *lac* operon—the effect of glucose—has not yet been discussed. Examination of this feature indicates that the *lac* operon is also subject to positive regulation.

β-galactosidase forms glucose by cleaving lactose. (The other cleavage product, galactose, is metabolized by enzymes of the galactose operon, described later.) Thus, if both glucose and lactose are present in the growth medium, the activity of the *lac* operon is not needed, so no β-galactosidase is formed until virtually all of the glucose in the medium has been consumed. β-galactosidase synthesis is regulated in part by the synthesis of *lac* mRNA. No *lac* mRNA is made in the presence of glucose, because in addition to an inducer to inactivate the repressor, another regulatory element is needed for initiating *lac* mRNA synthesis; the activity of this element is regulated by the concentration of glucose. However the inhibitory effect of glucose on the expression of the *lac* operon is quite indirect, as explained below.

The small molecule **cyclic AMP** (**cAMP**) is universally distributed in animal tissues, and in multicellular eukaryotes it is important in regulating the action of many hormones (Figure 14-5). It is also present in *E. coli* and many other bacteria. Cyclic AMP is synthesized by the enzyme adenylate cyclase, and its concentration is regulated indirectly by glucose. The entry of glucose into a cell makes the intracellular concentration of cAMP quite low. In a medium containing glycerol (or any other carbon source that enters the cell by a channel different from the type used by glucose), or under starvation conditions, the cAMP concentration in bacterial cells is high (Table 14-2). It is cAMP that links the *lac* operon activity with the intracellular concentration of glucose.

E. coli (and many other bacterial species) contain a protein called cAMP receptor protein (CRP) which is encoded in a gene called *crp*. Mutants of either *crp* or the adenylate cyclase gene are unable to synthesize *lac* mRNA, indicating that both CRP function and cAMP are required for *lac* mRNA synthesis. CRP and cAMP bind together to form a complex (denoted cAMP-CRP) that is a regulatory element in the *lac* system. The cAMP-CRP complex must be bound to a specific base sequence (called an activation site) in the promoter region in order for transcription of the *lac* operon to occur (Figure 14-6). Thus, the cAMP-CRP complex is a positive regulator, in contrast to the repressor, a

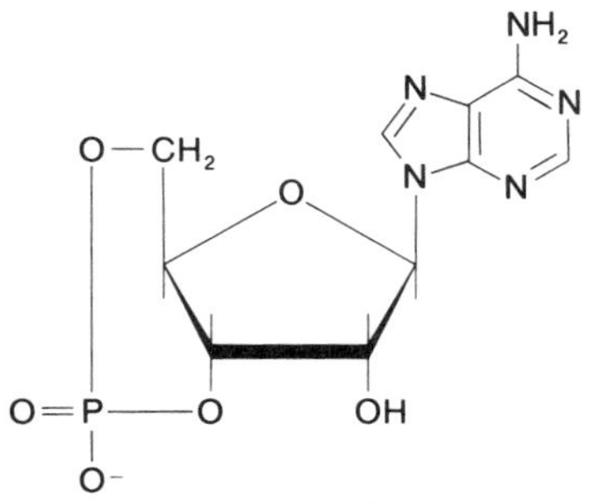

Figure 14-5
Structure of cyclic AMP.

Table 14-2

Concentration of Cyclic AMP in Cells Growing in Media Having the Indicated Carbon Sources

Carbon Source	cAMP Concentration
Glucose	Low
Glycerol	High
Lactose	High
Lactose + glucose	Low
Lactose + glycerol	High

negative regulator. The *lac* operon is therefore regulated both positively and negatively.

Using mixtures containing purified *lac* DNA, repressor, cAMP-CRP, and RNA polymerase, two key points have been established:

1. When there is not enough cAMP-CRP in the cell (due to high levels of glucose), the cAMP-CRP complex is not formed. In the absence of the cAMP-CRP complex, RNA polymerase binds weakly to the *lac* promoter, so transcription is rarely initiated. However, if cAMP-CRP is bound to the activation site in the *lac* promoter, it enhances the binding of RNA polymerase, and its subsequent initiation of transcription. When cAMP-CRP is bound to the activation site, it is believed to affect RNA polymerase activity by making direct contact with the enzyme. The cAMP-CRP complex has two domains—one binds to DNA, and the other contacts RNA polymerase (Figure 14-7).

2. Because there is partial overlap in the promoter and operator of the *lac* operon, when the *lac* repressor is bound to DNA at the operator

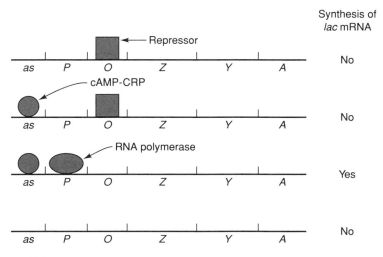

Figure 14-6

Four states of the *lac* operon showing that *lac* mRNA is made only if cAMP-CRP is present and repressor is absent. The symbol *as* represents the activation sites.

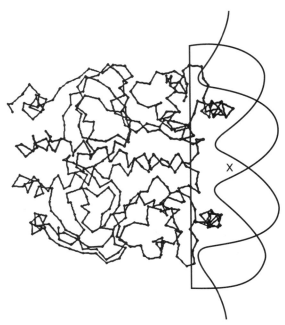

Figure 14-7
Fitting of the two protruding α-helical segments of a cAMP-CRP dimer to the adjacent major grooves of the DNA activation site. One subunit of CRP is shown in red and the other in pink. The structure of CRP is known from X-ray diffraction analysis of crystals of the pure protein. The center of the DNA dyad symmetry is marked by an X in the middle of the minor groove between the two major grooves. Recall that a similar binding phenomenon was described in Chapter 5 (Figure 5-13, page 90).

region, it physically blocks the binding of RNA polymerase to the promoter.

THE GALACTOSE OPERON: TWO PROMOTERS

AND DNA LOOPING

The galactose operon of *E. coli* is also an inducible system, regulated by cAMP-CRP and a repressor. The *gal* operon has three cistrons— *galE, galT,* and *galK,* which encode enzymes that metabolize galactose— and is transcribed into polycistronic mRNA molecules. The operon contains two promoters, *P1* and *P2*. The cAMP levels in the cell influence the *gal* operon in an interesting way. The cAMP-CRP complex binds to a non-coding DNA site in the operon and regulates transcription initiation from the two promoters in opposite fashions—cAMP-CRP **activates** transcription from promoter *P1*, but **inhibits** transcription from promoter *P2*. Thus, when the intracellular cAMP concentration is high, transcription of the operon is initiated at the *P1* promoter, but at a low cAMP level, it is initiated at the *P2* promoter. This mechanism of dual control insures that the galactose-metabolizing enzymes are syn-

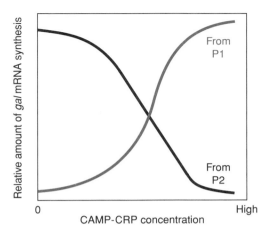

Figure 14-8
Regulation of transcription of the *gal* operon from the two promoters, *P1* and *P2*. In the absence of cAMP-CRP, *gal* mRNA is made mostly from the *P2* promoter. At high cAMP-CRP concentrations, *gal* mRNA is made principally from the *P1* promoter.

thesized whether the cells are growing in glucose, or in another carbon source. By using a mixture of purified components in an *in vitro* system and varying the cAMP concentration, it has been shown that the amount of synthesis of *gal* mRNA from the two promoters is inversely related to the level of cAMP (Figure 14–8). Note that the total amount of *gal* mRNA made at any level of cAMP is constant.

The negative regulator, the *gal* repressor, is the product of the *galR* gene. The *galR* gene, unlike the *lacI* gene (its counterpart in the *lac* operon), is located quite far from the operon it regulates. Gal repressor inhibits transcription from both promoters. To act, it must bind to two operator elements, designated O_E and O_I, present in the operon on opposite sides of the promoters (Figure 14–9). When cells are exposed to galactose, galactose acts as an inducer and causes derepression of the operon by preventing the binding of the repressor to O_E and O_I. Mutations of *galR* (a *trans*-acting mutation), or to either of the two operators (a *cis*-acting mutation), cause constitutive expression of the operon.

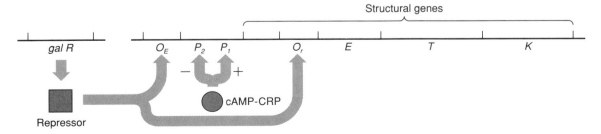

Figure 14-9
The regulation of the *gal* operon of *E. coli*. The two operators, O_E and O_I are shaded. The map is not drawn to scale.

The repressor of the *gal* operon functions differently than its counterpart in the *lac* system. Instead of physically blocking the binding of RNA polymerase, the two repressor molecules (one bound to each operator) interact with each other causing the formation of a loop in the DNA. This loop of DNA, in the region between the two operators, contains the promoters to which RNA polymerase is bound (Figure 14-10). This looping prevents the RNA polymerase from initiating transcription. While the exact mechanism of repression is not known, it is believed that the formation of DNA loops permits physical contact between regulatory protein molecules and RNA polymerase bound to areas spatially-separated on the DNA molecule; this contact somehow serves to regulate the rate and efficiency of transcription.

THE ARABINOSE OPERON

The arabinose operon in *E. coli,* which encodes enzymes involved in the utilization of the sugar arabinose, provides another example of variation in the behavior of regulatory proteins. The *ara* operon, like *lac* and *gal*, is inducible, and not normally expressed. The *ara* enzymes are made only when the cells are exposed to arabinose. The expression is regulated by a protein called AraC, which by itself acts as a repressor. When arabinose is present, the sugar binds to AraC and converts it into an activator. The arabinose-AraC complex activates transcription from the *ara* promoter. If the gene which encodes AraC is deleted, the *ara* operon is never expressed—even in the presence of arabinose. The arabinose operon thus provides an example of a regulatory protein that acts as a repressor under one condition, and as an activator under another.

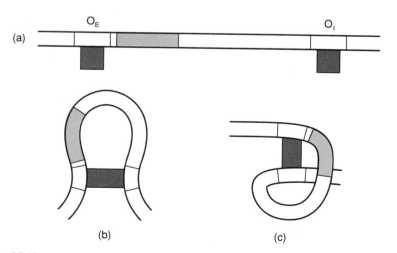

Figure 14-10

DNA looping in the *gal* operon. A repressor dimer binds to each operator (a) and the dimers associate together generating a loop of the intervening DNA by mechanisms (b) or (c). The *gal* promoter region is shaded.

THE TRYPTOPHAN OPERON AND ATTENUATION

In some operons, a single molecule does not by itself act as a repressor. The repressor for such an operon requires the presence of a small molecule (called a **corepressor**) in order to bind to the operator. In its absence, the repressor, called an aporepressor, is unable to bind to the operator, and the genes of the operon are expressed. In *E. coli*, the expression of enzymes involved in the synthesis of many amino acids and vitamins are regulated in this fashion. The cell represses the synthesis of the appropriate enzymes when an amino acid or vitamin is present in excess. This system of repression, which is present in the tryptophan operon, is highly economical in its use of energy.

There are five genes which encode the enzymes involved in tryptophan biosynthesis. The arrangement of the five cistrons (named *trpE, trpD, trpC, trpB,* and *trpA*), the promoter, and the operator is shown in Figure 14–11. Together, these components can form an operon, and following the definition of an operon, can be transcribed into a polycistronic mRNA molecule. In the *trp* operon, tryptophan acts directly in the repression system.

The regulatory protein of the repression system of the *trp* operon is called the Trp repressor, and is the product of the *trpR* gene (which is located elsewhere on the chromosome). Mutations in either the *trpR* gene (*trpR*-), or in the operator (0^c), cause constitutive initiation of *trp* mRNA synthesis, just as in the *lac* and *gal* operons. The tryptophan molecule (which acts as a corepressor) must bind to Trp repressor (the aporepressor) to achieve repression. The reaction scheme of this process is:

Trp repressor　+　Tryptophan　⇄　Trp Repressor　–　Tryptophan Complex
(Alone it does not　　　　　　　　　　(Binds to the operator, so
bind to the operator,　　　　　　　　transcription doesn't occur)
and transcription
occurs)

Thus, only when tryptophan is present does an active repressor molecule inhibit transcription. When the external supply of tryptophan is depleted (or at least reduced substantially), the equilibrium in the above equation shifts to the left, the operator is unoccupied, and transcription begins. This is the basic "on-off" regulatory mechanism of the *trp* operon.

Since the *trp* operon codes for a set of biosynthetic (rather than degradative) enzymes, neither glucose nor cAMP-CRP influences the activity of the operon. Furthermore, there is more to regulation of the *trp* operon than the simple on-off system described above. When a small amount of tryptophan is available (but not enough to allow cell growth without the concomitant synthesis of tryptophan) starvation is prevented by a modulating system in which the amount of transcription in the derepressed state is determined by the concentration of tryptophan. In the on state, a second level of regulation subjects the expression to

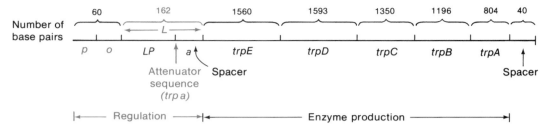

Figure 14-11
The *E. coli trp* operon. For clarity, the regulatory region is enlarged with respect to the coding region. The proper size of each region is indicated by the number of base-pairs. The regulatory elements are shown in red. *L* is the leader mRNA segment; *LP*, the leader peptide; and *a*, the attenuator.

a finer control, a control more sensitive to tryptophan concentration. This is provided by two factors:

1. Premature termination of transcription upstream of the first structural gene.

2. Regulation of the frequency of this termination by the intracellular concentration of tryptophan.

Note in Figure 14–11 that the *trp* operon has two additional regions, *LP* and *a*. At the 5′-end, upstream from the first cistron, the *trp* mRNA contains a 162 base segment called leader RNA (designated L in the figure). This leader sequence has several notable features (also see Figure 14-12):

1. An AUG codon and a downstream UGA stop codon in the same reading frame define a region encoding a polypeptide consisting of 14 amino acids, called the leader polypeptide (lp) (Figure 14-12(a)).

2. Two adjacent tryptophan codons are located in the *trp* mRNA at positions corresponding to the tenth and eleventh amino acids in the leader polypeptide. We will see the significance of these repeated codons shortly.

3. Four segments of the leader RNA (denoted 1, 2, 3, and 4) are capable of base-pairing in two mutually exclusive ways; by forming either the stem-loops 1–2 and 3–4 (Figure 14-12(b)), or just the stem-loop 2–3 (Figure 14-12(d)).

This arrangement enables premature termination of transcription to occur in the *trp* leader region by the following mechanism.

In the *trp* operon of wild-type bacteria, when tryptophan is present transcription terminates after making only a 140 base-pair mRNA, and before any of the structural genes are transcribed. In the absence of tryptophan, transcription continues, allowing the synthesis of *trp* enzymes. Mutants that have a deletion in part or all of a 28 base-pair long segment preceding position 160, make *trp* enzymes at a rate that is six to seven times greater than that of a wild-type cell growing in the absence of tryptophan—indicating that this region must have a regulatory role. This region (called an **attenuator**) has all the usual features of a transcription termination site, including a potential stem-loop configuration in the mRNA (stem-loop 3–4) followed by a sequence of several uridine residues. Transcription terminates following the seventh residue. Features of the leader RNA sense the level of tryptophan in the cell and modulate the amount of transcription accordingly. The mechanism for effecting this premature termination is based upon the potential of the RNA to form the alternative stem-loops 1–2 (protector), 2–3 (pre-emptor), and 3–4 (attenuator). The directive to terminate or read through depends upon whether or not the stem-loop 3–4 attenuator structure is formed.

Termination of transcription at the attenuator is mediated through translation of the leader peptide region. Because there are two tryptophan codons in this sequence, the translation of the sequence is sensitive to the concentration of charged $tRNA^{trp}$. That is, if the supply of tryptophan is inadequate, that amount of charged $tRNA^{trp}$ will be insufficient, and translation will be stalled at the tryptophan codons. Two points should be noted:

1. All base-pairing is eliminated in the segment of the mRNA that is in contact with the ribosome.

2. If the stem-loops 1–2 and 3–4 are formed simultaneously, then stem-loop 2–3 cannot be present.

Figure 14–12(c) shows that the end of the Trp leader peptide is in segment 1. Usually a translating ribosome is in contact with about ten bases in the mRNA past the codons being translated. Thus, when the final codons of the leader are being translated, segments 1 and 2 are partially covered by the ribosome, and so are not base-paired. Recall that in a coupled transcription-translation system, the leading ribosome is not far behind the polymerase. Therefore, if the ribosome is in contact with segment 2 when the synthesis of segment 4 by RNA polymerase is being completed, then segments 3 and 4 are able to form the stem-loop 3–4 without segment 2 competing for segment 3. The presence of the 3–4 stem-loop configuration allows termination to occur when the terminating sequence of seven uridine is reached. If there is very little tryptophan present, the concentration of charged $tRNA^{trp}$ becomes inadequate, and the translating ribosome stalls at the tryptophan codons (Figure 14–12(d)). These codons are located fourteen bases before the beginning of segment 2. Thus, segment 2 is free before segment 4 has been synthesized, so the stem-loop 2–3 can form. In the absence of stem-loop 3–4, termination of transcription does not occur, and the complete mRNA molecule is made, including the coding sequences for the *trp* structural genes. Hence, if tryptophan is present in excess, termination

Leader peptide

MetLysAlaIlePheValLeuLysGlyTrpTrpArgThrSer

pppAAGUUCACGUAAAAAGGGUAUCGACAAUGAAAGCAAUUUUCGUACUGAAAGGUUGGUGGCGCACUUCCUGAAACGGGCAGUGUAUUCACCA

1 2

TrpE Start code

AUG

Last base in attenuated mRNA

UGCGUAAAGCAAUCAGAUACCCAGCCCGCCUAAUGAGCGGGCUUUUUUUUUGAACAAAAUUAGAGAAUAACA**AUG**

3 4

(a)

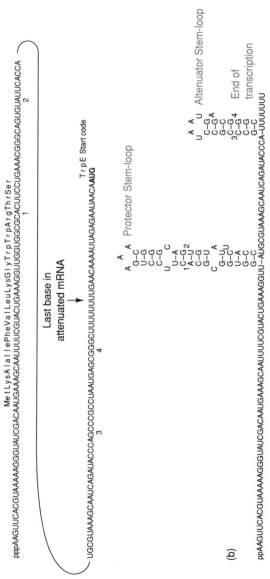

Protector Stem-loop

Altenuator Stem-loop

End of transcription

(b)

ppAAGUUCACGUAAAAAGGGUAUCGACAAUGAAAGCAAUUUUCGUACUGAAAGGUU–AUGCGUAAAGCAAUCAGAUACCCA–UUUUUUU

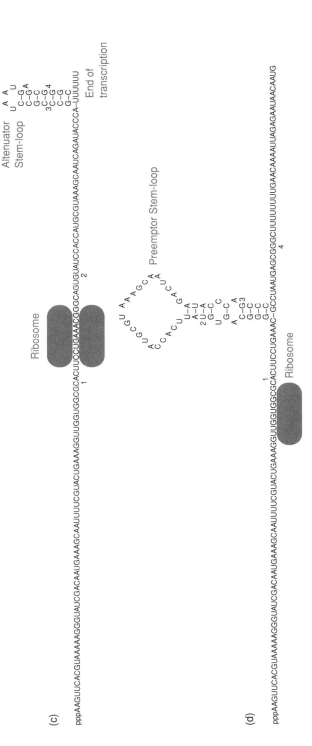

Figure 14-12

The leader region of *trp* mRNA and the accepted model for the mechanism of attenuation control in *E. coli trp* operon. (a) The 5' terminal region of *trp* mRNA, called the leader RNA. It shows the encoded 14 amino acid leader polypeptide with the 2 Trp codons, the chain terminating UGA codon, and the beginning **AUG** codon of the TrpE protein (bold face). If tryptophan is present, transcription terminates at RNA position 140 as indicated. Four different segments (designated 1, 2, 3, and 4), preceding the termination point potentially participate in alternative stem-loop structures as shown in (b), (c), and (d). (b) Free mRNA. It allows the formation of stem-loops 1–2 and 3–4. (c) mRNA structure in presence of high concentration of tryptophan. Ribosome reaches region 2, preventing stem-loop 1–2 but allowing stem-loop 3–4 formation. (d) Low concentration of tryptophan. Ribosome stalls in region 1, allowing formation of stem-loop 2–3 and preempting stem-loop 3–4.

of transcription occurs, and little enzyme is synthesized; if tryptophan is absent termination does not occur, and the enzymes are made. At intermediate concentrations, the fraction of initiation events that result in completion of *trp* mRNA synthesis will depend upon how often translation is stalled at the tryptophan codons, which in turn depends on the concentration of tryptophan.

Many operons responsible for amino acid biosynthesis (for example leucine, isoleucine, phenylalanine, and histidine operons) are regulated by attenuators equipped with the base-pairing mechanism for competition described in the tryptophan operon. In the histidine operon, which also has an attenuator system (that is, prematurely terminated mRNA), a similar base sequence encodes a leader polypeptide having seven adjacent histidine codons (Figure 14-13). In the phenylalanine operon, seven phenylalanine codons are also present in the leader, but they are divided into three groups (Figure 14-13).

SPECIFIC VERSUS GLOBAL REGULATORS

Regulatory proteins, both positive and negative, can be highly specific; for example, the Lac repressor controls only the *lac* operon and no other genes in *E. coli*. Alternatively, regulatory proteins have a more global effect on gene expression; cAMP-CRP for example, controls the expression of many genes in *E. coli,* and so does a regulatory protein, called LexA. LexA is a negative regulator of genes which encode enzymes involved in the repair of DNA damage induced by ultraviolet light. These genes are scattered around the *E. coli* chromosome, and LexA protein binds to the operator regions of all these genes. When the cells are exposed to a sublethal dose of ultraviolet light, LexA is inactivated, and the DNA-repair genes are expressed to carry out the repair job. A set of genes or operons which is regulated by a global regulator—positive or negative—is called a **regulon.** Specific and global regulators are frequent in bacterial cells.

AUTOREGULATION

The synthesis of some gene products, whose requirements in a cell vary greatly, is regulated by a mechanism called **autoregulation.** In the

(a)   Met Lys His Ile Pro Phe Phe Phe Ala Phe Phe Phe Thr Phe Pro Stop
      5' AUG AAA CAC AUA CCG UUU UUC UUC GCA UUC UUU UUU ACC UCC CCC UGA 3'

(b)   Met Thr Arg Val Gln Phe Lys His His His His His His His Pro Asp
      5' AUG ACA CGC GUU CAA UUU AAA CAC CAC CAU CAU CAC CAU CAU CCU GAC 3'

Figure 14-13
Amino acid sequence of the leader peptide and base sequence of the corresponding portion of mRNA from (a) the phenylalanine operon and (b) the histidine operon. The repeating amino acid is shaded in red.

simplest autoregulated system, the gene product is also a regulator: it binds to a DNA site at or near the promoter. The autoregulation could be positive or negative. When the concentration of the gene product exceeds what the cell can use, the product molecules occupy the DNA site and transcription is inhibited. At a later time when the concentration of unbound molecules decreases, the molecule bound to DNA leaves the site, allowing transcription by RNA polymerase from a free promoter. This is negative autoregulation. The synthesis of some regulatory proteins (e.g., CRP) are negatively autoregulated in this way. In a positively autoregulated system, the gene product acts as an activator of its own gene. An elegant example of positive autoregulation is the synthesis of a repressor protein of bacteriophage, λ. This system is described in Chapter 15.

Autoregulation at the level of translation has also been recognized. For example, when protein components of ribosomes become excessive, they negatively regulate their own synthesis by binding to mRNA and block the binding of intact ribosomes, and thus translation.

CONSTITUTIVE GENES

promoter

Many genes in bacteria have no on-off regulatory mechanism. They are expressed all the time, at rates predetermined by the nucleotide sequence of their **promoters** (promoter strength). Such constitutive genes generally code for the living cells house-keeping proteins, which are needed in a given amount. Examples of such genes are the *lacI* gene which encodes the Lac repressor, or genes encoding glucose-metabolizing enzymes, whose levels remain constant in a changing environment. The strength of promoters of constitutive genes, as well as of inducible or repressible genes, determines the maximum rate at which the cognate genes can be expressed.

Post-Transcriptional Control

Frequently, unequal amounts of proteins are made from different cistrons of an operon. For example, the ratios of the number of copies of β-galactosidase, permease, and transacetylase made from the *lac* operon are 100:50:20. These differences, which result from post-transcriptional regulation are achieved in two ways:

1. **Translational control.** The *lacZ* gene is translated first (Figure 14-14). The frequency with which it is translated is determined by the efficiency with which a ribosome attaches to a critical base sequence around the starting AUG codon in *lacZ* mRNA. Frequently the translating ribosome detaches from the mRNA molecule at the end of *lacZ* translation. The frequency at which the next cistron *lacY* is translated is also determined by the efficiency at which a ribosome (either dissociated from the end of the *lacZ* cistron or recruited from the cell's surplus

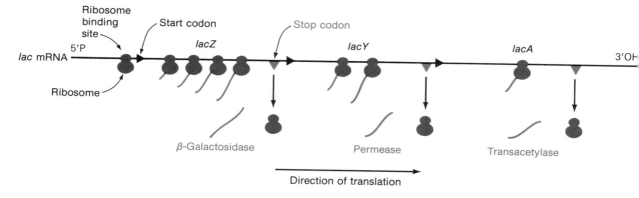

Figure 14-14
Translation of *lac* mRNA into proteins. All ribosomes attach to the mRNA at the ribosome binding site. At each stop codon some ribosomes detach.

supply) binds to the AUG codon of *lacY*. The same rule regulates the translation efficiency of the *lacA* cistron.

The efficiency at which the ribosome binds to the critical binding sequence on mRNA is also known to be modulated (inhibited) by formation of secondary structures involving this region and either another part of the mRNA or another regulatory RNA molecule with complementary base sequences.

2. **mRNA stability.** mRNA is unstable, the half-life is on the order of a few minutes in bacteria. We have discussed before that *lac* mRNA quickly disappears if the inducer (Figure 14–3) is removed from the cell. This instability helps cells quickly shift their spectrum of protein synthesis in response to radical changes in the environment. Degradation of *lac* mRNA is initiated from its 3′ terminus. Hence, at any given instant there are more copies of the *lacZ* portion of mRNA than of the *lacY* portion of mRNA, and more copies of the *lacY* mRNA than of *lacA* mRNA. Thus the amounts of the three *lac* enzymes made also depends upon the availability of the corresponding mRNA segments. In prokaryotes this mode of regulation occurs frequently.

In some systems, mechanisms also exist by which the degradation of mRNA from its 3′ end is substantially reduced by formation of RNA secondary structures resistant to degradation which enhances the amount of translation of the mRNA molecule.

Thus the overall expression of an operon is regulated by controlling transcription initiation of an mRNA molecule and premature transcription termination of the polycistronic mRNA (in some cases); whereas the relative concentration of the proteins encoded by the mRNA are determined by controlling both the degradation rate of the mRNA and the frequency of initiation of translation of each cistron. These general concepts of gene control have been firmly established by demonstrating that transcriptional or translational reactions with purified components are regulated by regulatory DNA elements and proteins. The chemical and physical basis of most of the gene regulatory macromolecular interactions are not well understood. Currently, they are the focus of intensive research among molecular biologists.

Control by Proteolysis

Unlike mRNA, most proteins are stable during normal growth unless they cease to be useful or become deleterious to a cell and must be removed for cell growth to succeed. An example of the second type of protein in *E. coli* is SulA, encoded by the *sulA* gene. When chromosomal DNA is damaged, the synthesis of SulA is induced along with DNA repair enzymes. The SulA protein inhibits cell division to provide time for DNA repair. When the repair process is completed, and SulA is no longer needed, it is quickly degraded by a protease called Lon. Mutant cells defective in Lon protease do not divide, and ultimately die when subjected to DNA-damaging agents.

Feedback Inhibition and Allosteric Control

When a large amount of an amino acid is available in a cell, the synthesis of the enzymes that make the amino acid is repressed because they are not needed any more, as previously explained. However, the cell contains a large amount of the enzymes before repression begins, and could continue to catalyze the amino acid biosynthetic reactions. But the cell has a mechanism by which the synthesis of the amino acid can be quickly stopped. Wasteful synthesis almost never occurs. For example, high concentrations of the amino acid isoleucine actually inhibit the activity of the first enzyme of the 5-enzyme pathway that catalyzes the synthesis of isoleucine from threonine (Figure 14-15). The bound enzyme is unable to catalyze the reaction. This stops isoleucine biosynthesis immediately. This process of very specific inhibition of catalytic activity of an enzyme in a biochemical pathway by binding to a product of a later enzyme in the same pathway, is called **feedback,** or **end product inhibition.** In feedback inhibition, isoleucine binds reversibly to a site different from the active site, and changes the structure of the enzyme protein, which keeps it from recognizing its substrate. Proteins, whose shape or conformation is changed by binding to a small molecule at a site different from the active site are called **allosteric proteins** (Figure 14-16). Small molecules that bring about the allosteric changes are called **allosteric effectors.** Effectors bind to the protein by weak bonds (hydrogen-bonds,

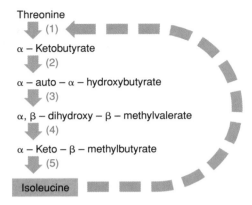

Figure 14-15
Feedback inhibition of the pathway for synthesizing Isoleucine. Isoleucine is boxed to indicate that it is the feedback inhibitor of the enzyme (1), as indicated by the dashed arrow.

ionic bridges, and van der Waal's attractions), and not by covalent bonds. This makes a reversal of the binding process possible, when the end product concentration falls to a low level, requiring the enzyme to be active again.

Some biosynthetic pathways are responsible for the synthesis of two products from a common precursor. A hypothetical example of such a **branched pathway** is the following:

$$A \xrightarrow{1} B \begin{array}{c} \nearrow^{4} E \xrightarrow{5} F \\ \searrow_{2} C \xrightarrow{3} D \end{array}$$

In this type of pathway it would be undesirable for a single product to inhibit enzyme 1, because both branches would be blocked. In general, the most economical kind of inhibition prevails—namely, D inhibits enzyme 2, and F inhibits enzyme 4. In this way, neither D nor F prevents the synthesis of the other. Conversion of A to B is wasteful if both D and F are present; therefore, in many branched pathways it is found that D and F together inhibit enzyme 1.

The elegance of feedback inhibition in a branched pathway is shown in Figure 14–17, which shows a well-studied multibranched pathway in which lysine, methionine, and threonine are synthesized from aspartate. Note how these amino acids inhibit enzymes 4, 6, and 8 immediately after the main branch in the pathway. Furthermore, three **isoenzymes** (different proteins that catalyze the same reaction)—1a, 1b, and 1c—are separately inhibited by lysine, homoserine, and threonine, respectively. Each of the three isoenzymes carries out the reaction at a different rate, each one synthesizing as much aspartyl phosphate as is needed to

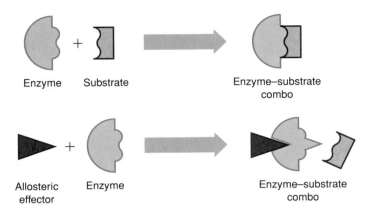

Enzyme   Substrate

Enzyme–substrate
combo

Allosteric   Enzyme
effector

Enzyme–substrate
combo

Figure 14-16
Conformational change of an enzyme protein by binding to an allosteric effector results in
the inability to bind to its substrate. Compare this conformational change with the models
for enzyme action illustrated earlier in Figure 4-12.

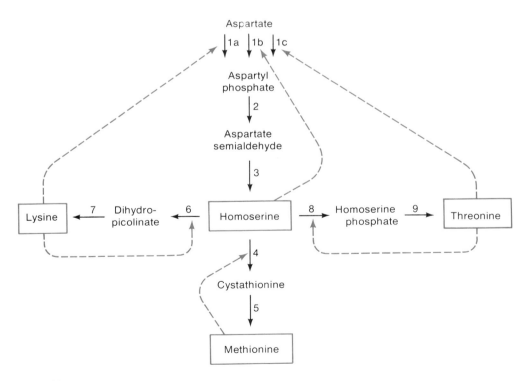

Figure 14-17
Feedback inhibition of the pathway for synthesizing lysine, methionine, and threonine. The boxed molecules are
inhibitors. The dashed arrows lead from an inhibitor to the enzyme that is inhibited.

form the appropriate amount of the product that inhibits the isoenzyme. Thus, when an isoenzyme is inhibited, the amount of aspartyl phosphate required to synthesize the needed amino acids is still made.

Feedback inhibition occurs in both eukaryotes and prokaryotes, but has been more fully described in the latter. It can occur in a variety of ways; for example, binding of the product (which acts as an inhibitor) to the enzyme causes a physical change in the shape of the enzyme and thereby alters its catalytic properties. The changes in the properties of the regulatory proteins we have discussed in this chapter are all examples of **allosteric conformational changes** induced by the binding of specific small molecule effectors—Lac and Gal repressors (from repressor to inactive repressor), CRP (from inactive activator to activator), AraC (from repressor to activator), and the Trp repressor (from inactive repressor to repressor). Note that allosteric inhibition or activation allow a very rapid response to changes in the cellular environment, whereas induction and repression provide an adjustment mechanism for longer-lasting changes in the environment.

Summary

The synthesis of particular gene products is controlled by mechanisms collectively called gene regulation. Gene regulation generally occurs at the level of transcription or translation. Transcriptional regulation can be characterized as positive or negative. In negative regulation an inhibitor (repressor) bound to the DNA must be removed before transcription can occur. In positive regulation an activator molecule must bind to the DNA to allow transcription. The *E. coli* lactose system was first explained by the operon model, and is an example of both negative and positive regulation. In this system, lactose is an inducer which alters the shape of a repressor (lacI) so the repressor can no longer bind to its operator. However, transcription of the lactose operon also requires the presence of the cAMP-CRP activator complex. Other relatively well understood operons include the galactose operon, the arabinose operon, and the tryptophan operon. Each has its own set of components, and each works in a unique fashion.

Regulatory proteins, both positive and negative can be highly specific (lacI), or may have a global effect (CRP) and control the expression of several genes. Some gene products can also regulate their own expression (autoregulation). The overall expression of an operon is regulated by controlling transcription initiation and premature transcription termination; whereas the relative concentration of protein is determined by controlling the degradation rate of the mRNA, the frequency of initiation of translation, and the degradation rate of the protein itself. The activity of the protein can also be under the control of allosteric interactions or feedback inhibition.

Drill Questions

1. What distinguishes negative from positive regulation? How would you distinguish between negative and positive autoregulation?

2. What are the main features of a repressor, co-repressor, and aporepressor?

3. What do the terms induced and derepressed mean in relation to gene functions?

4. What is the genetic evidence that *lac* O^c mutations are *cis*-dominant?

5. Explain how the regulatory protein AraC can be both a repressor and an activator.

6. Why does attenuation not occur in eukaryotes?

7. List two mechanisms that a bacterial cell uses to control the amount of mRNA present inside the cell.

8. List three mechanisms a bacterial cell uses to control the amount of protein present inside the cell.

Problems

1. Suppose *E. coli* is growing in a growth medium containing lactose as the sole source of carbon. The genotype is $I^- \ Z^+ \ Y^+$. Glucose is then added. Which one of the following will happen?
   a. nothing
   b. lactose will no longer be utilized by the cell
   c. lac mRNA will no longer be made
   d. the lac repressor will bind to the operator

2. Imagine a mutant lactose repressor (denoted *lac* I^x) that could interact with its operator and prevent transcription initiation, but could not interact with its inducer (lactose or IPTG). What would be the lac Z phenotype (i.e., expressed or not expressed), in the presence and absence of inducer, corresponding to the following genotypes?
   a. $I^x \ O^+ \ Z^+$ $^+$IPTG
                                    $^-$IPTG

   b. $I^x \ O^+ \ Z^-/I^+ \ O^+ \ Z^+$ $^+$IPTG
                                             $^-$IPTG

   c. $I^x \ O^+ \ Z^+/I^+ \ O^+ \ Z^-$ $^+$IPTG
                                             $^-$IPTG

3. Ribonuclease P is an enzyme needed for the proper maturation of tRNAs. It has been shown that some mutations in ribonuclease P cause increased expression of the *trp* biosynthetic operon. Explain.

4. An *E. coli* mutant is isolated that renders the cell simultaneously unable to utilize a large number of sugars, including lactose, zylose, and sorbitol. However, genetic analysis shows that each of the operons responsible for utilization of these sugars is free of mutation. What are the possible phenotypes of this mutant?

Conceptual Questions

1. Imagine an enzyme which must be synthesized in response to an environmental stimulus (e.g. lacZ in response to lactose). What are the advantages and disadvantages for regulating this enzyme at the transcriptional vs. post-transcriptional level?

2. Genetic analysis has been fundamental in identifying and characterizing the individual components of a regulatory system. For example, the lactose operon was initially characterized by making hundreds of *lac*$^-$ mutants and looking at their genetic interactions. Once a regulator protein (lacI, galR, AraC, trpR, CRP) has been identified, how might the genetic consequences of inactivating that regulatory protein be used to discriminate between negative and positive control?

3. Why is the autogenous type of regulation common for abundant proteins that are incorporated into macromolecular assemblies? (For example, ribosomal proteins in prokaryotes are autogenously regulated at the level of translation.)

David Parma (left)
Senior Research Associate
Department of Molecular, Cellular, and Developmental Biology

Birthday: **10 May 1940**

Birth Place: **Santa Barbara, California**

Undergraduate Degree: **University of California, Davis**
Major: Biological Sciences, B.S. 1964; Genetics, M.S. 1965

Graduate Degree **University of Washington, Ph.D. 1968**

Postdoctoral Training: **Carnegie Institute of Washington, 1968–1970**

Present Position: **University of Colorado, Department of Molecular, Cellular, and Developmental Biology**

Address: **Boulder, Colorado**

Larry Gold (right)
Professor of Biochemistry
Department of Molecular, Cellular, and Developmental Biology

Birthday: **16 August 1941**

Birth Place: **Schenectady, New York**

Undergraduate Degree: **Yale University**
Major: Biochemistry 1963

Graduate Degree: **University of Connecticut, Ph.D. 1967, Biochemistry**

Postdoctoral Training: **Rockefeller University, 1967–1969**

Present Position: **University of Colorado, Department of Molecular, Cellular, and Developmental Biology**

Address: **Boulder, Colorado**

What is the nature of your current research?

WE ARE INTERESTED in gene expression, macromolecular assemblies, and cell growth and physiology. We are interested as well in anything not included within those categories. Translation initiation and its regulation has been an area of both historical and recent work; we continue to study the mechanisms by which RNA binding proteins can alter translation in a message-specific manner. Other RNA-related problems under study include the action of an endonuclease that cuts target messages in their translation initiation regions and the mechanism by which bacteriophage T4 group 1 introns are mobile.

Another project seeks the biochemical role of the lambda *rex* proteins. The *rexB* protein appears to be a proton and monovalent cation channel of some kind, perhaps even having a homologue in the membrane ATPase operons of many organisms. The *rex* proteins provide bacteriophage lambda with a primative immune system.

A new project now involves a large number of people. The work began with Craig Tuerk's development of a technology called SELEX (for Systematic Evolution of Ligands by Exponential Enrichment) and the emergent development of a companion technology called SPERT (for Systematic Polypeptide Evolution by Reverse Translation). In short, one can evolve either nucleic acid or polypeptide ligands aimed at any target, in much the same way that antibodies can be developed toward nearly any target. Both SELEX and SPERT are in-vitro techniques with a deep capacity to be useful in many areas of basic science. We hope to answer, experimentally, questions related to the functional possibilities of nucleic acids and also to help answer two big and related evolutionary questions: Why are there proteins in the present biosphere? and, What are the real chemical capacities of RNA?

15

Bacteriophage

Bacteriophage, or phage, have played an important role in the development of molecular biology. Because of their relative simplicity, the availability of enormous numbers of mutants, and ease of propagation, phage have been extraordinarily useful in the study of basic processes such as replication, transcription, translation, and regulation. In this chapter, we shall examine some basic features of phage biology and provide a few examples of reproductive strategies by looking at particular stages of the life cycles of certain phage.

A bacteriophage is a bacterial parasite. By itself, it can persist, but a phage can reproduce only within a bacterial cell. Although phage possess genes encoding a variety of proteins, all known phage use the ribosomes, protein-synthesizing factors, amino acids, and energy-generating systems of the host cell.

Phage must perform four minimal functions for continued survival. These are:

1. Protect their nucleic acid from environmental chemicals which could alter the molecule (for example, break the molecule or cause a mutation).

2. Deliver their nucleic acid to the inside of a bacterium.

3. Convert an infected bacterium to a phage-producing system which yields a large number of progeny phage.

4. Release phage progeny from an infected bacterium.

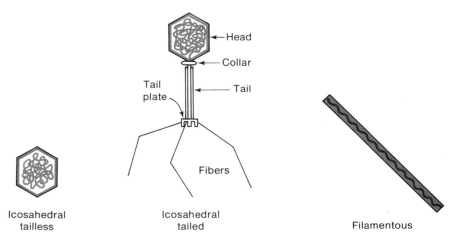

Figure 15-1
Two-dimensional view of three basic phage structures. The nucleic acid is shown in red.

These problems are solved in a variety of ways by different phage species.

Phage particles also differ in their physical structures from species to species, and often certain features of their life cycles are correlated with their structure. The three basic structures are shown in Figure 15-1.

Table 15-1

Properties of the Nucleic Acid of Several Phage Types

Phage	Host	DNA or RNA	Form	Molecular Weight, $\times 10^6$	Unusual Bases
φX174	E	DNA	ss, circ	1.8	None
M13, fd, f1	E	DNA	ss, circ	2.1	None
PM2	PB	DNA	ds, circ	9	None
186	E	DNA	ds, lin	18	None
B3	PA	DNA	ds, lin	20	None
Mu	E	DNA	ds, lin	25	None
T7	E	DNA	ds, lin	26	None
λ	E	DNA	ds, lin	31	None
N4	E	DNA	ds, lin	40	None
P1	E	DNA	ds, lin	59	None
T5	E	DNA	ds, lin	75	None
SPO1	B	DNA	ds, lin	100	HMU for thymine
T2, T4, T6	E	DNA	ds, lin	108	Glucosylated HMC for cytosine
PBS1	B	DNA	ds, lin	200	Uracil for thymine
MS2, Qβ, f2	E	RNA	ss, lin	1.0	None
φ6	PP	RNA	ds, lin	2.3, 3.1, 5.0*	None

Note: Abbreviations used in this table are: E, *E. coli*; B, *Bacillus subtilis*; PA, *Pseudomonas aeruginosa*; PB, *Ps. aeruginosa BAL-31*; PP, *Ps. phaseolica*; ss, single-stranded; ds, double-stranded; circ, circular; lin, linear; HMU, hydroxymethyluracil; HMC, hydroxymethylcytosine.

* φ6 contains three molecules.

The most common type of nucleic acid in phage is double-stranded linear DNA; however, double-stranded circular DNA, single-stranded linear and circular DNA, and single- and double-stranded linear RNA are also found (Table 15-1). The molecular weight of phage nucleic acid has a hundredfold range of variation among species, in sharp contrast with bacteria, for which the weight rarely varies more than about 10%. Phage with larger nucleic acid molecules possess more genes, and consequently can have more complex life cycles and are less dependent on bacterial enzymes for their reproduction. An unusual feature of the DNA of some phage is the presence of bases other than the standard A, T, G, C, as shown in the table. For example, T4 contains glucosylated 5-hydroxymethylcytosine instead of cytosine, and SP01 has hydroxymethyluracil instead of thymine. Phage nucleic acid is always isolated from the extracellular environment by an enclosing protein shell called either the **coat** or the **capsid** and is thereby protected from harmful substances.

Stages in the Lytic Life Cycle of a Typical Phage

Phage life cycles fit into two distinct categories—the **lytic** and the **lysogenic** cycles. A phage in the lytic cycle converts an infected cell to a phage factory, and many phage progeny are produced. A phage capable only of lytic growth is called **virulent.** The lysogenic cycle, which has been observed only with phage containing double-stranded DNA, is one in which no progeny particles are produced; the phage DNA usually is inserted into the chromosome of the host bacterium. A phage capable of such a life cycle is called **temperate.**★ In this section the lytic cycle is outlined.

There are many variations in the details of the life cycles of different phage. There is, however, what may be called a basic lytic cycle, which is the following (Figure 15-2):

1. *Adsorption of the phage to specific receptors on the bacterial surface.* These receptors serve the bacteria for purposes other than phage adsorption. (For example, the receptor for phage T6 is involved in transporting nucleosides into the bacterial cell.)

2. *Passage of the DNA from the phage through the bacterial cell wall.* Some types of tailed phage use an injection sequence shown schematically in Figure 15-3.

★ Most temperate phage also undergo lytic growth under certain circumstances.

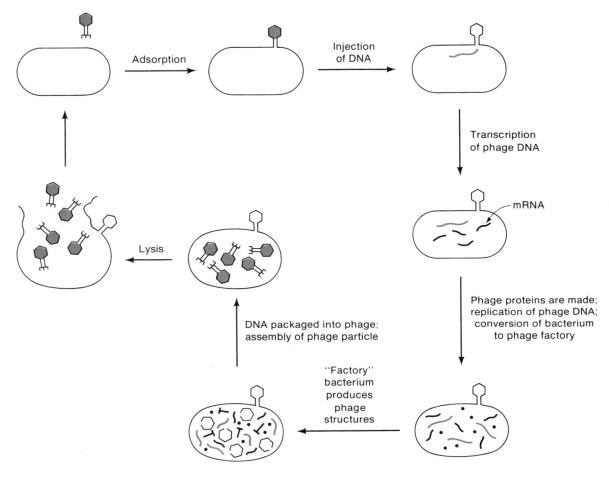

Figure 15-2
Schematic diagram of the life cycle of a typical phage.

3. *Conversion of the infected bacterium to a phage-producing cell.* Following infection by most (but not all) phages, a bacterium loses the ability either to replicate or to transcribe its DNA; sometimes it loses both. This shutdown of host DNA or RNA synthesis is accomplished in many different ways depending on the phage species.

4. *Production of phage nucleic acid and proteins.* The phage redirects the biosynthetic pathways of the infected cell to make copies of phage nucleic acid and synthesize phage proteins. This reprogramming is accomplished by transcribing and then translating phage genes. Transcription of phage genes is almost always initiated by the bacterial RNA polymerase, but after the first set of transcriptions, the bacterial poly-

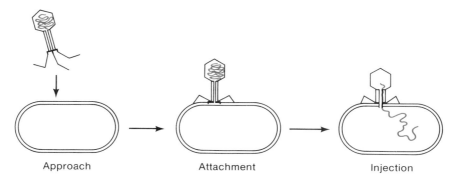

Figure 15-3
Injection sequence of a tailed phage. In the injection stage the tail sheath contracts and drives a core protein tube through the cell wall like a hypodermic syringe.

merase is usually modified to recognize phage-specific promotors, or a phage-specific RNA polymerase is synthesized. Transcription is regulated, and phage proteins are synthesized sequentially in time as they are needed. Distinct classes of mRNA are made at various times after infection; the major division is between early mRNA and late mRNA. Early mRNA usually encodes the enzymes required for takeover of the bacterium for phage DNA replication, and for the synthesis of late mRNA. Late mRNA encodes the components of the phage particle, proteins needed for packaging of nucleic acid in the particles, and enzymes required to break open the bacterium. The more complex phage usually synthesize several classes of early mRNA, which are made in a particular time sequence.

5. *Assembly of phage particles.* This process is often called **morphogenesis.** Two types of proteins are needed for the assembly process: **structural proteins,** which are present (possibly in modified form) in the phage particle, and **catalytic proteins** which participate in the assembly process, but do not become part of the phage particle. A subset of the latter class consists of the maturation proteins, which convert intracellular phage DNA to a form appropriate for packaging in the phage particle. Usually 50–1000 particles are produced; the number depends on the phage species.

6. *Release of newly synthesized phage.* With most phages a phage protein called a **lysozyme** or an **endolysin** is synthesized late in the cycle of infection. This protein causes disruption of the cell wall, so that phage are released to the surrounding medium. This disruptive process is called **lysis.**

lysis

The events just described occur in an orderly sequence, as exemplified by the life cycle of *E. coli* phage T4 listed below (times in minutes at 37°C):

$t = 0$	Phage adsorbs to bacterial cell wall. Injection of phage DNA probably occurs within seconds of adsorption.
$t < 1$	Synthesis of first phage mRNA begins.
$t = 2-3$	Synthesis of host DNA, RNA, and protein is turned off.
$t = 5$	Degradation of bacterial DNA begins. Phage DNA synthesis is initiated.
$t = 9$	Synthesis of late mRNA begins.
$t = 12$	Completed heads and tails appear.
$t = 13$	First complete phage particle appears.
$t = 25$	Lysis of bacteria; release of about 300 progeny phage.

The total duration of the cycle is typical, most phage having a life cycle of 20–60 minutes, which is comparable to the generation times of most bacteria. The life cycles of animal viruses are considerably longer—24 to 48 hours—but this is again comparable to the life cycle of an animal cell.

Specific Phage

On the scale of biological complexity, phage are relatively simple forms of life. However, phage are sufficiently complex that there is not a single phage type for which all of the molecular details of its life cycle are completely understood. The best known are a small number of phage species that grow in *E. coli, Bacillus subtilis,* and *Salmonella typhimurium.* Special features of each phage have resulted in particular phage being more suited to the study of certain processes. For example, regulation of transcription is best understood in phage λ and T7, the study of morphogenesis has been more successful with phage T4 and λ than with other phage, phage P1 and P22 gave the first clues to understanding transduction. T5 has yielded the best information about DNA injection. T4 has been most profitable in the study of DNA synthesis, and the study of T7 has shown clearly how a phage takes over a bacterium.

In the following sections several well-understood features of a few phage will be described.

E. COLI PHAGE T4

The *E. coli* phage T4 is large by bacteriophage standards (Table 15-1). Its genome contains 166,000 base-pairs of DNA, sufficient coding capacity for about 200 average-sized genes. The nucleotide sequences of 119 genes of known function, and 81 open reading frames of unknown

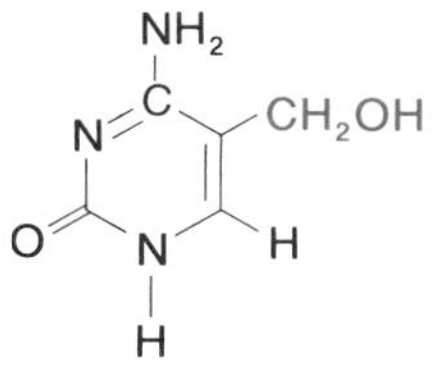

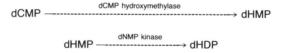

5-Hydroxymethylcytosine (HMC)

Figure 15-4
5-Hydroxymethylcytosine. If the CH₂OH (red) in HMC were replaced by hydrogen, the molecule would be cytosine.

function (approximately 90% of the genome) have been determined. The large genome of T4 allows it to be less dependent on host functions and have a more complex life cycle than is possible for smaller phages. In this section we will consider examples of increased independence and complexity in T4 DNA metabolism.

Unlike other phage, T4 DNA replication proceeds normally in thymine-requiring hosts in the absence of added thymine, because T4 encodes its own thymidylate synthetase. In fact, T4 specifies a number of enzymes involved in nucleotide metabolism (dihydrofolate reductase, nucleotide reductase, thioredoxin reductase). These enzymes do more than duplicate *E. coli* functions; they are responsible for the high rate of replication in infected cells.

A second complexity of T4 DNA metabolism is that T4 DNA contains a modified form of cytosine, called **5-hydroxymethylcytosine (HMC),** which base pairs with guanosine (Figure 15-4). This base is further modified by glucosylation—that is, a glucose molecule is coupled to the -OH group of HMC. As we shall see, HMC plays an important role in the T4 life cycle.

E. coli does not encode enzymes for HMC synthesis; but rather, its biosynthesis is directed by two T4 enzymes: dCMP hydroxymethylase and dNMP kinase.

$$dCMP \xrightarrow{\text{dCMP hydroxymethylase}} dHMP$$

$$dHMP \xrightarrow{\text{dNMP kinase}} dHDP$$

The *E. coli* enzyme, nucleoside diphosphate kinase, forms all nucleoside triphosphates in *E. coli,* then converts dHDP to dHTP, the immediate precursor of HMC in DNA (Figure 15-5).

As indicated in Figure 15-5, T4 also encodes nucleases that attack cytosine containing DNA and thus degrade the host chromosome to mononucleotides. If T4 DNA did not contain HMC, it too would be degraded by these nucleases.

Most dCMP (whether produced *de novo* or from degradation of the host DNA) is converted to dHMP as described above. However, some will be converted to dCTP, which can be incorporated into replicating T4 DNA. Of course, this DNA would be susceptible to attack by the phage-encoded nucleases. To prevent dCTP incorporation, another phage-specified enzyme, dCTPase, hydrolyzes both dCTP and dCDP to dCMP, the substrate for dCMP hydroxymethylase.

Although HMC containing DNA is resistant to many DNases (including most restriction enzymes), *E. coli* does possess an endonuclease that attacks certain nucleotide sequences containing HMC. To avoid this damage, HMC is glucosylated. This is accomplished by two phage enzymes, α-glycosyl transferase (α*gt*) and β-glucosyl transferase (β*gt*),

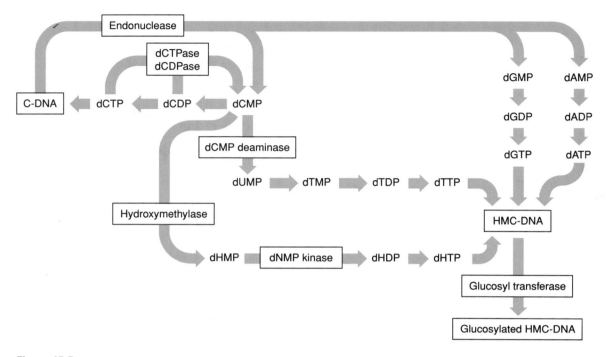

Figure 15-5
Pathway for biosynthesis of hydroxymethylcytosine (HMC) DNA induced by phage T4. Boxed elements are macromolecules; thin boxes indicate enzymes and heavy boxes indicate DNA. Substances found only in phage-infected cells are shown in black. C-DNA is cytosine-containing DNA.

each of which transfers a glucose from uridine diphosphoglucose (UDPG) to HMC that is already in DNA. Glucosylation is an example of postreplicative modification.

The *E. coli* endonuclease is inactive against glucosylated DNA. A simple genetic experiment shows that the protection of HMC DNA from endonucleolytic attack is the only essential function of glucosylation. A T4 αgt^- βgt^- double mutant cannot carry out glucosylation, so its newly synthesized DNA is attacked by the *E. coli* HMC endonuclease. However, T4 αgt^- βgt^- double mutants do grow on an *E. coli* mutant ($rglB^-$), which lacks this nuclease, even though the phage DNA is nonglucosylated.

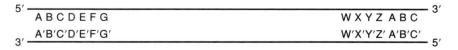

Figure 15-6
A terminally redundant molecule.

An important property of T4 DNA molecules packaged in phage particles is that they are terminally redundant (Figure 15-6)—that is, a sequence of bases (about 1% of the total) is repeated at both ends of the molecule. Terminal redundancy is a feature of the DNA of many phage species and can be generated in several ways, as will be described below.

Another feature of T4 DNA is that although each phage particle contains one DNA molecule, the molecules differ from phage to phage, even if the population was produced by replication of a single phage particle and its progeny. A sample of T4 DNA molecules is **circularly permuted,** the meaning of which is shown in Figure 15-7. The figure indicates schematically that the termini of the DNA molecules in the population can be found at many different bases within an overall sequence—in reality, probably at any point in the base-sequence. Note that circular permutation is a property of a phage population, whereas terminal redundancy is a property of an individual phage DNA molecule. Both terminal redundancy and circular permutation are a consequence of the mechanism by which DNA is packaged into the phage head.

The mechanism of packaging of T4 DNA in a phage head is not yet fully understood. The basic problem is how the long strand of DNA is tightly folded so that it will fit in the capsid. It is thought that packaging begins by the attachment of one end of a DNA molecule to a protein contained in a precursor of the phage head. Then either a condensation protein or a small, very basic molecule (such as a polyamine) induces folding (Figure 15-8). The best-understood feature of the process is that the molecule is cut from long concatemers. The cuts are not made in unique base sequences in the DNA, because if they were, T4 DNA could not be circularly permuted. Instead, the cuts are made at positions that are determined by the amount of DNA that can fit in a head.

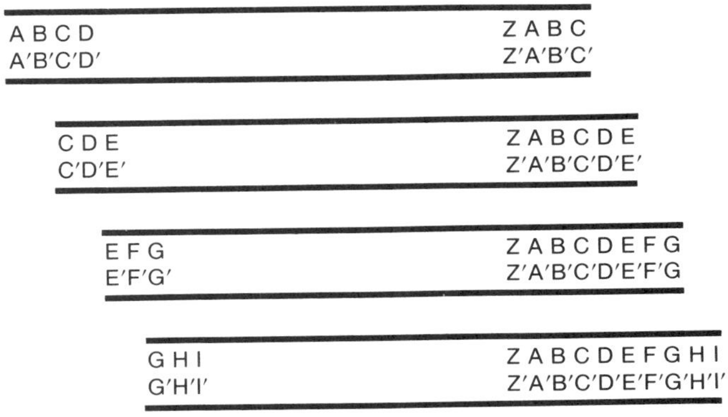

```
A B C D                              Z A B C
A'B'C'D'                             Z'A'B'C'


  C D E                             Z A B C D E
  C'D'E'                            Z'A'B'C'D'E'


    E F G                         Z A B C D E F G
    E'F'G'                        Z'A'B'C'D'E'F'G


      G H I                     Z A B C D E F G H I
      G'H'I'                    Z'A'B'C'D'E'F'G'H'I'
```

Figure 15-7
A circularly permuted collection of terminally redundant DNA molecules.

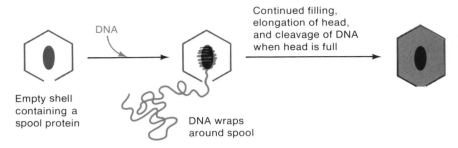

Figure 15-8
Proposed model for filling a T4 head. Cleavage and rearrangement of head proteins is known to occur at several stages of the process.

Presumably a free end of the DNA molecule enters the head, and this continues until there is no more room: then the concatemer is cut. This is known as the **headful mechanism** and it explains how both terminal redundancy and circular permutation arise (Figure 15-9). The essential point is that the DNA content of a T4 particle is greater than the length

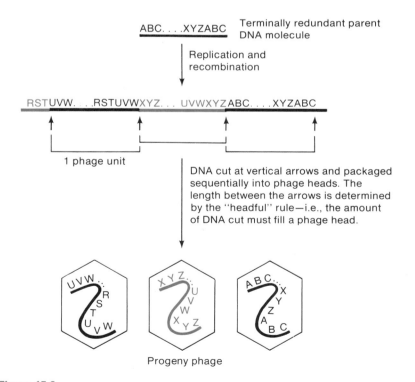

Figure 15-9
Origin of circularly permuted T4 DNA molecules. Alternate units are shown in different colors for clarity only.

Figure 15-10
Electron micrograph of
phage T7. Note the very
short tails. (Courtesy of
Robley Williams.)

of DNA required to encode the T4 proteins. Thus, when cutting a headful from a concatemeric molecule, the sequence at the end of the packaged DNA is a duplicate of the sequence packaged first—that is, the packaged DNA is terminally redundant. The first segment of the second DNA molecule that is packaged is not the same as the first segment of the first phage. Furthermore, since the second phage must also be terminally redundant, a third phage-DNA molecule must begin with still another segment. Thus, the collection of DNA molecules in the phage produced by a single infected bacterium is a circularly permuted set.

E. COLI PHAGE T7

Phage T7 is a midsized phage (Table 15-1). Its DNA has a molecular weight of about 26×10^6 and has no unusual bases. It has a terminal redundancy of 160 base-pairs, but it is not circularly permuted. The DNA molecule is encased in a head that is attached to a very short tail (Figure 15-10).

T7 has fewer genes than T4 (about 50), which simplifies analysis of its life cycle. Furthermore, most of the gene products have been identified by gel electrophoresis, and many have been sequenced by chemical means. The complete sequence of 39,930 base-pairs of T7 DNA has been determined; this information allows direct inspection to locate all of the promoters, termination sites, regulatory sites, spacers, leaders, initiation codons, termination codons, and genes. A great deal is known about the molecular biology of T7; however, we shall only outline how it regulates transcription.

T7 transcription occurs in three temporal stages (Figure 15-11), with Class I transcripts occurring first. Class I transcripts originate at three promoters (pI) located very near the left end of the T7 DNA molecule. These transcripts end at the termination site, T_E, and are made by *E. coli* RNA polymerase. As expected, the pI sequences strongly resemble those of *E. coli* promoters.

Two class I gene products are essential for the transition to class II and III transcription. One of those is a protein kinase enzyme which phosphorylates and thereby inactivates *E. coli* RNA polymerase (though inactivation is not complete). The second enzyme is T7 RNA polymerase. Only T7 RNA polymerase recognizes the promoters (pII and $pIII$) for class II and III transcripts. The sequences of these T7 specific promoters are very different from those recognized by *E. coli* RNA polymerase.

T7 RNA polymerase does not recognize the terminator, T_E. Hence, class II transcripts proceed past the site to the terminator T_ϕ (Figure 15-11). Class II transcripts encode replication proteins, the major structural protein of the phage head, and a second inhibitor of *E. coli* RNA polymerase. This latter protein completely inhibits the remaining activity of *E. coli* RNA polymerase.

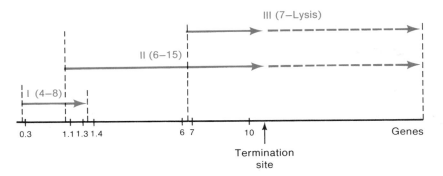

Figure 15-11
A simplified transcription map of T7. The DNA and gene numbers are in black; the three classes of transcripts are in red. The red numbers indicate the time interval (in minutes at 37°C) during which the transcript is made. The termination site indicated by the black arrow is about 90% efficient; thus 10% of the class II and III molecules are extended further, as indicated by the dashed lines.

Termination at T_ϕ is only 90% efficient, meaning that 10% of the transcripts proceed to the right end of the DNA molecule. The extended region encodes the tail, maturation, and lysis proteins, which are required in small quantities compared to the head proteins.

Since DNA replication proteins are enzymes, they are needed in only very small amounts compared to the head and tail proteins. Late in infection when there is no need to continue making these enzymes, class II transcription shuts off, and class III transcription begins.

Class III transcripts are initiated at seven *pIII* promoters to the left of T_ϕ and encode both structural and lysis proteins. These transcripts, like those of class II, terminate at T_ϕ with 90% efficiency. However, there are three additional *pIII* promoters to the right of gene 12 (Figure 15-11), so that only gene 11 and 12 expression is entirely dependent on read-through.

The mechanism of temporal regulation of class II and III transcription is unusual and rather unsophisticated in comparison to phage λ. Shutoff of class II transcription depends on promoter strength and the synthesis of a transcription inhibitor. Class II promoters are weak compared to those of class III. One of the proteins encoded in class II transcripts reduces all transcription. When this protein is made, transcription is initiated only at strong (*pIII*) promoters.

The mechanism that delays class III expression is quite uncommon. Most phage species inject their DNA within a few seconds of adsorption, injection of T7 DNA takes about 10 minutes. Thus there is a period when *pII* promoters are the only promoters available to T7 RNA polymerase. Entry of *pIII* promoters into the cell occurs around seven minutes after adsorption.

Let us now summarize how the three classes of transcripts are regulated. At first, *E. coli* RNA polymerase makes class I transcripts. An inhibitor, translated from class I transcripts, prevents further synthesis of these transcripts. T7 RNA polymerase, translated from class I transcripts, makes class II transcripts. An inhibitor translated from class II transcripts prevents their further synthesis. Slow injection delays entry of *pIII* promoters into the cell; however, once *pIII* promoters are available, class III transcripts are made. The net result is early synthesis of the DNA replication proteins and late synthesis of the structural proteins of the phage particle.

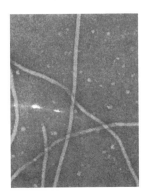

Figure 15-12
Electron micrograph of phage M13.

E. COLI PHAGE M13

E. coli phage M13 (Figure 15-12) differs from other bacteriophage discussed in this chapter in many ways. It is one of the smallest known phage (Table 15-1). Its genome is composed of only 10 genes (Figure 15-13), all of which are essential. The phage particles contain a circular DNA molecule (the viral or (+) strand) that is only 6407 nucleotides long. M13 is a rod-shaped, filamentous phage that can package circular DNA molecules that are several times longer than its viral strand. The phage adsorbs only to male strains of *E. coli,* but does not lyse infected cells. The infected cells continue to grow and divide, producing 100–200 phage per cell, per generation.

In this section, we will consider M13 DNA replication and its role in recombinant DNA technology.

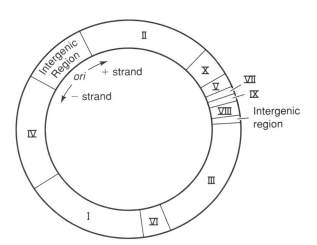

Figure 15-13
Genetic map of M13. Genes are designated by roman numerals: II. replication protein; V. single-stranded DNA binding protein; VIII. major coat protein; IX. minor coat protein. The origins of replication are in the intergenic space between genes II and IV.

Table 15-2

Replication Cycle of M13

Stage	Time, Minutes at 33°C	Event
ss → RF	0–1	Adsorption and penetration; viral single strand converted to parental RF; RF transcribed
RF → RF	1–20	Parent RF replicates and 100–200 progeny RF are formed; RF transcription continues
	25	Replication of RF molecules stops
RF → ss	20–30	Rolling circles form progeny viral DNA; from these, about 500 ss molecules are produced; DNA is packaged into phage particles which are excreted

Note: Abbreviations used are: ss, single-stranded phage DNA; RF, circular replication form.

The M13 replication cycle can be divided into three stages (see Table 15-2). In the first stage, the infecting viral (+) strand is converted to a double-stranded circular DNA molecule (replicating form or RF), ss → RF. In stage two, the double-stranded circles produce daughter double-stranded circles, RF → RF. In the final stage, double-stranded circles produce viral (+) single strands, RF → ss.

Although there are three stages of replication, there are only two mechanisms; one for producing (−) strands and one for (+) strands. The complementary or (−) strand serves two important functions. It is the template for all M13 transcription, and for (+) strand synthesis. The origins for both (−) and (+) strand replication are located in the intergenic space between genes II and IV (Figure 15-13). M13 proteins are not required for the synthesis of (−) strands, as might be expected in view of the (−) strand's role as a template for transcription.

The first step of (−) strand replication is the synthesis of an RNA primer by *E. coli* RNA polymerase at the (−) origin on a single-stranded viral DNA molecule. The primer is 30 nucleotides long, and is extended from its 3′-OH end by *E. coli* DNA polymerase III around the circle until it reaches the 5′-P end of the primer. *E. coli* DNA polymerase I then degrades the primer and replaces it with deoxynucleotides. The two ends of the (−) strand are immediately sealed by *E. coli* DNA ligase. The resulting supercoiled, double-stranded molecule is known as **RFI (replication form I)**.

M13 gene II protein is required for viral strand synthesis. These strands are made from RFI templates by a variation of the rolling-circle model (Figure 15-14). Synthesis is initiated when M13 gene II protein nicks the (+) strand at the (+) origin. Using the *E. coli* replication proteins (Rep, ssb, and DNA polymerase III), the (+) strand is extended from its 3′-OH end at the nick. As the old (+) strand is displaced by the elongating new (+) strand, it is coated with ssb. When the new (+) strand reaches the origin, it is nicked again by gene II protein, freeing the old (+) strand whose 3'-OH and 5'-P ends are joined by gene product II.

Note that replicating RF's do not directly produce double-stranded daughter molecules. The products of one round of replication are a single-stranded (+) circle, and a double-stranded RF molecule.

At early times during infection the (+) strands serve as templates for the synthesis of (−) strands, accounting for RF → RF replication. During this period RFs are also transcribed, producing a coat protein (gene VII product) that associates with the cytoplasmic membrane; and gene V product, a single-stranded DNA binding protein, which is necessary for the final stage of replication.

The switch from RF → RF to RF → ss occurs when sufficient gene V protein has accumulated to bind the new viral strands, and prevent them from serving as templates for (−) strand synthesis. The gene V protein and viral DNA complex then move to the cell membrane where the gene V protein is displaced by coat protein. The completed virus particles are excreted through the cell membrane without lysing the cell.

If the M13 DNA molecule is enlarged by inserting foreign DNA at a site near the minus origin, no phage functions are altered. The enlarged molecule replicates and is packaged into a larger coat (unlike T4, the size of the DNA molecule determines the size of the phage particle). Thus M13 is a very useful tool in DNA cloning. Using recombinant DNA techniques, foreign DNA can be inserted into double-stranded RF DNA. The inserted DNA is replicated as a part of the M13 molecule. The DNA packaged in phage particles is an abundant source of single-stranded DNA for sequencing, site-directed mutagenesis, and other procedures.

E. COLI PHAGE λ: THE LYTIC CYCLE

λ is a temperate E. coli phage (Figure 15-15). It can engage in both a lytic cycle in which progeny phage are produced, and a lysogenic cycle in which a phage-DNA molecule is inserted into the bacterial chromosome. Special features of the elegant regulatory systems of λ determine which pathway is used in any given infection.

The DNA of λ has two components: a double-stranded segment containing 48,489 nucleotide pairs of known sequence, and a 12 nucleotide single-strand extension at the 5'-P terminus of each strand. The two terminal single-strand extensions have complementary base sequences, and are called **cohesive ends.** Immediately following injection of the DNA, the cohesive ends base-pair, yielding a circle (Figure 15-16). The adjacent 5'-P and 3'-OH groups are quickly sealed by E. coli DNA ligase, and the circle is then supercoiled. No phage gene products are involved in this process.

The arrangement of the genes of phage λ is shown in Figure 15-17. As is typical for phage, the genes are clustered according to function.

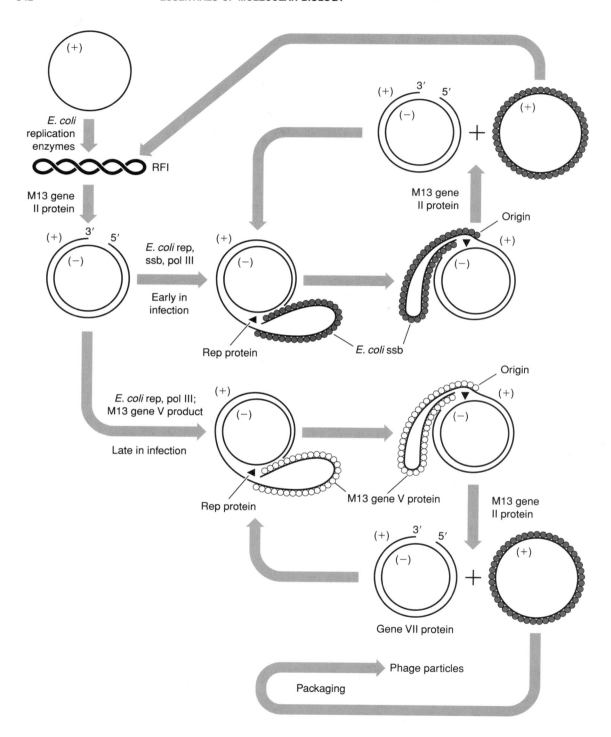

Figure 15-15
Electron micrograph of
phage λ. (Courtesy of
Robley Williams.)

For example, the head, tail, replication, and recombination genes form four distinct groups. Clustering and transcription of the members of a cluster from the same DNA strand allow λ to regulate transcription with a small number of regulatory sites (Figure 15-17).

In this section we will consider the regulatory events of the lytic cycle and replication and maturation of λ DNA. In the next section we will examine the regulatory events of the lysogenic pathway and the factors that determine which pathway will be followed.

Transcription of λ occurs in three temporal stages: very early, early, and late. Very early and early transcription precede the decision to follow the lytic or lysogenic pathway. Hence they are the same in both types of infections. Very early mRNA encodes regulatory proteins, early transcripts encode replication and recombination enzymes, and late lytic mRNA encodes the structural proteins of the phage particle and lysis proteins.

In marked contrast to T7, λ uses *E. coli* RNA polymerase to transcribe all of its mRNA. Qualitative control of λ transcription is accomplished by regulating both initiation and termination. Initiation is regulated by λ-encoded activators (*cI*, *cII*) and repressors (*cI*, *cro*) that control promoter accessibility. Termination is controlled by phage-specified **antiterminators** (*N* and *Q*), proteins that allow RNA polymerase to transcribe beyond specific termination sites.

Temporal (i.e., over time) regulation is achieved by sequential expression of regulatory proteins. Consider the following example. A DNA segment contains the sequence

$$p\ A\ B\ t1\ C\ t2$$

in which p is a promoter, *A B C* are genes, and *t1* and *t2* are terminators. Initially, RNA polymerase will attach to p and transcribe *A* and *B*. Assume that *B* encodes an antiterminator so that RNA polymerase ignores *t1*. In this case, before the *C* product can be made, the *B* product must be made; thus the *C* product will be synthesized at a later time than the *A* product.

Figure 15-18(a) shows a portion of the genetic map of λ, with three regulatory genes (*cro*, *N*, and *Q*), three promoters (*pL*, *pR*, and *pR′*),

Figure 15-14
A diagram of the replication of phage M13. *E. coli* replication enzymes convert the single-stranded phage DNA into the RFI form, which is then nicked by the M13 gene II protein. Rolling circle replication (see Figure 7-2 in Chapter 7) ensues to generate a daughter strand (red) and a displaced (+) single strand that is coated with *E. coli* ssb (filled circles) early in infection, or M13 gene V protein (open circles) late in infection. When the entire (+) strand is displaced, it is cleaved from the daughter (+) strand and circularized by the joining activity of the M13 gene II protein. The cycle is ready to begin anew. Note that the (−) strand is never cleaved. Viral (+) strands coated with M13 gene V protein are packaged into phage particles after the gene V protein is displaced by gene VII protein.

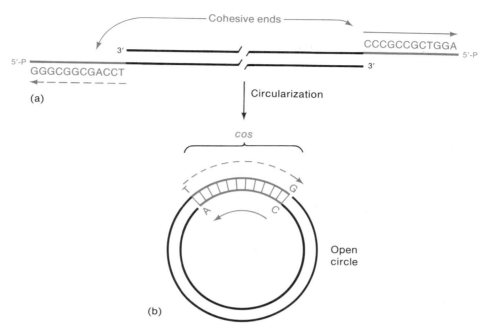

Figure 15-16
(a) A diagram of a λ DNA molecule showing the complementary single-stranded ends (cohesive ends). Note that 10 of the 12 bases are G or C. (b) Circularization by means of base-pairing between the cohesive ends. The double-stranded region that is formed is designated *cos*.

two operators (*oL* and *oR*), and four termination sites (*tL1*, *tR1*, *tR2* and *tR3*). Immediately after infection, *E. coli* RNA polymerase initiates transcription at the strong promoters (*pL*, *pR*, and *pR'*). These transcripts terminate at *tL1*, *tR1*, and *tR3*. Only two genes are encoded in these very early transcripts, *N* and *cro* (control of repressor operator); both are crucial for regulation. The *N* gene product is an antiterminator. It binds to λ DNA at the *nut* (*N* utilization) sites (Figure 15-18(a)), allowing RNA polymerase to transcribe beyond the terminators *tL1* and *tR1*.

The longer (early) transcripts from *pR* terminate at *tR2*, expressing the replication genes *O* and *P*, and the antiterminator *Q*, as well as *cro*. The early transcripts, extended beyond *tL1*, express λ recombination genes. At this point, if the infection follows the lytic pathway, the *Q* antiterminator binds to the *qut* (*Q* utilization) site permitting transcripts initiated at *pR'* to continue beyond *tR3* into the late region (head, tail and lysis genes).

Late in the lytic cycle, only structural proteins need to be synthesized. Transcription from *pL* and *pR* is repressed by the Cro protein

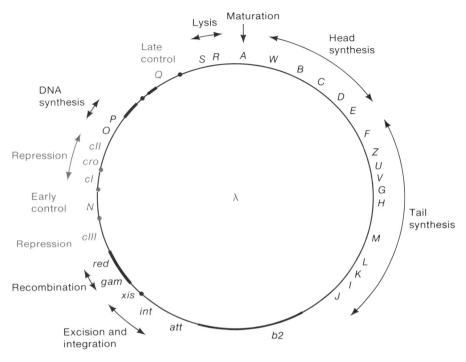

Figure 15-17
Genetic map of phage λ. Regulatory genes and functions are given in red. All genes are
not shown. Major regulatory sites are indicated by black solid circles. Regions nonessential for both the lytic and lysogenic cycles are denoted by a heavy line.

which binds to *oL* and *oR* (Figure 15-18(a)), preventing RNA polymerase
from binding to *pL* and *pR*. In order to understand how the Cro protein
inhibits transcription, we must examine the structure of *oR* and *pR* more
closely (*oL* and *pL* have analogous structures).

Actually *oR* is tripartite (Figure 15-18(b)). The three parts of *oR*
(*oR3*, *oR2*, and *oR1*) are similar but not identical sequences, and each
can bind a molecule of the Cro repressor. Since *oR1* and the right half
of the *oR2* overlap *pR*, the binding of RNA polymerase to *pR* is prevented by Cro protein bound to either *oR2* or *oR1*. Since *oR3* does not
overlap *pR*, *cro* repressor bound to *oR3* has no effect on *pR*.

Early in infection Cro protein is present at low concentrations and
binds only to *oR3*, which has a much higher affinity for Cro protein
than *oR2* and *oR1*. Since transcription from *pR* and expression of the
cro gene continue, the concentration of *cro* product increases. Late in
infection the concentration of Cro protein is sufficiently high that it
binds to the low affinity operators *oR2* and *oR1*, shutting off further

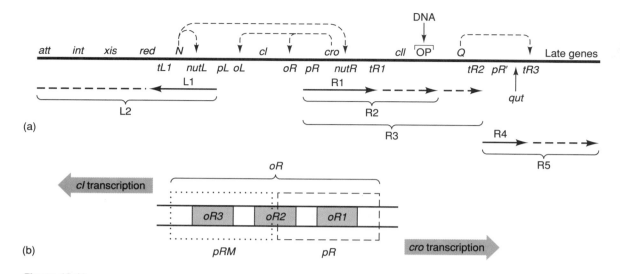

(a)

(b)

Figure 15-18
(a) A genetic map of the regulatory genes of phage λ. Genes are listed above the line; sites are below the line. The mRNA molecules are red. The dashed black arrows indicate the sites of action of the N, Cro, and Q proteins. (b) A blowup of the *oR* region showing its tripartite structure. The sub-operators, *oR3*, *oR2*, and *oR1* (shown in red) are each 17 nucleotides long. Each is capable of binding *cI* repressor or Cro protein. *pR* (dashed line) overlaps *oR1* and the right half of *oR2*, while *pRM* (dotted line) overlaps *oR3* and the left half of *oR2*.

transcription from *pR*. Thus *pR'* is the only active promoter at late times. Note that the delay in production of late proteins in comparison to early proteins is primarily a result of the time required to transcribe and translate the *N* and *Q* genes.

It was mentioned earlier that λ DNA circularizes shortly after infection and that the cohesive ends base-pair to form the double-stranded *cos* region. Clearly, at some stage of the life cycle the single-stranded termini must be regenerated, since the DNA is linear within a phage head. This is accomplished by a cutting system, called **terminase** (or **Ter**), which cuts within the *cos* region of progeny DNA. However, the cuts are not made in progeny circles; instead, DNA replication and cutting are coordinated in a way that is found with many phage species. The earliest stage of replication of λ DNA is the bidirectional θ mode seen in Figures 7-4 and 7-25. Progeny circles do not continue to replicate in this way, but switch to the rolling circle mode, yielding a long linear branch (Figure 15-19). Progeny λ DNA molecules are cut from the branch of the rolling circles by terminase. The circular portion of the rolling circle is not cut (this would prevent further replication), because terminase requires two *cos* sites or one *cos* site plus one free, single-stranded terminus. Thus, excision of λ units occurs by sequential cleav-

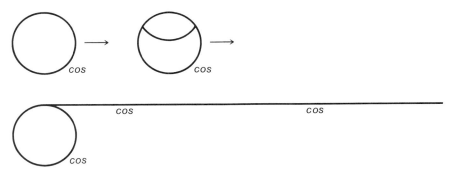

Figure 15-19
The replication sequence of phage λ.

age from the free end of the branch, as shown in Figure 15-20. This mechanism for determining the amount of DNA to be packaged is fundamentally different from the T4 headful mechanism. For λ, like M13, the length of the phage chromosome determines how much DNA is packaged.

E. COLI PHAGE λ: THE LYSOGENIC LIFE CYCLE

There are two types of lysogenic cycles. Phage λ is the prototype for the more common one, the essential features of which are shown in Figure 15-21.

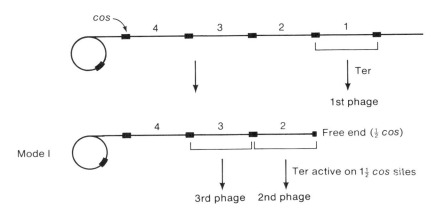

Figure 15-20
The mode of cutting λ units from a branch of a rolling circle. The units are cut sequentially from the free end by terminase.

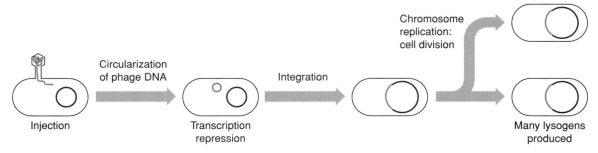

Figure 15-21
The general mode of lysogenization by insertion of phage DNA into a bacterial chromosome.

1. A DNA molecule is injected into a bacterium.
2. After a brief period of transcription, needed to synthesize an integration enzyme and repressor of phage transcription, transcription is turned off.
3. A phage DNA molecule is integrated into the DNA of the bacterium forming a **prophage.**
4. The bacterium continues to grow and divide and the prophage is replicated as part of the bacterial chromosome.

The less common type, for which *E. coli* phage P1 is the prototype, differs from the preceding one in that the phage DNA molecule becomes a plasmid (an independently replicating circular DNA molecule), rather than a segment of the host chromosome. In this section we will consider only the λ lysogenic cycle.

Two important properties of lysogens are the following:

1. A lysogen is *immune* to infection by a phage of the type that lysogenized the cell. Although the DNA of the superinfecting phage is injected and circularized it is neither replicated nor transcribed.

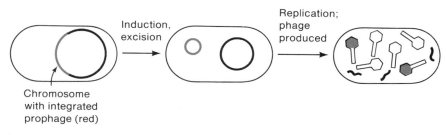

Figure 15-22
An outline of the events in prophage induction. The prophage DNA is in red. The bacterial DNA is omitted from the third panel for clarity.

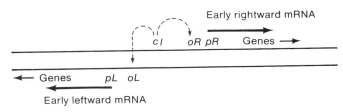

Figure 15-23
The repressor-operator system of λ, showing the two early mRNA molecules. Symbols: *cI*, repressor gene; *p*, promoter; *o*, operator; *L*, left; *R*, right.

induction

2. Even after many cell generations, a lysogen can initiate a lytic cycle; in this process (called **induction**) the prophage is excised as a circular DNA molecule (Figure 15-22) and initiates lytic growth.

In order to understand **immunity** and **induction,** we must examine how the *cI* gene product maintains lysogeny. The *cI* gene product, commonly called λ repressor, represses transcription from *pL* and *pR* by binding to *oL* and *oR* (Figure 15-23). In addition it activates and represses its own synthesis from a weak promoter, *pRM* (promoter for repressor *m*aintenance). The *pRM* overlaps *oR3* and the left half of *oR2* (Figure 15-18(b)), and is so weak that it must be activated to bind RNA polymerase. All three sub-operators (*oR3*, *oR2*, *oR1*) can bind *cI* repressor, but with different affinities and consequences.

A λ repressor binds first to *oR1*, the highest affinity operator, inhibiting transcription from *pR*, and facilitates binding repressor to *oR2* through protein-protein interactions. Repressor bound at *oR2* does not inhibit transcription from *pRM*, rather it activates *pRM* (binding of RNA polymerase to *pRM* is facilitated by protein-protein interactions between the repressor and RNA polymerase). Repressor binds to *oR3* with a low affinity, and when bound shuts off transcription from *pRM*.

At low concentrations, *cI* binds *oR1* and *oR2*, blocking transcription from *pR* and activating *pRM*. At high concentrations it binds to oR3 repressing transcription from *pRM*. In a lysogen, this circuit regulates λ repressor synthesis to maintain about 100 molecules of unbound repressor per cell in addition to those bound to *oL* and *oR*.

Immunity results from the λ repressor-operator system (Figures 15-18 and 15-23). When a λ phage infects a lysogen, unbound repressor molecules immediately bind to *oL* and *oR* of the injected DNA molecule, preventing RNA polymerase from binding to *pL* and *pR*. Thus λ repressor inhibits very early transcription directly, and early and late transcription indirectly. Since there is no transcription from the superinfecting DNA molecule, the bacterium grows and divides normally, and the superinfecting DNA molecule is progressively diluted out.

Induction is also based on the operator-repressor interaction, and results from the inactivation of the *cI* repressor. The repressor may be inactivated by high temperature in mutants that make a temperature labile *cI* protein or by chemical or physical agents that damage DNA. Damaged DNA activates the *E. coli recA* protein (which then functions as a protease) cleaving the cI repressor. The cleaved repressor is thus rendered inactive; *pRM* is no longer activated and *pL* and *pR* are no longer repressed, allowing the expression of *N* and *cro*. In addition to all of the gene products that are made in the lytic cycle, an enzyme called **excisionase** is made. This enzyme cuts the prophage DNA from the bacterial chromosome and forms a circle; thus, the λ DNA is in the same state as in the beginning of a lytic cycle.

In the lysogenic cycle a λ DNA molecule is integrated at a preferred position in the *E. coli* chromosome; this site is between the *gal* operon and the *bio* (biotin) operon and is called the λ attachment site or, genotypically, *att* (Figure 15-23). Integration occurs as a physical breakage and rejoining of phage and host DNA—precisely, between the bacterial DNA attachment site and another attachment site in the phage DNA that is located near the center of the phage DNA molecule (Figure 15-24). A phage protein, **integrase** (its gene designation is *int*), recognizes the phage DNA and bacterial DNA attachment sites and catalyzes the physical exchange. This results in integration of the λ DNA molecule into the bacterial DNA. As a consequence of the circularization and integration, the linear order of the genes in the infecting λ DNA molecule is rearranged, as shown in Figure 15-24.

A question that arose after this model was proposed was: why doesn't the integrase excise the prophage shortly after integration occurs, since the prophage is flanked by two attachment sites? Furthermore, excision would seem to be the preferred reaction since, kinetically, two attachment sites in the same DNA molecule (that is, the host chromosome) ought to interact with each other more rapidly than two attachment sites in different DNA molecules (phage and bacterial DNA). Why this does not happen became clear when it was discovered that, inasmuch as the DNA attachment sites in the phage and the bacterium are not the same, they recombine to form prophage attachment sites that, as a consequence, also differ from the sites joining the bacterial and phage DNA.

All of the λ attachment sites have three different components. One of these is common to all sites and is denoted by the letter *O*.⋆ The phage attachment site is written *POP′* (P for phage) and the bacterial attachment site is written *BOB′* (B for bacteria). Thus, in the integration reaction two new attachment sites, *BOP′* and *POB′*, are generated (Figure 15-25). These are often written *attL* and *attR* to designate the left and right prophage-attachment sites, respectively. Integrase cannot catalyze a reaction between *BOP′* and *POB′*, so that the reaction

---

⋆ Another common notation is a raised dot, ·, so that the bacterial attachment site, which we write *BOB′*, may be written *B · B′*. In this book we prefer the *O*.

$$BOB' - POP' \xrightarrow{\text{integrase}} BOP' + POB'$$

Bacterium Phage   **Prophage**

is irreversible if integrase is the only enzyme present.

The result of the integration reaction is that the λ DNA is linearly inserted between the *gal* and *bio loci* and henceforth is replicated simply as a segment of *E. coli* DNA (Figure 15-24). The first evidence for insertion came from genetic experiments. Final proof for insertion came from the following physical experiment. A prophage was formed in a circular plasmid whose molecular weight was 105×10^6 and which contained *BOB'*. The plasmid remained circular but its molecular weight increased by the amount of one λ molecule (31×10^6) to 136×10^6.

Finally, we must ask what determines whether a λ infection will by lytic or lysogenic. Although *cI* expression is essential to lysogeny, *pRM* dependent expression cannot establish lysogeny because activation of *pRM* transcription requires prior synthesis of the *cI* gene product. In λ, the solution for the initial transcription of *cI* is to use a second promoter, *pRE* (promoter for repressor establishment), that does not require prior *cI* expression. However, *pRE* is a weak promoter and must be activated by the *cII* gene product.

The *cII* gene is expressed early in infection along with *cro, O, P* and *Q*, as a consequence of antitermination of the *pR* transcript (Figure 15-26). The *cII* product then activates transcription of *cI* from *pRE* and *int* from *pI*. The first *cI* repressor molecules bind to *oL* and *oR*, inhibiting transcription from *pL* and *pR* and activating *pRM*. As lysogeny is established, transcription of *cI* from *pRE* ceases because the *cII* protein is unstable and is no longer synthesized once transcription from *pR* is inhibited. However, *cI* transcription continues from the activated *pRM*. Transcription from *pI* produces high levels of *int* protein which is necessary for integration of λ DNA into the host chromosome (Figure 15-26).

If the initial stages of the lytic and lysogenic pathways are identical, what determines which pathway will be followed? The principal determinant seems to be the level of *cII* protein. High levels of *cII* product

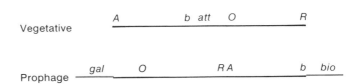

Figure 15-24
The order of the genes on the DNA molecule in the phage head (vegetative order) and in the prophage (prophage order). The genes have been selected arbitrarily to provide reference points.

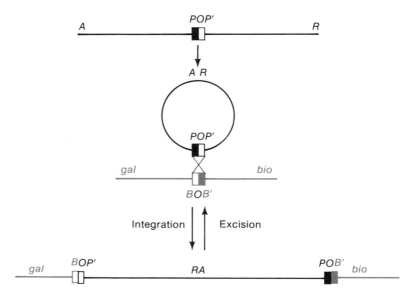

Figure 15-25
The Campbell model for the mechanism of prophage integration and excision of phage λ. The phage attachment site has been denoted *POP'* in accord with subsequent findings. The bacterial attachment site is *BOB'*. The prophage is flanked by two new attachment sites denoted *BOP'* and *POB'*.

favor lysogeny while low levels favor the lytic pathway. The instability of the *cII* protein is due to degradation by proteases. Under favorable growth conditions proteases are plentiful, the *cII* protein is degraded and hence *pRE* is not activated; the lytic cycle is favored. Under poor growth conditions proteases are less abundant and there is less degradation of *cII* protein, so *pRE* is activated, favoring lysogeny.

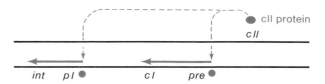

Figure 15-26
The role of the *cII* protein in stimulating synthesis of *cI* and *int* mRNA from *pre* and *pl*, respectively. The mRNA molecules are drawn as red arrows which indicate the direction of RNA synthesis. The arrows are located nearer the DNA strand that is copied. Dashed arrows point to the sites of binding of the cII protein (red dots).

Transducing Phage

Some phage species, known as **transducing phage,** are able to package bacterial DNA as well as phage DNA. Particles containing bacterial DNA, called **transducing particles,** are not always formed and usually constitute only a small fraction of a phage population. Such particles can arise by two mechanisms, aberrant excision of a prophage and packaging of bacterial DNA fragments. We begin with a discussion of the first mechanism.

One mechanism by which transducing particles form is shown in Figure 15-27, which depicts the formation of the galactose- and biotin-transducing forms of phage λ—namely, λ*gal* and λ*bio*.

When λ prophage is induced, an orderly sequence of events ensues in which the prophage DNA is precisely excised from the host DNA (Figure 15-27). In the case of phage λ, this is accomplished by the combined efforts of the *int* and *xis* genes acting on the left and right prophage attachment sites. At a very low frequency—namely, in about one cell per 10^6–10^7 cells—an excision error of two incorrect cuts is made—one within the prophage, and the other cut in the bacterial DNA. The pair of abnormal cuts will not always yield a length of DNA that can fit in a λ phage head—it may be too large or too small. However if the spacing between the cuts produces a molecule between 79% and 106% of the length of a normal λ phage-DNA molecule, packaging can

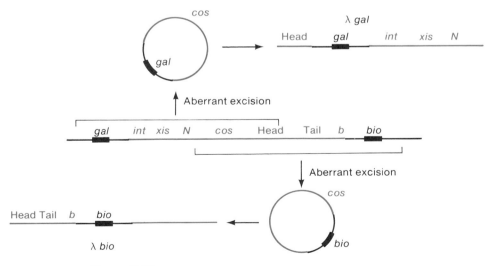

Figure 15-27
Aberrant excision leading to production of λ*gal* and λ*bio* phages.

occur. Since the prophage is located between the *E. coli gal* and *bio* genes and because the cut in the host DNA can be either to the right or the left of the prophage, transducing particles can arise carrying the *bio* genes (cut to the right), or the *gal* genes (cut to the left). Formation of the λ*gal*- and λ*bio*-transducing particles entails loss of λ genes. The λ*gal* particle lacks the tail genes, which are located at the right end of the prophage; the λ*bio* particle lacks the *int, xis, . . .* genes from the left end of the prophage. The number of missing phage genes of course depends on the position of the cuts that generated the particle and thus correlates with the amount of bacterial DNA in the particle. The missing phage genes come from the prophage ends, but because of the permutation of the gene order in the prophage and the phage particle, the deleted phage genes are always from the central region of the phage DNA, as shown in Figure 15-27.

Since these transducing particles lack phage genes, they might be expected to be nonviable. This is indeed true of the λ*gal* type, which lack genes essential for synthesis of the phage tail and, in the case of the larger deletion–substitutions, the genes for the phage head also. Thus, these particles are incapable of producing progeny; they are defective and this is denoted by the symbol d—thus, a *gal*-transducing particle is written λd*gal* (and sometimes λd*g*). The λd*gal* particle contains the cohesive ends and all of the information for DNA replication and transcription; it therefore goes through a normal life cycle, including bacterial lysis. In fact, if the head genes are not deleted, the concatemeric branch produced by rolling circle replication is cleaved by the Ter system. However, tails are not added to the filled head, no viable particles are produced in the lysate, and plaques are not formed on a host lawn. Note that there is no discrepancy between the formation of a λd*gal* transducing particle and its ability to reproduce itself. A λd*gal* particle fails to reproduce only because it lacks the tail genes, but such a particle arises from a normal prophage having a full set of genes.

The situation is quite different with λ*bio* particles for these usually lack only nonessential genes—*int, xis,* etc. These genes are needed for the lysogenic cycle, but not for the lytic cycle, so these particles are able to replicate and to form plaques. To denote this, the letter p, for plaque-forming, is added; that is, the particle is called λp*bio*.

Transducing phage, like λ, able to form transducing particles containing bacterial genes from selected regions of the bacterial chromosome, are called **specialized transducing phage.** Other phage species are able to package fragments of bacterial DNA from all parts of the chromosome. These are called **generalized transducing phage.** The specialized transducing phage producing transducing particles only come from a lysogen. In contrast, the generalized transducing phage form transducing particles from infected cells. The mechanism by which these particles are produced is straightforward. Some phage species, of which T4 is an example, degrade bacterial DNA early in infection. With T4 the degradation is complete, but in other species, of which *E. coli* phage P1 and *Salmonella typhimurium* phage P22 are examples, fragments of chromosomal DNA comparable in size to phage DNA are present at the time phage DNA is packaged. The packaging system of λ is a site-specific one, requiring *cos* sites. However, some phages, like T4, either merely fill a head, or like P1 and P22, are able to package fragments having a range of sizes. The latter phage cannot distinguish bacterial DNA from phage-DNA and occasionally package a piece of bacterial DNA. In a P1 population about one percent of the particles contain *only* a fragment of bacterial DNA.

Transducing particles are quite useful in genetic experiments. For example, they can be used to transfer a bacterial gene from one cell to another. They have also been used as means of isolating particular segments of a bacterial chromosome.

Summary

A bacteriophage, or phage, is an extracellular nucleoprotein particle that can infect and replicate within a bacterium. A typical phage consists of a protein shell in which is enclosed a molecule of nucleic acid. Phages contain either single- or double-stranded DNA or single- or double-stranded RNA. The most common phage structure is an icosahedral head to which a tail is attached. Less common are tailless heads and filamentous structures.

Phages have two distinct life cycles—the lytic and lysogenic cycles. In the lytic cycle progeny phage are made during the infection and released from the cell shortly after their assembly. A complete lytic cycle generally occurs in 20–60 minutes at 30–37°C. In the lysogenic cycle a replica of a phage DNA molecule becomes incorporated into the bacterial chromosome. The bacterium survives the infection and the phage DNA replicates as part of the bacterial genome. At a later time, if the bacterium is damaged, a lytic cycle ensues. Both life cycles begin by adsorption of a phage particle to the bacterium. Following adsorption tailed phages inject their nucleic acid into the cell; tailless phages release the nucleic acid on the cell surface, and the nucleic acid is taken into the cell in a manner that is not understood. Filamentous phages penetrate the cell wall, and the coat is digested during the process.

Once phage DNA is in a cell, a variety of strategies are used to produce progeny particles depending on the phage species and the life cycle selected. The earliest phage mRNAs are made by the bacterial polymerase; later mRNAs may be made by a newly made phage RNA polymerase or by various modified forms of the bacterial enzyme. There are three types of phage proteins made: catalytic, regulatory, and structural. The catalytic proteins (enzymes) are responsible for DNA replication, further transcription, preparation of DNA for packaging into a coat, cell lysis, and, occasionally, production of DNA precursors and destruction of bacterial DNA. Regulatory proteins may inactivate bacterial enzymes, but are primarily responsible for causing the various stages of phage production to occur in an orderly and efficient way. Structural proteins form heads, tails, and other components of the finished phage particle. If the lytic cycle is selected, the cycle ends with the bacterium being broken up by lysing enzymes and release of the progeny phage to the surrounding medium.

Mammalian viruses have been extensively studied, in ways that are similar to phage studies. The supplement entitled "The Molecular Biology of Mammalian Viruses" describes several of the key features of this study (see page 422).

Drill Questions

1. How many DNA molecules are contained in a typical phage particle?

2. What are the three morphological forms of phage particles?

3. What kind of nucleic acids are found in phages of different species?

4. What kind of nucleic acids is found in the following phages: T4, T7, λ, M13, MS-2? Tell about circularity also.

5. Why can a particular phage not absorb to any bacterial species?

6. What class of molecules are found in the phage tails?

7. What structural feature is present at the termini of λ DNA that is absent in the DNA of T4, T7?

8. A phage capable only of lytic growth is called _____.

9. Name the modified form of cytosine found in the T4 phage.

10. Phages able to package bacterial DNA are called _____ phages.

Problems

1. A suspension of a phage is exposed for a long time to a DNase. Will the treatment alter the number of infective phages?

2. List the main stages of a phage life cycle.

3. Name the minimal (essential) structural components of the simplest phage particle.

4. What is meant by a nonessential gene? Can a phage gene be truly nonessential?

5. What is meant by the term phage-host specificity, and what is the most frequent cause of this specificity?

6. A particular bacterial mutant cannot utilize lactose as a carbon source. If a phage adsorbs to such a bacterium and the infected cell is put in a growth medium in which lactose is the sole carbon source, can progeny phage be produced?

7. What is the source of the RNA polymerase that makes the first phage mRNA in the life cycle of a phage containing double-stranded DNA?

8. Considering the enzymes used for transcription, how does transcription of T7 differ from λ?

9. What DNA intermediate seems to be obligatory in the replication of single-stranded DNA phages?

10. Following injection, what structural transitions does phage λ undergo, prior to transcription?

Conceptual Questions

1. Since infection of a bacterium by a phage is usually lethal to the bacterium, why have bacteria not evolved to lose their phage receptors?

2. What would you expect to happen if, after virulent phage infection (such as with T4), all the phage genes were transcribed and translated at once following phage DNA infection?

3. How would you show whether or not a phage-encoded gene product is required throughout the infectious cycle or only at a unique time?

4. A new protein X appears in infected cells. Describe various experiments that you might perform to prove that the gene coding for X is phage-encoded and is not encoded in host DNA.

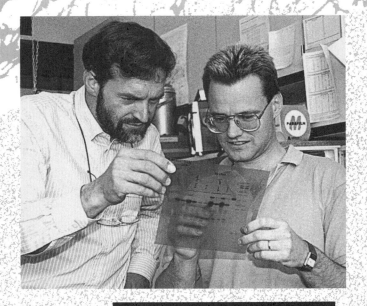

Walter Schaffner *(left)*
Professor of Molecular Biology
Institute of Molecular Biology (II)

Birthday: **26 October 1944**

Birth Place: **Aarau, Switzerland**

Undergraduate Degree: **University of Zurich**
Major: Zoology 1970

Graduate Degree: **University of Zurich, Ph.D. Molecular Biology, 1975**

Postdoctoral Training: **University of Zurich, 1975–1976;**
Medical Research Council, Cambridge, U.K., 1976;
Cold Spring Harbor Laboratory, New York, 1977–1978

Present Position: **Director, Institute of Molecular Biology (II), University of Zurich**

Address: **Zurich, Switzerland**

Hans-Peter Müller *(right)*
Postdoctoral Fellow
Department of Immunology and Microbiology

Birthday: **2 September 1960**

Birth Place: **Grabs, Switzerland**

Undergraduate Degree: **University of Zurich**
Major: Biology 1985

Graduate Degree: **University of Zurich, Ph.D. 1990**

Postdoctoral Training: **University of California, San Francisco, 1992**

Present Position: **University of California, San Francisco, Department of Immunology and Microbiology**

Address: **San Francisco, California**

What motivated you to decide to pursue a career in molecular biology?

OUR MOTIVATION COMES from a deep interest in general biology, coupled with the wish to understand more and more aspects of cellular behavior at the molecular level and to hopefully come closer to the answer of what life is in terms of chemistry and physics. Besides that, we like the daily lab work that affords a challenging interplay of experimental and intellectual skills. Molecular biology is still a young and growing field, and we find it particularly exciting to be able to discuss the rapid progresses with colleagues from all over the world. Moreover, molecular biology becomes of increasing importance in everybody's daily life. As with almost every powerful technology, molecular biology might help to solve some of today's most urgent problems, for example AIDS; but on the other hand, it may also intensify problems of our society. As molecular biologists we hope to always be aware of both types of developments.

16

Regulation of Gene Activity in Eukaryotes

More than two hundred cell types have been identified in the human body including those for muscle, liver, nerve, lung, epithelium, bone marrow (B), and thymus (T) lymphocytes. These cell types differ dramatically in both morphology and function. However, with some exceptions, the genetic information within a given organism is virtually the same. These different cell types seem to arise mainly from the differential expression of an identical genome, and are due to the expression of a characteristic subset of relatively few proteins and mRNAs in any given cell type. The majority of the proteins and mRNAs, however, are produced in all cell types, and their levels do not differ by more than a few fold from one cell to another (i.e., they are responsible for so-called housekeeping functions). Eukaryotes have evolved complex regulatory mechanisms to express a specific gene in one cell type, but not in another, without changing the sequence of the gene itself. In principle, gene expression can be regulated at each step in the circular synthesis/modification pathway from DNA to RNA to protein to DNA (Figure 16-1). Potential control levels include:

1. Transcription
2. RNA processing (splicing or other types of processing/modification)
3. mRNA transport
4. mRNA degradation and storage
5. Translation
6. Posttranslational modulation of protein activity

The expression of a given gene is likely to be regulated at several of these levels.

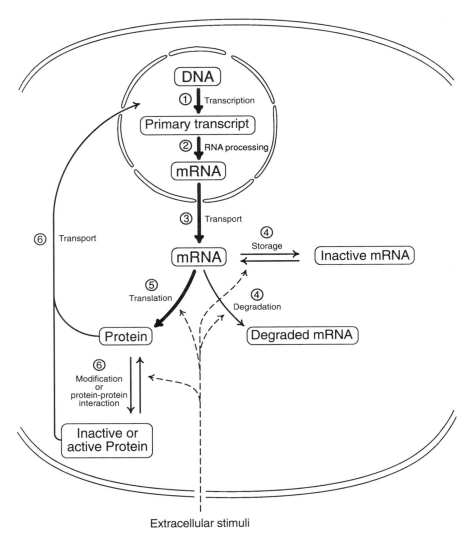

Figure 16-1
Potential control levels for the regulation of gene expression in eukaryotes. The levels 1 and 2 are localized in the nucleus and 4 and 5 in the cytoplasm; 3 and 6 indicate translocations between the two compartments. Note that with 6 only one route of protein translocation (from the cytoplasm to the nucleus) is indicated.

Important Differences in the Genetic Organization of Prokaryotes and Eukaryotes

Numerous differences exist between prokaryotic and eukaryotic cellular and genomic organization. The most striking differences are the following:

1. The eukaryotic cell contains a membranous nucleus which includes the chromosomes as carriers of genetic information. In addition to the nucleus, the eukaryotic cell contains many other subcellular compartments not found in prokaryotic organisms.

2. The chromosomal DNA of eukaryotes is packaged into a highly regular nucleoprotein complex, called **chromatin.**

3. In eukaryotes (particularly those with a large genome) many DNA segments are repeated hundreds, thousands, or even millions of times. The function of these highly repeated sequences—if there is any—has yet to be elucidated. In contrast, prokaryotic genomes contain essentially no repetitive sequences, except for a few repeats of ribosomal RNA (rRNA) and transfer RNA (tRNA) genes.

polycistronic mRNA

4. A typical eukaryotic gene is **monocistronic** (i.e. one transcription unit usually encodes one translation unit). Prokaryotes, however, often have several translation units organized in a single operon which comprises one transcription unit (**polycistronic** gene). The translation units of one operon usually encode different proteins involved in the same regulatory pathway. In contrast, eukaryotic genes, even if they belong to the same regulatory pathway, are usually not grouped closely on the same part of the chromosome.

intron

5. A characteristic feature of most eukaryotic genes is that they are split (i.e. the amino acid sequence of the gene product is usually not colinear with the genetic information). These **intron** sequences interrupt the eukaryotic genes and are also present in their primary transcripts. These introns have to be "spliced out" in a complicated process that generates mature messenger RNA (mRNA).

In the following sections we will see how eukaryotes have incorporated some of these features into particular modes of regulation.

The Regulation of Transcription Initiation

Whether a gene is expressed or not is often determined at the level of transcription. This is not so surprising, since it would seem to be the most economical way to control the expression of a gene. Nature sometimes defies the obvious expectation (e.g. by transcribing a specific gene in all cell types, but converting the transcript into an active form only in a subset of cell types).

THREE DIFFERENT RNA POLYMERASES TRANSCRIBE

EUKARYOTIC GENES

In prokaryotes, all genes are transcribed by the same RNA polymerase, and gene classes are recognized by different accessory proteins (sigma factors). In eukaryotes three different RNA polymerases fulfill the different transcriptional tasks. Although functionally and structurally distinct, these three multicomponent enzymes seem to share a few of their many subunits. RNA polymerase I transcribes only the repeated genes encoding ribosomal 18S, 5.8S, and 28S RNAs. RNA polymerase III transcribes some small RNA genes such as those encoding tRNAs, ribosomal 5S RNA and the U6 small nuclear RNA (U6 snRNA). Transcripts from these genes are produced in all cell types and therefore can be regarded as housekeeping genes. RNA polymerase II transcribes all the protein-coding genes, as well as most small nuclear RNA (snRNA) genes.

During the differentiation of a cell type, specific sets of genes are turned on, while others are turned off. Regulatory genes which control the establishment and maintenance of a specific cell type and genes which encode cell-type-specific proteins (which may be needed in massive amounts) are among those genes that are activated. Examples of the latter type of proteins are the hemoglobins of erythroid cells, insulin of pancreatic β cells, immunoglobulins of B lymphocytes, and actin and myosin of muscle cells.

DNA-BOUND PROTEINS THAT STIMULATE TRANSCRIPTION

BY RNA POLYMERASE II

Regulatory molecules which are decisively involved in the establishment of different cell types are generally proteins. Therefore, the genes encoding such regulatory proteins, as well as the mechanisms by which those regulatory proteins control transcription of other genes have drawn much attention from molecular biologists over the last few years. We will focus in the subsequent paragraphs entirely on the protein encoding genes—those that are transcribed by RNA polymerase II.

Two types of regulatory sequences which control transcription by RNA polymerase II have been described: **promoters** and **enhancers** (Figure 16-2). Both regulatory sequences are recognized by specific DNA binding proteins. These proteins (**transcription factors**) mark the DNA

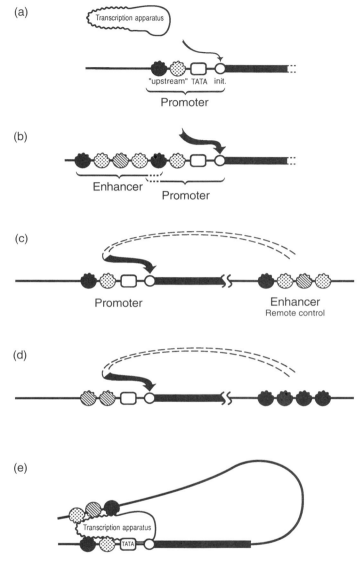

Figure 16-2
Two types of regulatory sequences recognized by DNA-binding proteins (transcription factors) can influence transcription initiation by RNA polymerase II: enhancers and promoters.

for RNA polymerase II—unlike prokaryotic RNA polymerase, eukaryotic RNA polymerase II cannot recognize "naked" regulatory sequences.

The promoter consists usually of a sequence around the transcription initiation site (cap-site or initiator box), an AT-rich sequence (TATA box), and one or more upstream elements which are each about 10 base-pairs in length (Figure 16-2(a)). The initiator box (init.) and the TATA box, together with the bound TATA box factor, direct the transcription apparatus (RNA polymerase II and accessory factors) to begin transcription at the initiation site located about 25 base-pairs further downstream from the TATA box. While all promoters recognized by RNA polymerase II contain one or more upstream elements, they do not always contain a TATA box. Many housekeeping genes, for example, lack TATA boxes and tend to have scattered transcription start sites. The upstream promoter elements and their associated transcription factors control the frequency of transcription initiation and therefore the amount of transcription of the corresponding gene. An increased number of upstream promoter elements usually induce a greater level of transcription.

Enhancers increase the rate of transcriptional initiation. They are located either upstream or downstream of the start site—close to the promoter (Figure 16-2(b)) or thousands of base pairs away (Figure 16-2(c)). These regulatory DNA sequences contain sequence motifs similar to upstream promoter regions and thus can bind some of the same transcription factors. In several cases, when a single upstream promoter element is multimerized experimentally, the resulting synthetic segment behaves as a strong enhancer, i.e., it activates transcription from a remote position (Figure 16-2(d)). This indicates that at least some transcription factors stimulate transcription via a similar mechanism, regardless of whether they are bound to promoter sequences or remote enhancer. Therefore, promoters and enhancers can overlap both physically and functionally (Figure 16-2(b)).

The mechanism by which enhancers exert their effect on transcriptional initiation by RNA polymerase II is still unclear. Most likely, transcription factors bound to a remote enhancer interact with factors bound to the promoter, while the intervening DNA is looped out. According to the so-called looping model (Figure 16-2(e)), transcription factors binding to enhancer and promoter sequences interact indirectly through one or more components of the general transcription machinery, such as the TATA box factor or the RNA polymerase II.

In recent years, a number of genes encoding DNA-binding transcription factors have been cloned, and much has been learned about their structure and function. Interestingly, the overall structure of transcription factors is very similar in species as phylogenetically distant as yeast and humans. Indeed, it has been shown that several yeast and insect transcription factors work well in human cells. **Transcription factors** consist of at least three different domains which specify DNA binding, allow translocation of the protein to the nucleus, and interact with the general transcription apparatus—i.e., the TATA box factor or RNA polymerase II itself. Quite often a transcription factor contains an additional domain for interaction with one or more regulatory proteins that can substantially increase or decrease the activity of the factor (for further details see the section on "Regulation of Protein Activity" later in this chapter).

Some transcription factors are found exclusively in one or a few cell types and help to control the establishment and/or maintenance of specific cell types (e.g. Myo D, found in muscle cells). The experimental expression of the transcription factor Myo D in fibroblasts for example, induces these cells to differentiate into muscle cells. The Myo D gene in conjunction with a few other genes appears to regulate a large set of genes which are necessary to induce such a cell-type transition. Other transcription factors are involved in the production of the above-mentioned abundant cell type-specific proteins. The transcription factor Oct 2A, for example, is found mainly in B-lymphocytes and is involved in the cell type-specific transcription of immunoglobulin genes, which encode antibodies.

There is also a large body of evidence that indicates that transcription factors (especially the **homeobox-containing** proteins) are responsible for many crucial steps in the embryogenesis of invertebrates and vertebrates. It came as a surprise to find that at least some of the morphogens, which were postulated decades ago to establish gradients of tissue or organ formation within the embryo, are transcription factors. For example, the product of the maternally transcribed *bicoid* gene is a transcription factor. This homeobox-containing protein forms an anterior-posterior gradient in an early stage of embryogenesis in the fruitfly, *Drosophila melanogaster*.

GENE TRANSCRIPTION CORRELATES WITH DECONDENSED CHROMATIN AND DNA HYPOMETHYLATION

As mentioned earlier in this chapter, the genome of eukaryotes is packaged into a highly organized nucleoprotein complex, called **chromatin** (see also Chapter 5). This complex, however, is not homogenous throughout the entire genome. Chromatin containing activity transcribing genes is at least partially devoid of histones, the major protein components of chromatin. It is packed much more loosely than inactive chromatin. As a consequence, regions of active chromatin typically exhibit an increased overall sensitivity to experimental treatment with DNase. In addition, these DNase-**hypersensitive sites** often are associated with active genes that are even more sensitive to DNase. These sites may be located in front of, behind, and sometimes within the gene, and they often coincide with promoter and enhancer sequences. Transcription factors bound to these regulatory sequences might prevent tight chromatin packaging and thus allow more easy access by DNase. In addition, some of these transcription factors have been shown to induce DNA to bend, which would make it more vulnerable to nuclease attack.

The characteristic hypersensitive sites of a given gene appear in a cell- and stage-specific manner. They may precede the actual process of transcription, but they are not found in tissues where the gene is always inactive. Hemoglobin genes, for example, have hypersensitive sites in both immature and mature erythroid cells but not in nonproducer cells such as those found in the liver, brain, or muscles. Most likely, hypersensitive sites are formed as soon as the transcription factors that specifically bind to these sites appear in the cell at a given stage of differentiation. The details of this transition are, however, still unknown.

Gene activity correlates not only with a decondensed chromatin structure but also with hypomethylation of the gene. The principle of DNA modification by **methylation** of specific cytosines is exploited by both prokaryotes and eukaryotes. The specific methylation pattern of a given gene (or even a large chromosomal region) can be maintained during many rounds of cell division (Figure 16-3). In bacteria, methylation of the adenine in GATC and cytosine in CCA/TGG sequences can occur. In higher eukaryotes, only the modified DNA base 5-methyl cytosine has been detected thus far. It is typically present in the CG dinucleotide sequence (and also as CNG in plants). In invertebrates (for example, sea urchins), methylation seems to be used as a means to inactivate highly repetitive DNA, which is also constitutively packaged into dense **heterochromatin.**

In vertebrates, cytosine methylation appears to be used in a more sophisticated manner, since the same DNA sequence can be methylated in one cell type, but not in another. The β-globin gene, for example, is methylated in nonproducer cells, but unmethylated in erythroid cells where it is transcribed. Also, the housekeeping genes transcribed by RNA polymerase II are unmethylated even though they are usually particularly rich in CG dinucleotides. These CG rich sequences, usually containing enhancer or promoter sequences, are called **CpG islands.** They are unmethylated in both autosomes and one X chromosome in all cell types. Mammalian females usually contain two X chromosomes. One of these X chromosomes, however, is always packaged into a dense chromatic structure (**Barr body**) and the CpG island promoters on this chromosome are fully methylated. In cell culture, rare variants can be selected that re-express a given gene on the inactive X chromosome. Invariably, such reactivated genes have also lost the methyl groups in the CpG island, consistent with the correlation of methylation and inactivity of the gene.

Differential methylation of single genes or entire chromosome regions can occur in a cell type-specific (e.g. β-globin) or in a chromosome-specific manner (e.g. X chromosome). Most interestingly, in the phenomenon called genomic imprinting, differential methylation patterns of chromosomal regions seem to be established in a germline-specific manner. In other words, the methylation pattern and activity of certain genes can differ, depending on whether the genes are inherited from the

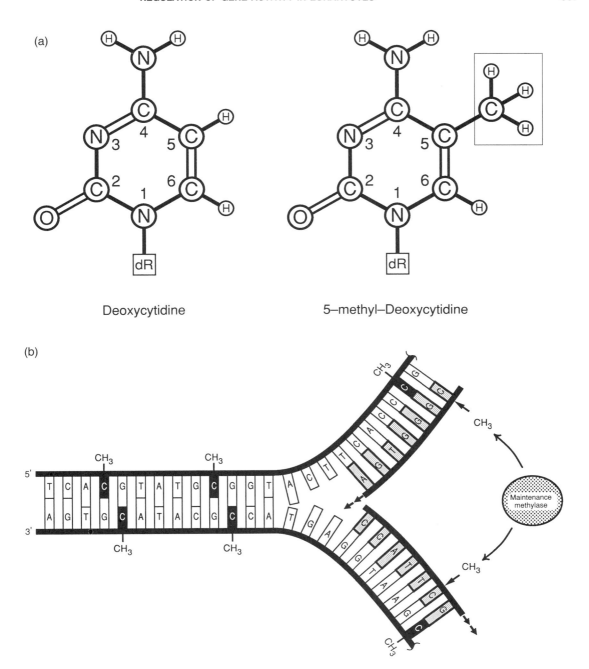

Figure 16-3
(a) Deoxycytidine and 5-methyl-deoxycytidine are indicated (dR, deoxyribose). (b) The methylation, in which the hydrogen atom at carbon number 5 of deoxycytidine is replaced by a methyl group, can be catalyzed by maintenance methylase which recognizes hemimethylated CpG dinucleotides and fully methylates them shortly after DNA replication.

father or from the mother. We are only beginning to understand the meaning of this type of differential methylation. It might explain the lack of parthenogenesis (embryonic development from fatherless eggs which have undergone duplication of the maternal genome) in mammals. Similarly, it is also impossible to create embryos from a duplication of the paternal set of chromosomes. One simple explanation would be that a 1:1 mix of differentially methylated genes inherited from both sperm and egg is required for the proper gene balance and correct development of the embryo. Therefore, differential methylation may prevent parthenogenesis in mammals.

Which regions of a given inactivated gene are methylated? It is obvious that the position of methylation is more crucial than the overall amount of methylation. For example, methylation of a promoter region can abolish activity, while extensive methylation in the protein-coding part of the gene may have no effect on gene expression. Therefore, methyl groups do not act as a "roadblock" to transcribing RNA polymerase II. However, it is not clear if methylation of a promoter generally abolishes binding of transcription factors in a direct manner, or indirectly by inducing tighter chromatin packaging of this sequence. While there is no question about a strict correlation between gene inactivity and DNA methylation, the phenomenon has eluded a thorough analysis for years. Part of the confusion can be attributed to the abnormal methylation patterns often obtained with immortalized cells grown in culture.

The question of whether DNA methylation is a cause or an effect of gene inactivation remains a matter of debate. In several cases, it was found that gene inactivity preceded the addition of methyl groups to the DNA. Therefore, it was argued that methylation was just perpetuating, rather than initiating, the inactive state of a gene. However, model experiments have shown that methylation of the promoter region, and even of binding sites for a single transcription factor, can result in a complete loss of gene activity. The mechanism that controls gain or loss of a methylation pattern during cell type-specific differentiation is one of the most challenging problems of contemporary research on vertebrate development.

The Regulation of RNA Processing

ALTERNATIVE SPLICING

The discovery of split eukaryotic genes in 1977 was completely unexpected, since previous studies with bacterial genes had revealed strict colinearity between a gene and its product. In most genes of higher *exon*

eukaryotes, the coding sequences (**exons**) are interrupted by non-coding intervening sequences (**introns**). Therefore, the primary transcripts of such a gene have to be processed in order to obtain the mature messenger RNAs containing a reading frame that is colinear to the gene product. While introns can be found in most of the genes in higher eukaryotes, they are rare in lower eukaryotes, e.g., in yeast. Are introns a relatively recent development in the genes of higher eukaryotes? It is generally believed that the opposite is true: introns are probably almost as old as life itself, although their structure might have changed through time. Since, in general, organisms with a short generation time tend to have fewer and shorter introns, it seems most likely that introns have been widely lost in prokaryotes and lower eukaryotes as an adaptation to their short generation time and pressure to minimize genome size.

RNA splicing

Splicing, i.e. the removal of intron RNA from the primary transcript to generate the mature mRNA, is an extremely precise process. Eukaryotes have evolved a splicing machinery (called a **spliceosome**), consisting of several protein/RNA complexes (called **small nuclear ribonucleoprotein complexes** or **snRNPs**) as shown in Figure 16-4. The cell cannot afford to miss any of the splice junctions by a single nucleotide, because this would result in an interruption of the correct reading frame, leading to a truncated protein. Skipping of exons, such as direct splicing from the 5′ site of exon 1 to the 3′ site of exon 3, also has to be avoided. It came as a surprise, therefore, to find that whenever a correct 5′ or 3′ splice site was eliminated by mutation, splicing nevertheless occurred, albeit to previously undetectable, cryptic splice sites within the same transcript. These cryptic splice sites seem to be secondary choices due to a lower match to the splicing consensus sequence, or to an unfavorable RNA structure. Obviously the splicing machinery is able to distinguish between stronger and weaker splice sites and chooses the best one available. In a number of human diseases, loss of the function of a necessary protein is a consequence of a mutation in a correct splice site. This is best documented in several thalassemias, where there is an underrepresentation of hemoglobin α or β chains. In some of these genetic diseases globin production is reduced or eliminated by mutation of a splice site, resulting in an RNA that undergoes erroneous splicing to cryptic sites and can thus no longer be translated to the correct protein.

Even though splicing in principle has to be very precise, a given set of splice sites is not used in all cells under all circumstances. In mammalian cells, alternative modes of splicing (frequently cell type-specific) are often used to produce several protein species of a given gene which have overlapping, but distinct functions. Even though the general splicing machinery is the same in all cells, the eukaryotic cell has evolved a strategy to produce **differentially spliced** versions of a primary transcript. This does not involve just one mechanism, but rather several, depending on the mode of **alternative splicing.** Different modes of alternative splicing are summarized in Figure 16-5:

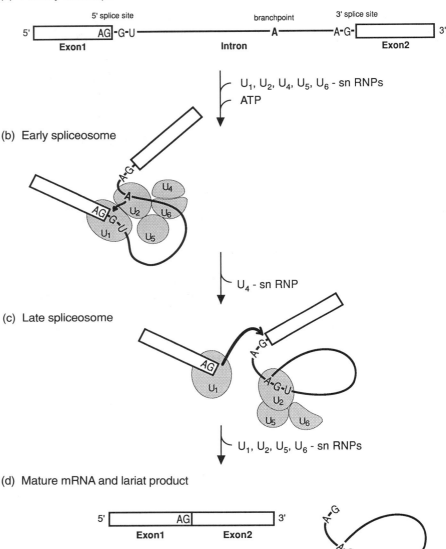

(a) Primary transcript

5' splice site   branchpoint   3' splice site

5' [Exon1 AG]-G-U————————A————A-G-[Exon2] 3'

Exon1   Intron   Exon2

U_1, U_2, U_4, U_5, U_6 - sn RNPs
ATP

(b) Early spliceosome

A-G
A
AG-G-U U_4
U_1 U_2 U_6
U_5

U_4 - sn RNP

(c) Late spliceosome

AG
U_1
A-G
A-G-U
U_2
U_5 U_6

U_1, U_2, U_5, U_6 - sn RNPs

(d) Mature mRNA and lariat product

5' [Exon1 AG | Exon2] 3'

Exon1   Exon2

A-G
A-G-U

Figure 16-4
The production of mature messenger RNA (mRNA) by splicing of the primary transcript. RNA splicing occurs in large ribonucleoprotein complexes known as spliceosomes, which are assembled from abundant small nuclear ribonucleoprotein particles (snRNP) which recognize specific sequences on the primary transcript (a) and (b). The first step in splicing (b) involves cleavage at the conserved 5' splice site sequence with the concomitant covalent joining of a conserved guanosine (G) residue at the 5' end of the intron via a 2',5'-phosphodiester bond to a site within the intron known as the branch point sequence, yielding exon 1 and intron-exon 2 intermediates. In the second step of splicing (c), cleavage at the conserved 3' splice site occurs, and the two exons are joined by ligation, creating the mature mRNA and the excised intron which is also called a lariat product because of its shape (d). Subsequently the spliced (mature) mRNA is transported to the cytoplasm, while the lariat product remains in the nucleus and is degraded.

(a) Alternative selection of promoters

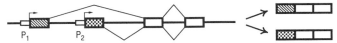

(b) Alternative selection of cleavage/polyadenylation sites

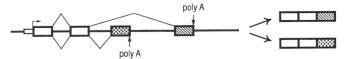

(c) Intron retaining mode

(d) Exon cassette mode

Figure 16-5
Different modes of alternative splicing.

(a). *Differential promoter selection* due to cell type-specific transcription factors can dictate the splicing pattern. In the longer primary transcript, which has been started from promoter P1, the (stronger) 5' splice site overrides the second one. Genes for a myosin light chain as well as an amylase are regulated in this manner.

(b). *Differential cleavage/polyadenylation site selection* can also determine the splicing pattern. The selective usage of a particular poly(a) site, in a cell type-specific manner, creates a longer or shorter primary transcript. Accordingly, the splicing apparatus uses the most downstream (stronger) 3' splice site, as long as it is available. In this way, a differential choice of poly (A) sites results in an alternative splicing of immunoglobulin gene transcripts. The product is an antibody molecule that is either membrane-bound or is secreted into the serum. The same mode of alternative splicing is found in α and β tropomyosin (cytoskeleton proteins) and in calcitonin/CGRP genes.

(c). An example of the **intron retaining mode** is the cell type-specific splicing of primary transcripts of the P element transposase gene in *Drosophila*. Transposition of this P element significantly contributes to the mutational load of the organism. Perhaps for this reason, transposition in somatic cells is abrogated. The transposase gene is transcribed in somatic cells, but the third intron remains part of the mature mRNA, which therefore codes only for a truncated protein (the third intron contains a stop codon; therefore the reading frame is shorter). A mature mRNA with a full-length open reading frame is only produced in germ cells by also splicing out the third intron. This leads to the production of competent transposase exclusively in the germline and therefore a low level of transposition activity in the fly.

(d). In the **exon cassette mode,** some exons can be included or excluded independently of other exons, and usually the same reading frame is maintained whether the exon is spliced out or not. This mode was found in the genes for neural cell adhesion molecules (N–CAMs) and also in troponin-T genes.

In (a) and (b), alternative splice patterns are obtained not by modification of the constitutive splicing activity, but by differential promoter or polyadenylation site selection. However, the intron retaining and the exon cassette modes (c and d) must involve cell type-specific factors that modify the constitutive splicing activity on certain transcripts. It is conceivable that a protein factor binds at a splice site and thus prevents splicing by steric hindrance. The splicing machinery, in an opportunistic manner, would then just choose the next available splicing site. The converse situation, namely a factor changing the structure of the transcript such that a previously inaccessible site becomes available for splicing, is also conceivable.

In mammals, more and more splicing variants of a great number of proteins are being discovered. Most recently, it has become obvious that some DNA-binding transcription factors can also occur as a family of multiple splice variants. Some of these variants seem to serve overlapping but distinct functions. Alternative splicing can create two factors while both bind to the same DNA sequence, one harbors an activating domain while the other lacks it and thus works as an antagonist of the activator, i.e. as a repressor of transcription. Sex determination in *Drosophila,* and possibly other organisms, is also controlled by alternative splicing. During embryogenesis, there is a cascade of gene activations, each leading to correct splicing of the mRNA of the next gene.

TRANS-SPLICING

RNA splicing occurs usually in *cis,* i.e. exons of the same transcript are joined. When **trans-splicing,** namely the splicing of an exon of one transcript to another exon that is located on a different transcript, was shown to be possible in model experiments, it was considered a curiosity, rather than a natural occurrence. It was originally argued that if *trans*-splicing were possible under natural circumstances, it would wreak havoc among cellular transcripts, leading to hundreds of thousands of new protein combinations. However, as was seen to be the case with surface antigen proteins of trypanosomes, orderly *trans*-splicing is possible by selective matching of two specific RNAs yielding a mature mRNA. *Trans*-splicing has also been found in nematodes but not as yet in higher eukaryotes. It may well be a remnant of an ancient condition where new proteins could be generated from different exons by exon shuffling at the *trans*-splicing (RNA) level as well as the genome (DNA) level.

RNA EDITING

Viruses and lower eukaryotes have been found to exhibit several unexpected mechanisms for the control of gene expression. Perhaps the most spectacular one is **RNA editing,** which was first found in the mitochondria of the unicellular trypanosomes, the blood parasite that causes sleeping sickness. For several of their mitochondrial genes it was found that the corresponding primary transcripts were extensively modified by insertion (or deletion) of uridine nucleotides, to yield the final messenger that could be translated into protein.

Another type of RNA editing has been discovered in paramyxoviruses (e.g. in measles viruses). In this case stuttering of the RNA-dependent RNA polymerase at a particular region results in the addition of non-coded G residues, which shifts the reading frame, and yields mRNAs encoding one or more additional proteins.

A cell type-specific form of editing has been found in mammalian cells. The messenger RNA for intestinal apolipoprotein B is modified at a single base-position by specific replacement of a C by a U residue in intestinal (but not liver) cells. This converts a glutamine codon (CAA) to a stop codon (UAA) and thus leads to a truncated protein. This base change may be caused by a deaminase recognizing some structural feature of the mRNA in a cell type-specific manner. However, it remains to be determined if this post- or co-translational modification takes place in the nucleus or the cytoplasm. A similar type of editing might occur in some transcripts of plant mitochondria. In these transcripts, editing changes the genetic code for arginine (CGG) into the code for tryptophan (UGG).

Regulation of Nucleocytoplasmic mRNA Transport

One of the most prominent features of eukaryotic cells is the existence of a membranous nuclear envelope that creates distinct nuclear and cytoplasmic compartments. The local separation of transcription and mRNA processing in the nucleus and translation in the cytoplasm greatly increases the regulatory potential of eukaryotes. The nucleocytoplasmic translocation of macromolecules such as the RNAs and proteins itself provides a further level of gene regulation. We are just beginning to understand the importance of this process.

The nuclear envelope consists of two concentric membranes, whereby the outer membrane can be regarded as an extension of the **endoplasmatic reticulum (ER). Nuclear pore complexes,** which are generally believed to be involved in the translocation of different types of molecules in both directions, are evenly spread over the nuclear envelope (2000–4000 per nucleus). Each nuclear pore complex is highly organized in an eight-fold symmetric structure. It consists of a large number of different proteins with a total molecular weight of $25-100 \times 10^6$ daltons, and is about 100 nm in diameter. The pore itself is an aqueous channel and has a diameter of about 10 nm and a length of about 15 nm. It allows free diffusion of molecules of up to about 20,000 daltons. However, larger molecules have to be actively transported through these pores, probably by receptor-mediated gating of the otherwise too-narrow channels. Specific transport occurs in both directions: some proteins and RNAs are unidirectionally transported, while others seem to be able to shuttle between the two compartments, depending on the type of macromolecule and the physiological state of the cell.

During and immediately after synthesis, the nuclear RNA is covered by RNA binding proteins forming **hnRNPs (heterogeneous nuclear ribonucleoprotein particles**). It has been postulated—although not proven—that hnRNPs bind to receptors thought to be located on the nucleoplasmic side of the nuclear pore complexes through which they are exported to the cytoplasm. During translocation, at least some of the proteins might get stripped off the RNA and remain in the nucleus. Studies performed primarily with yeast have revealed a relation between splicing and transport. Primary transcripts without introns leave the nucleus immediately after synthesis (which includes capping and polyadenylation). Intron containing transcripts, however, need to be spliced first. The stronger the affinity of a pre-mRNA for splicing factors, the less efficient the transport of the unspliced RNA out of the nucleus seems to be. Splicing factors which bind to the RNA appear to prevent export until splicing is completed. RNA mutants which are not recognized by splicing factors, however, are easily exported.

In spite of the inverse correlation between spliceosome assembly and nuclear export of intron-containing RNAs, the cell has ways to export mRNAs which still contain one or several introns. This is documented by the intron retaining mode of alternative splicing (Figure 16-5(c)). The regulation of export of unspliced RNA might happen in at least two ways: Specific RNA binding proteins might directly block the spliceosome assembly, or some regulatory protein might bind to an RNA sequence distinct from splice sites and then transport the pre-mRNA to another subcompartment of the nucleus where no splice factors are available. So far there is no cellular or viral protein known which would function according to the first mode. However, studies of the

endoplasmatic reticulum (ER)

human immunodeficiency virus (HIV) have revealed the existence of a viral protein called Rev which likely functions according to the latter mode. Rev is a phosphorylated protein of 19,000 daltons. Its expression increases the cytoplasmic level of unspliced viral RNA. Rev binds to the RRE (rev-responding element) of the HIV RNA, but does not directly regulate splicing. Instead, Rev seems to activate nuclear export of the viral pre-mRNA by translocating it first into another subcompartment of the nucleus which might be deficient of splice factors. This model is particularly appealing since the subnuclear localization of Rev (mainly in the nucleolus) and the spliceosomes (in splice islands all over the nucleus) are clearly distinct. Eukaryotic viruses have often provided the first glimpses of novel cellular regulatory mechanisms. It remains to be seen if Rev is the harbinger of a new class of cellular regulatory proteins.

Regulation of mRNA Stability

Most prokaryotic mRNAs are short-lived, i.e., they are degraded within minutes. This allows for rapid responses to changes in the environment by transcription of another set of genes. In eukaryotes, the longevity of individual mRNA species can be very different, but in general eukaryotic mRNAs are more stable than prokaryotic mRNAs. The half-life (or time required for 50% of the starting level to disappear) of β-globin mRNA, for example, is more than 10 hours. On the other hand, many mRNAs encoding regulatory proteins such as growth factors or the products of the protooncogenes (e.g., *fos* and *myc*), are rather short-lived with a half life between 15 and 30 min. Such short-lived mRNAs share an evolutionarily conserved AU-rich sequence of about 50 bases in their 3' nontranslated region. When such a sequence motif is genetically engineered into the 3' nontranslated region of the β-globin messenger, this mRNA becomes short-lived. Interestingly, eukaryotes have exploited mechanisms to differentially regulate the stability of a given mRNA in response to environmental stimuli or to changes of the physiological state.

Steroid receptors, such as the estrogen or the glucocorticoid receptor, are well-known as hormone-dependent transcription factors, which stimulate transcription of their target genes upon addition of hormone. Surprisingly, steroid receptors can also increase the stability of certain mRNAs in a manner which is not understood. For example, the half-life of the mRNA for vitellogenin (the major protein in eggs of the frog *Xenopus laevis*), increases about 30 times upon addition of estrogen.

Conversely, the addition of iron to cells decreases the stability of the transferrin receptor mRNA (Figure 16-6(a)). As a consequence, less transferrin receptor is made. This is reasonable from a physiological point of view, since the transferrin receptor is an iron scavenging protein which increases the intracellular concentration of iron. The regulation of the stability of the transferrin receptor mRNA is mediated by an iron-responding factor (IRF). At a low intracellular iron level, this protein binds to a stem and loop structure at the 3′ end of the mRNA. However, upon addition of iron, the factor is released and unmasks, most probably a site which then is attacked by an unknown nuclease. Interestingly, the same factor also controls translation of ferritin mRNAs (Figure 16-6(b)). Ferritin binds free iron within the cell. With a low level of intracellular iron, IRF binds the 5′ end of the mRNA and blocks translation (for further details see the section on "Regulation of Translation" later in this chapter).

The control of mRNA stability has also been studied extensively for the mRNA that encodes histones (Figure 16-6(c)). During the S-phase (DNA replication phase) of the cell, there is a great demand for histone proteins. Not unexpectedly, therefore, histone mRNA stability (and also transcription) is cell cycle regulated. In the S-phase, the half life of the histone mRNA is about one hour. As soon as replication is completed or artificially blocked with chemicals, the histone mRNA is degraded within minutes. The regulation of histone mRNA stability depends upon a short 3′ stem and loop structure that is specific for these RNAs and replaces the poly(A) tail found in other mRNAs. Replacing the 3′ region of the β-globin with the 3′ region of a histone mRNA is sufficient to make the β-globin mRNA stability DNA replication-dependent. The exact mechanism of this replication-dependent mRNA-stability regulation is not yet understood. Interestingly, degradation is dependent on ongoing translation, as is also the case for the above-mentioned instances. If translation stops more than 300 bases upstream of the actual stop codon (due to the experimental insertion of a stop codon), the histone mRNA is no longer rapidly degraded in non-S-phases. Therefore, it has been suggested that a ribosome-associated nuclease is somehow involved in the degradation. The hypothesis would explain why most unstable mRNA is selectively stabilized when cells are treated with the protein synthesis inhibitor cycloheximide.

The regulation of α and β tubulin mRNA stability is also intimately linked to translation (Figure 16-6(d)). The α and β tubulin proteins are the principal subunits of microtubules. The degradation of their mRNA is auto-regulated since it is dependent on free α and β tubulin proteins. As shown for β tubulin mRNA, the sequence of the mRNA necessary for auto-regulated instability is located, unexpectedly, in the first 13 bases to be translated. Introduction of premature translation termination codons reveals that auto-regulated mRNA destabilization requires β tubulin mRNA to be translated beyond codon 41. In the current model, some cellular component (presumably tubulin itself) binds to the peptide encoded by the first nucleotides of the open reading frame, after which an unknown nuclease is activated which degrades the β tubulin mRNA.

(a) Regulation of transferrin receptor mRNA stability

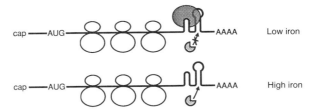

Low iron

High iron

(b) Regulation of ferritin mRNA translation

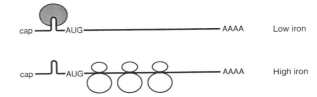

Low iron

High iron

(c) Regulation of histone mRNA stability

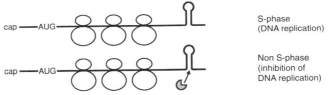

S-phase
(DNA replication)

Non S-phase
(inhibition of
DNA replication)

(d) Regulation of β tubulin mRNA stability

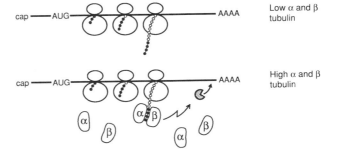

Low α and β
tubulin

High α and β
tubulin

Figure 16-6
Different examples for the regulation of mRNA stability are shown in (a), (c), and (d). (b) displays the regulation of the translation of ferritin mRNAs which involves most probably the same factor (IRF) as used for the regulation of transferrin receptor mRNA stability indicated in (a). Open double circles represent ribosomes; filled circles the IRF protein; notched circles, various nucleases; α and β, the subunits of tubulin.

Regulation of Translation

In contrast to the few well-defined factors required to start translation in *E. coli,* there are a great number of eukaryotic initiation factors (eIFs; see Chapter 10). In prokaryotes, genes of related function, such as enzyme chains for a specific metabolic pathway, are often grouped into operons and transcribed as one contiguous, polycistronic transcript. This transcript is then subject to translational control to produce the appropriate amount of each gene product. Therefore, translational control is of obvious importance in prokaryotes. Nevertheless, translational control is also widely used in eukaryotes, even though one might have thought that eukaryotic genes, which typically are monocistronic, could be sufficiently regulated at the transcription/processing level.

A very well-studied case of negative translational control is the ferritin system (see also the section on "Regulation of mRNA Stability" earlier in this chapter; Figure 16-6(b)). As described on pages 376 and 377, ferritin is an iron storage protein. Its intracellular concentration is tightly controlled at the level of translation, and is dependent on the intracellular iron concentration. If there is a low intracellular level of iron, the iron responsive protein factor (IRF) binds to a stem and loop structure at the 5′ end of the ferritin mRNA. In this way IRF blocks translation, most likely by physically blocking the binding of the translation machinery to the initiation codon AUG. However, upon iron addition, IRF no longer binds to the RNA and translation of the ferritin mRNAs is increased 100 fold.

Negative translational control is very important in embryogenesis of higher eukaryotes. A great number of mRNA species which are synthesized during development of the oocyte is deposited as so-called maternal mRNA in the eggs of higher eukaryotes. These messengers are stored in an inactive form and, upon fertilization, are activated for translation by an unknown mechanism.

The translational controls discussed so far influence the rate of translational initiation. Usually, protein synthesis goes on automatically once translation is initiated. However, nature also has exploited exceptions to this rule (e.g., with the phenomenon of translational frameshifting found in retroviruses and in a coronavirus). In *Rous sarcoma* virus, for example, 95% of the protein translated from the *gag-pol* mRNA is the major glycoprotein antigen (*gag*). However, a few percent are translated into a longer combined protein (*gag-pol*) which encodes the characteristic retrotranscriptase of the virus. How is this 20:1 ratio of *gag* to retrotranscriptase (*gag-pol*) regulated? A ribosome that encounters a specific sequence shortly before the UAG terminator triplet of the gag coding sequence is kicked out of its reading frame and moves back by one nucleotide. From there it continues translation of the *pol* (retrotranscriptase) moiety in a "minus 1" reading frame. A similar process also has been observed with the human immunodeficiency virus HIV. This **trans-**

lational frameshifting, as it turns out, is only one way to ensure the creation of two polypeptides from one mRNA. In another retrovirus (Moloney leukemia virus) the ribosome reads across a UAG stop codon in 5% of the cases, inserts glutamine instead of terminating, and continues translation into the polymerase moiety which, in this case is in the same reading frame.

Regulation of Protein Activity

TWO PRINCIPAL PATHWAYS FOR INTRACELLULAR PROTEIN TRANSLOCATION

The regulation of protein activity comprises mainly two types of processes, namely the translocation of proteins into different cellular compartments and reversible or irreversible enzymatic modifications. These apparently different cellular functions are intimately related with each other.

The subcellular translocation of newly synthesized polypeptides occurs via two major pathways. In the cytosolic pathway the entire protein is translated in the cytoplasm (membrane-free translation). Subsequently, the product is either taken up by one of four different organelles (nucleus, mitochondria, peroxisomes, or chloroplasts), or it remains in the cytoplasm. The uptake into the organelles occurs most probably via organelle-specific receptors which recognize short peptide sequences of the proteins (called translocation signals). The proteins which remain resident in the cytoplasm do not contain a specific translocation signal.

The second pathway is through the endoplasmic reticulum or ER. The decision for this pathway is made during translation. A protein destined for this pathway usually contains a hydrophobic signal sequence at its N-terminus. This signal sequence is recognized by the signal recognition particle (SRP), a protein/RNA complex, which in turn docks the polysome (mRNA with the translation machinery) to the SRP receptors of the ER membrane. During ongoing translation (now membrane-bound), the newly-synthesized polypeptide is translocated into the lumen of the ER. The proteins are either completely secreted into the ER lumen or, if they are transmembrane proteins, they remain bound to the ER membrane. Subsequently, most proteins are transported to the Golgi apparatus and from there either to the lysosomes, secretory vesicles, or directly to the plasma membrane. These translocations typically occur via vesicles which fuse with the membrane of their subcellular target and release the protein content into the lumen of the new compartment, or into the extracellular space. All plasma membrane proteins (e.g., receptors of peptide hormones) or enzymes and proteins that are secreted (e.g., digestive enzymes of the intestinal tract) are translocated via the ER pathway.

polyribosome (polysome)

POSTTRANSLATIONAL MODIFICATIONS AND THE REGULATION
OF PROTEIN STABILITY

A great number of posttranslational modifications of amino acid side-chains have been described in proteins. For most of these modifications the function is not known. Some of them may be unimportant chemical side reactions, but others may have important functions, since they are linked to specific enzymes. Any given modification occurs preferentially or exclusively in one or the other of the two translocation pathways and in specific cell compartments. Some modifications are permanently required and are irreversible (e.g., attachment of prosthetic groups to some enzymes). Other covalent modifications are reversible and allow rapid responses to extra- and intra-cellular stimuli. The most prominent reversible modification appears to be protein phosphorylation. Other frequently found modifications are acetylation, methylation, and glycosylation. The importance of protein phosphorylation was discovered through its ability to regulate enzymes involved in intermediary metabolism. However, it is clear that phosphorylation also is exploited extensively in other systems, such as signal transduction or cell-cycle regulation. Phosphorylation can either increase or decrease the activity of the target protein.

Posttranslational modification processes are also evident in the regulation of protein stability. Some cytosolic proteins are stable for several days, and then are randomly degraded and replaced by new ones. On the other hand, rate-limiting enzymes of a metabolic pathway often have half-life times of only a few minutes. Other short-lived proteins are the products of the protooncogenes such as *fos* or *myc*, which probably play important roles in the control of cell growth. Since these proteins are degraded so rapidly, their intracellular concentration can be changed very quickly by changing the rate of synthesis according to extracellular stimuli. Usually these types of proteins also have rather unstable mRNA (see the section on "Regulation of mRNA Stability" earlier in this chapter). Proteins seem to bear not only a translocation signal, but also sequences which determine their half-life. The half-life of a given protein appears to be a consequence of some destabilizing sequences and general features of the secondary structure. Interestingly, in some model experiments, a strict correlation between the instability of a given cytosolic protein and the first amino acid at its N-terminus has been found. The amino acids Met, Ser, Thr, Ala, Gly, Val, and most probably also Cys and Pro, have a stabilizing effect on the protein when located at its N-terminus. All other amino acids make the protein vulnerable to a protease that rapidly degrades it.

The proteolytic machinery responsible for selective protein degradation is ubiquitin-dependent. **Ubiquitin** is a small protein of 76 amino acids. When the ubiquitin-dependent proteolytic machinery recognizes a destabilizing amino acid at the N-terminus of a protein, a ubiquitin molecule is covalently attached either to the N-terminus itself, or to a nearby lysine. Subsequently, a series of additional ubiquitins are added producing a multiubiquinated, branched protein. A specific ATP-consuming protease recognizes such ubiquinated proteins and degrades them while recycling the ubiquitin polypeptides.

As already mentioned, ubiquitin–dependent protein degradation is specified by the first amino acid at the N-terminus of each protein. Although in all eukaryotic protein genes the first codon (AUG) codes for a methionine, very few mature protein species have a methionine at their N-terminus. Usually the N-terminus is cotranslationally modified by trimming and end-addition activities which are not well understood. Interestingly, proteins entering the ER pathway often bear an N-terminus, making them highly unstable in the cytosol, but not in the ER. It might be very important in allowing the cell to be able to specifically degrade proteins that have entered the cytosolic pathway by accident.

POSTTRANSLATIONAL REGULATION OF TRANSCRIPTION FACTOR ACTIVITY

A transcription factor binds to an enhancer or a promoter in a sequence-specific manner. Subsequently, transcription of the linked gene is stimulated (see the section on "The Regulation of Transcription Initiation" earlier in this chapter). Recently, a number of transcription factors and their corresponding genes have been studied. While some transcription factors, especially those stimulating the transcription of housekeeping genes, are constitutively active, several transcription factors are now known whose activity is regulated at the posttranslational level. The regulation involves differential subcellular translocation, stable interactions with other proteins, and enzymatic modifications.

Perhaps the best characterized case is the **glucocorticoid receptor (GR)**, which is a steroid-(glucocorticoid) inducible mammalian transcription factor (Figure 16-7). In the absence of hormone, GR is complexed to the 90kD heat shock protein (hsp90), and is located in the cytoplasm. Therefore, hsp90 could be regarded as a cytoplasmic anchorage protein of GR. When the steroid hormone glucocorticoid diffuses into the cell, it binds to the hormone binding domain of GR. This causes the release of GR from hsp90, most probably because of a conformational change induced by binding of the hormone. The release leads to the exposure of a nuclear translocation signal of GR. The transcription factor is translocated into the nucleus via nuclear pore complexes. In the nucleus, GR binds to its target sequence and stimulates transcription of the GR-responsive genes.

A similar type of regulation has been found with studies of NF-κB, a mammalian transcription factor which induces transcription of genes involved in immune defense or inflammation. It is found in a number of different mammalian cell types, often in the cytoplasm in an inactive form. Similar to GR, inactive NF-κB forms a complex with a cytoplasmic anchorage protein, designated as I-κB (Inhibitor of NF-κB).

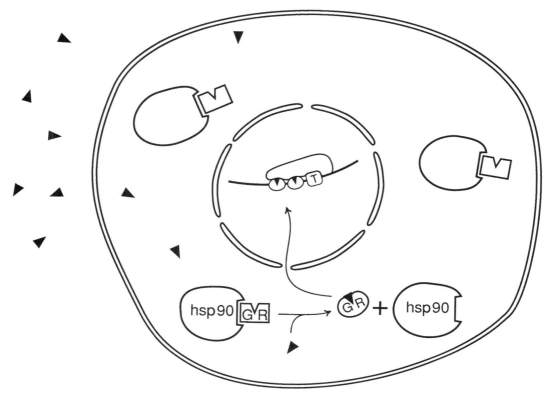

Figure 16-7
The glucocorticoid receptor: a paradigm for the posttranslational regulation of a transcription factor. Solid triangles represent glucocorticoid molecules.

NF-κB is posttranslationally activated upon addition of physiological agents which impair cellular functions (e.g., agents such as UV, DNA-damaging chemicals; viruses, parasites, or lipopolysaccharides of bacteria). This activation probably involves a cascade of regulatory processes which finally result in the phosphorylation of I-κB. Following this, NF-κB is released and is translocated into the nucleus, where it stimulates transcription of its target genes.

Other transcription factors can be located in the nucleus but nevertheless remain inactive. The regulation of the activity of the yeast transcription factor Gal4 is an example. Gal4 has the potential to stimulate genes that encode proteins involved in galactose metabolism. Gal4 always binds to its DNA target, but in the absence of galactose it cannot activate transcription. This is due to the Gal80 protein, which stably interacts with Gal4 and covers its transcription-activating domain. Upon galactose addition, Gal80 is released and Gal4 can stimulate transcription. In contrast, the activity of the mammalian transcription factor Oct-1 can be specifically augmented by the protein VP16, which is encoded by herpes simplex virus and contains a very strong activation domain. VP16 can form a complex with Oct-1 and thus stimulate transcription by virtue of its own very strong activation domain. Neither Gal80 or VP16 bind to DNA by themselves.

The transcriptional activity of a gene can also be influenced by direct enzymatic modification. The yeast heat-shock transcription factor HSF can stimulate transcription of genes which are essential for heat-shocked cells. HSF always binds to its target HSE (heat shock responsive element). However, HSF is only active when the cells have been incubated at a high temperature. The activation of HSF strongly correlates with its phosphorylation. So far it is not clear if this modification helps to expose an activating domain of HSF, or if the phosphorylated region itself acts as an activating domain.

Gene Rearrangement: Joining of Coding Sequences in the Immune System

The life of a multicellular organism is under constant threat by all kinds of microorganisms and viruses. Most of these are warded off by the skin, or by secretions of the digestive tract, such as hydrochloric acid in the stomach, and digestive enzymes. Quite often, however, a virus enters the body via a sensitive mucus epithelium or via small lesions in the skin. Once inside the body, the invader might initiate a deadly race with the immune system of the infected organism. The immune system of vertebrates is designed to offer a selective defense against all possible infectious agents, and to be ready even against those that might appear on earth in the future. How is this possible? Several million different protein molecules, **immunoglobulins** or **antibodies** (see Chapter 4), are produced, whereby each can bind a distinct infectious target molecule or **antigen** in a specific manner. If each antibody molecule were encoded by a distinct gene, a significant fraction of the entire genome would have to be used for antibody synthesis. The number of antibody genes (about 10^8) would even exceed the estimated total number of human genes (about 10^5). However, the immune system has devised another way to produce such a huge variety of antibodies. In general, a mature antibody-producing cell (B-lymphocyte) makes only one type of antibody. Hence, there must exist some mechanism that is responsible for programming each cell. The mechanism involves specific gene rearrangements and is therefore very different from all other modes of gene regulation discussed so far, none of which change the genome. However, the mechanism for generation of antibody diversity is not the same for all vertebrates. For example, birds have evolved a system which is rather different from the one in mammals. In this section we will focus on the mammalian system.

DIFFERENT IMMUNOGLOBULIN GENE SEGMENTS

Immunoglobulin G (IgG) is one of several classes of immunoglobulins. It is a tetrameric protein containing two L chains and two H chains (for the structure of IgG see Chapter 4). Both the L and H chains consist of two regions, the constant and the variable region. The variable region is responsible for recognizing and binding a specific antigen. Note that the antigen-binding site consists of amino acid residues from both heavy and light chains. There are two types of L chains, called κ and λ. We will only describe how the κ type is produced; the basic mechanism for the λ type is similar, differing only in detail.

Three different classes of gene segments are used to form a κ-type L chain, namely V, J, and C segments, which are all located on the same chromosome (Figure 16-8(a)). There are about 300 different V gene-segments (which are responsible for the synthesis of the first 95 amino acids of the variable region), 4 different J gene-segments (which encode the final 12 amino acids of the variable region and join the V and C regions), and 1 copy of the C gene-segment (which encodes the constant region). In an embryonic cell the V segments form a tight cluster, the J segments form a second tight cluster far downstream from the V segment cluster, and the C segment follows not far after the J segment cluster. Note that each V gene-segment is preceded by a promoter from which transcription can potentially be initiated, and that an enhancer is located between the J_4 and the C segment (Figure 16-8(a)).

Regions encoding particular IgG molecules have been cloned by recombinant DNA techniques from various mouse cell lines, each producing a particular IgG molecule. For each clone obtained from an antibody-producing cell line, it has been found that a large segment of the embryonic DNA sequence is absent, and that the missing DNA is always a sequence between the particular V gene-segment and the J gene-segment. This is explained by a gene rearrangement in which DNA is deleted between the V and J segments chosen for recombination. Many different genomic sequences, each encoding a particular κ chain, have been cloned, and each clone lacks a different DNA segment. For example, in Figure 16-8(b) the entire DNA between V_3 and J_2 is absent, while in another clone there might be a V_{210}-J_1 junction instead.

Any of the approximately three hundred V segments can be joined to any of the 4 J segments, so that at least 1200 (300 × 4) different variable regions (or V/J combinations of the κ chain) can be potentially encoded. The H (heavy) chain genes are organized in a similar, but not identical way. In the mouse, for example, there are three types of gene segments: 1000 V, 12 D, and 4 J segments. Therefore, about 48,000 (1000 × 12 × 4) heavy chain variable regions can be potentially encoded upon appropriate DNA rearrangements. These rough calculations (the exact number of V segments in these pools of segments is not known) give about 5 × 10^7 different antigen-binding sites, assuming both the heavy and light chain variable regions equally contribute to the antigen-

(a) Germline DNA

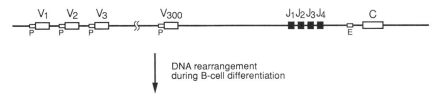

DNA rearrangement
during B-cell differentiation

(b) Rearranged DNA of specific B-lymphocyte

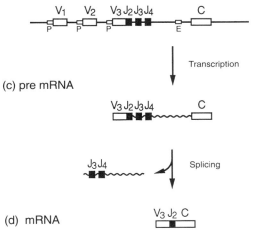

(c) pre mRNA

Transcription

Splicing

(d) mRNA

Figure 16-8
The production of the mature κ-type L chain mRNA of a particular IgG molecule involves
processes of the DNA and the RNA level. During B-cell differentiation, the DNA between
a V and a J segment (in our example between V_3 and J_2) is excised, i.e., the genomic
sequences are specifically changed (compare (a) and (b)). With this step, the enhancer,
E, is moved closer to the promoter, P, of V_3 and the given B-lymphocyte is able to
produce a specific κ-type L chain pre mRNA (c). This pre-mRNA undergoes splicing
whereby the C segment is joined to V_3J_2. This results in the mRNA (d), which is ready for
translation to produce the specific κ-type L chain protein.

binding site. The number of potential antigen–binding sites is even
higher, since the junctions produced by two particular gene segments
(e.g., by linking a given V with a given J segment) is variable. A variable
number of nucleotides often is lost (or sometimes added) from the ends
of the two recombining segments (junctional diversification). This mechanism increases the possible number of different antigen–binding sites
once more by a factor of about 1000. Following such addition or deletion
of nucleotides, only in one of three cases will the correct reading frame
be restored, to yield a functional gene.

HOW ARE IMMUNOGLOBULIN GENE REARRANGEMENTS CONTROLLED?

What is the enzymatic machinery that creates all of these combinatorial variants of immunoglobulin genes? As shown in Figure 16-9(a), the signal sequences for the V-J joining are a pair of specific base sequences—3' of each V segment (heptamer/spacer/nonamer) and 5' of each J segment (nonamer/spacer/heptamer). Presumably a site-specific **recombinase** enzyme recognizes these sequence motifs, cuts the DNA, and joins randomly selected V and J segments (Figure 16-9(b)). The same sequence motifs are used in the heavy chain pool to join V, D, and J segments. Perhaps the most striking aspect of these gene rearrangements is their precision. In order to avoid a scrambling of the entire genome, the recombinase has to be targeted specifically to the immunoglobulin locus. Furthermore, the time window of enzyme activity is quite narrow, being restricted to a specific phase in B-lymphocyte development.

Figure 16-8(b) indicates that the joining of segments can not fully generate an L chain reading frame, since the spacer between the J and C gene segment remains, and the actual L chain has amino acids derived from only one V gene segment and one J segment, and recombined DNA still usually contains many V and J segments. The recombination process however, has brought the enhancer (E) much closer to the promoter of V_3 (P), allowing transcription to start. This leads to the production of primary transcripts (pre mRNA, see Figure 16-8(c)). The correct reading frame is obtained by a final RNA splicing event, as shown in Figure 16-8(d). Note how the particular V-J joint determines the splicing pattern. The RNA removed by splicing is always between the C segment and the right end of the J segment which has been joined to the V segment.

ALLELIC EXCLUSION, CLONAL SELECTION AND AFFINITY MATURATION

Whether the site-specific rearrangement processes happen in advance of, or in response to the occurrence of a specific antigen has been a major question. Today it is generally accepted that the events occur at random in the course of development, and when differentiation is complete, there are several million different antibody-producing B lymphocytes. Each of these cells produces only a single type of antibody, due to the strict feedback-regulation of the site-specific DNA rearrangements. The feedback mechanism designated as **allelic exclusion** allows the production of only one type of rearranged H and L chain gene per cell. However, very little is known about this phenomenon at the molecular level.

How does an organism know when to produce a huge amount of a particular antibody in response to a particular antigen? Each antibody-producing cell makes a small amount of a specific antibody. Some of

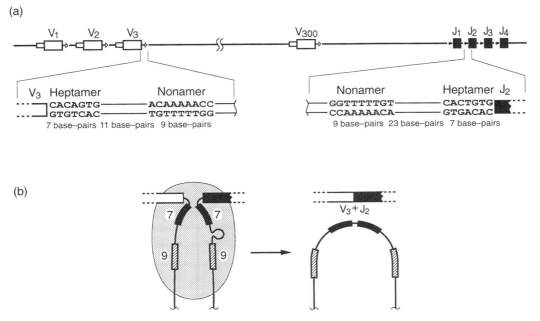

Figure 16-9
(a) The joining of V and J DNA segments during B-cell differentiation is a highly precise process which involves conserved DNA sequences, called heptamer (7) and nonamer (9) sequences. (b) A site-specific recombinase is thought to recognize these sequences, after which the randomly selected V segment and J segment are joined.

this antibody becomes bound to the cell surface. When such a cell is exposed to the specific antigen that can react with its antibody, a complex antigen-antibody network forms. This event stimulates cell division, and extensive production of specific antibodies. This process is called **clonal selection.**

Another interesting mechanism that allows the immune system to increase the affinity of antibodies to a given antigen, again by modifying genomic sequences deserves mentioning. Repeated immunization of an animal with the same antigen results in the production of antibodies which have, on average, a higher affinity to the antigen. This phenomenon is called **affinity maturation.** By somatic hypermutation (the molecular details of which are not understood), specific sequences in the V region accumulate point mutations with repeated exposure to a given antigen. Since B-lymphocytes are stimulated to proliferate in response to the binding of antigens, those cells which bear hypermutations yielding an antibody with higher affinity to the antigen will proliferate and outgrow the cells which produce antibodies with lower affinity. The mutation frequency over an immunoglobulin variable region can be as high as 10^{-5}/cell cycle/base-pair. This frequency would completely inactivate a genome if it were not restricted to the variable regions of active immunoglobulin light and heavy chain genes.

Summary

Regulation of genetic systems in eukaryotes is accomplished in a variety of ways, most of which are quite different from the mechanisms observed in prokaryotes. In eukaryotes, three different RNA polymerases fulfill transcriptional tasks which in prokaryotes are carried out by a single RNA polymerase. It is at this level of transcription where decisions are often made with regard to gene expression. Sequences upstream of eukaryotic genes known as promoters and enhancers are required for initiation of transcription. Activation of transcription is carried out by binding of transcription factors to these sites. Once mRNA is transcribed, further regulation can occur during splicing events that remove eukaryotic introns and create mRNA sequences which can be translated into functional proteins. For example, alternate splice sites can be chosen resulting in several different mRNA species encoded in one gene. The nucleocytoplasmic translocation of these RNAs can then provide a further level of gene regulation. Once the mRNA has arrived in the cytoplasm the variety of their half-lives as compared with the short-lived nature of prokaryotic mRNA allows for further regulation. Eukaryotes have also developed mechanisms whereby the stability of mRNA can be changed in response to environmental stimuli or physiological states. Expression regulation can also be controlled at the protein level through translational regulation, posttranslational modification, and protein stability.

Drill Questions

1. What RNA polymerase(s) transcribe eukaryotic genes? Name the polymerase(s) and the type of gene(s) it transcribes.

2. In prokaryotes, regulatory elements are in fixed positions with respect to the gene(s) regulated. How does the situation differ in eukaryotes?

3. Do the V and J segments of an antibody gene join to form the variable or the constant region of an IgG molecule?

4. List several mechanisms a cell uses to increase the concentration of a particular mRNA molecule to a very high value.

5. How does the organization of chromatin differ with regard to gene activity? What experimental technique can be used to find such regions of activity?

6. Give the two types of splicing that can occur in the processing of eukaryotic mRNA, and state the differences between them.

7. How might a cell be signaled to synthesize a particular protein but no other?

8. Outline the mechanism for generating antibody diversity.

Problems

1. a. If you wanted to redesign the mouse genome to increase antibody diversity, but by adding the least amount of DNA, which component would you increase?

   b. If an organism has 150 different V gene segments, 12 J gene segments, and 3 possible V-J joints for L chains, and can make 5000 different H chains, how many different antibodies can be made?

2. If an enhancer were moved from a position near gene *A*, where it has strong enhancing activity, to a position 50 nucleotides upstream from gene *B* which is constitutively transcribed, would transcription of gene *B* be increased?

3. In prokaryotes, positive regulatory elements often provide a binding site for RNA polymerase, which strengthens the binding of the polymerase to the transcription start site. What is the major evidence that suggests this is not the function of all eukaryotic regulatory elements?

4. A substance (Q) is made in a cell culture in response to an external effector (E). In the absence of E, a nuclear RNA consisting of about 800 nucleotides can be found that hybridizes with the cloned gene encoding Q. The RNA is capped, has a poly(A) tail, and in an *in vitro* protein-synthesizing system directs synthesis of a tripeptide. If E is added to the growth medium, very little of this nuclear RNA is found and an mRNA molecule containing 770 nucleotides is found in the cytoplasm. This mRNA directs synthesis of Q, but not the tripeptide. How might E regulate synthesis of Q?

5. Several cases are known in which a single effector molecule regulates the synthesis of different proteins encoded in distinct mRNA molecules 1 and 2. Give a brief possible molecular explanation for each of the following observations made when the effector is absent.
   a. Neither nuclear nor cytoplasmic RNA can be found that hybridizes to either of the genes encoding molecules 1 and 2.
   b. Nuclear (but not cytoplasmic) RNA can be found that hybridizes to the genes encoding molecules 1 and 2.
   c. Both nuclear and cytoplasmic RNA (but not polysome-associated RNA) can be found that hybridize to the genes encoding molecules 1 and 2.

6. Two proteins are synthesized initially at the same time because the two RNAs encoding each protein are made in response to the same signal. At a later time, when the transcription signal is no longer present, one of the proteins is still made at nearly the same rate, and the other is not detected. Suggest a means for this temporal programming.

7. Does the presence of DNase-hypersensitive sites in DNA always indicate promoter and enhancer sequences? If not, what else might these sites indicate?

Conceptual Questions

1. Since there are a number of diseases resulting from loss of function of a gene product due to a mutation in a correct splice site, how might such diseases be treated using molecular biology approaches? What, if any, negative consequences could these approaches have?

2. Almost nowhere in the biological kingdom are there mechanisms for regulating gene expression which involve permanent changes in the nucleotide sequence of DNA observed. Why do you suppose such mechanisms do not exist?

3. Phosphorylation of proteins is another method used by eukaryotic cells to regulate gene expression. Proteins called kinases phosphorylate amino acids such as tyrosine and serine on other proteins to cause either increased or decreased activity. Similarly, many kinases control their own activity level through autophosphorylation. Two such kinases are c-src and its oncogenic counterpart, v-src. The src kinase phosphorylates tyrosine residues. A c-src kinase contains two tyrosine residues, while v-src contains only one. It is known that the presence of c-src in cells does not lead to a high level of cancer transformation, while the presence of v-src is highly transforming. Propose an explanation for this observation.

Judith A. Jaehning
Assistant Professor of Biology
Department of Biology

Birthday: **14 October 1950**

Birth Place: **Yakima, Washington**

Undergraduate Degree: **University of Washington, Seattle Major: Chemistry 1972**

Graduate Degree: **Washington University, St. Louis, Ph.D. 1977, Biological Chemistry**

Postdoctoral Training: **University of California, Berkeley, 1977–1978, Biochemistry; Stanford University, 1978–1980, Biochemistry**

Present Position: **Indiana University, Biology Department**

Address: **Bloomington, Indiana**

MY LABORATORY STUDIES RNA polymerases, the enzymes responsible for copying portions of DNA into RNA. We study the structure of these complex, multisubunit proteins, and we analyze how they interact with DNA. We are particularly interested in the signals (sequences) in the DNA that direct the RNA polymerases to start and stop in precise positions. We know that additional regulatory proteins are required to "read" the DNA sequences and to tell the RNA polymerase where to start and stop copying. We are also interested in the mechanisms that coordinate the expression of genes in the nucleus and in the mitochondria of eukaryotic cells. We are studying nuclear and mitochondrial RNA polymerases to learn if these enzymes are involved in communication between the organelles.

What were the major influences or driving forces that led you to a career in molecular biology?

I decided to be a molecular biologist when I was about 14, after reading books from the library about viruses. The fact that a single molecule of nucleic acid could reproduce itself was fascinating to me, and I wanted to understand this process of self-propagation. I was fortunate to have excellent science teachers in high school, who encouraged my interest in these areas and introduced me to the complexities of chemistry and biology.

Appendix

Chemical Principles
Important for Understanding
Molecular Biology

The complex structures and functions of the molecules described in this text are a consequence of the chemistry of their component parts. To understand how these molecules function, we need to briefly review some of the fundamentals of chemistry. In particular, we need to understand the unique features of carbon, hydrogen, oxygen, nitrogen, phosphorus, and sulfur, which enable these elements to form the compounds essential for life. We will begin with an introduction to atomic structure and chemical bonds.

Structure of the Atom

Atoms are composed of a nucleus, which contains protons (positively-charged particles) and neutrons (particles with no charge), and electrons, which are negatively-charged particles. The number of protons in the nucleus defines the atomic number and specifies the identity of an element. For example, an atom with six protons has an atomic number of six and has been given the name "carbon." (See a Periodic Table of the Elements). The electrons are added to shells around the nucleus.

Isotopes of the same element all have the same number of protons, but differ in the number of neutrons they contain. In an uncharged atom, the number of electrons and protons is equal.

Chemical Bonds

When individual atoms form chemical bonds between one another, they create molecules. Chemical bonds are important for the structure and properties of small molecules, such as water, and simple salts, acids, and bases; they are equally important for the structure and properties of large molecules, such as DNA, and proteins (see Chapters 2–4). Chemical bonds can be of several different kinds, which are described in the sections below.

COVALENT BONDS

Covalent bonds are formed when electrons are shared between two atoms. The electrons occupy shells that optimally contain the number of electrons shown in Figure A-1. Elements with completely filled outer shells do not readily form bonds with other atoms. For example, helium with two electrons is unreactive. These elements are found in the far right column of a Periodic Table of the Elements. In contrast, the elements found in biological compounds (hydrogen, carbon, oxygen, nitrogen, sulfur, and phosphorus) are uniquely suited to share electrons in covalent bonds, thereby forming very stable structures. A simple example of a covalent bond is shown in Figure A-2, where an atom of hydrogen (atomic number 1), with a single electron occupying its outer shell, combines with another atom of hydrogen to form molecular hydrogen. The two electrons are equally shared, giving each atomic nucleus a filled shell.

As shown in Figure A-1, the second shell can hold eight electrons. This means that oxygen (atomic number 8) has only six electrons in its outermost shell and needs to attract two electrons to share to fill its outer shell. Molecular oxygen, shown in Figure A-3, consists of two oxygen nuclei sharing two pairs or four electrons (dotted circles in Figure

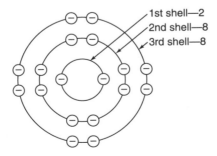

1st shell—2
2nd shell—8
3rd shell—8

Figure A-1
Model of an atom showing electrons in shells around the nucleus.

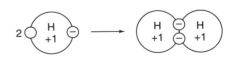

Figure A-2
Hydrogen atom and molecular hydrogen, H_2.

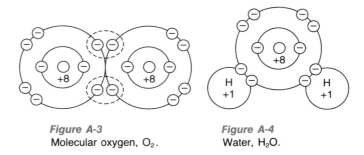

Figure A-3
Molecular oxygen, O₂.

Figure A-4
Water, H₂O.

A-3). This is known as a double bond. In this way, each oxygen gains two electrons and satisfies its need for a total of eight electrons. Oxygen can also share electrons with many other types of atomic nuclei, as can be seen in the molecules described throughout this text. When oxygen shares electrons to form covalent bonds with two hydrogen atoms, water is formed, as shown in Figure A-4. These covalent bonds are very strong, and they have unique properties as described further below.

In Table A-1, we see the atomic numbers and the number of electrons required for a filled outer shell for each of the major elements found in biological molecules. The number of electrons needed to fill an atom's outermost shell reflects the number of bonds in which that atom can engage. Therefore, all the elements listed in Table A-1 except hydrogen can form multiple bonds such that large, complex molecules are possible. Furthermore, the smaller an atom is, the easier it is to share electrons. Since the elements found in biological compounds all have relatively low atomic numbers, they form very strong covalent bonds. The most common bonds are single bonds (in which only one pair of electrons is shared), but double bonds (in which two pairs of electrons are shared) are also important for biological function. Carbon, oxygen, nitrogen, and phosphorus can all form a variety of double bonds. Carbon and nitrogen are also found in rare triple bonds (Figure A-5).

Table A-1
Atomic Configuration of Biologically Important Atoms

Element	Atomic Number	Electrons in Full Shell	Number Needed to Complete Shell
H	1	2	1
C	6	10	4
N	7	10	3
O	8	10	2
P	15	18	3
S	16	18	2

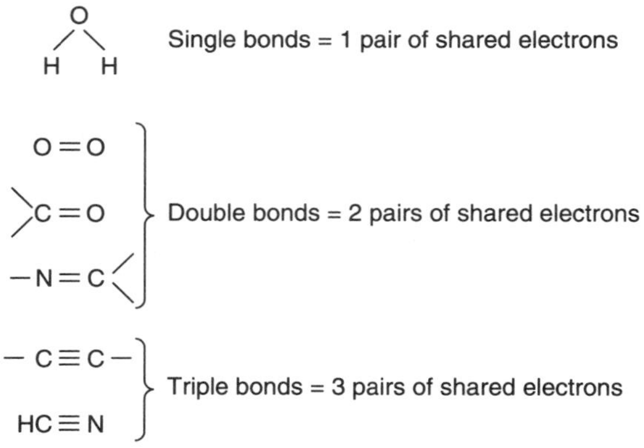

Single bonds = 1 pair of shared electrons

Double bonds = 2 pairs of shared electrons

Triple bonds = 3 pairs of shared electrons

Net charge +1   Net charge −1

Figure A-5
Some examples of single, double, and triple bonds.

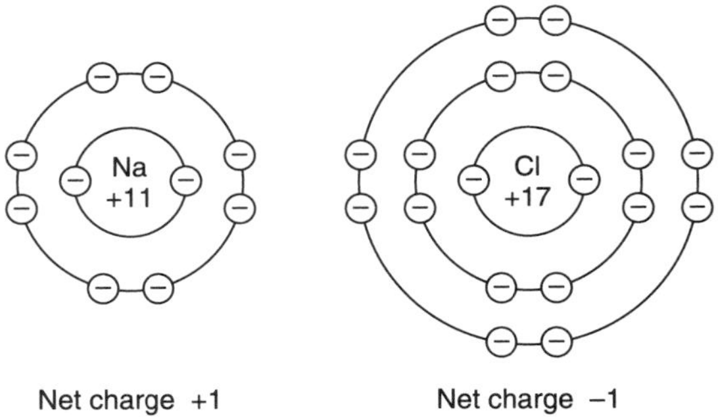

Figure A-6
The ionic bond of sodium chloride, NaCl.

IONIC BONDS

In some compounds, the electrons are not equally shared, but are effectively transferred from one atom to the other. The result is two ionized species that associate due to their opposite charge. The salt that is formed from one atom of sodium and one atom of chlorine is an example of an ionic bond. In sodium chloride, an electron from sodium (atomic number 11) is transferred to chlorine (atomic number 17). The atomic nuclei in the resulting ionic compound each have a completely full outer shell of electrons and opposite charges (Figure A-6). The strength of ionic bonds varies more widely than does the strength of covalent bonds (Table A-2). In biological compounds, we will see many ionized species. In particular, oxygen, nitrogen, and phosphorus are frequently involved in ionic bonds.

POLAR COVALENT BONDS

Many bonds in biological compounds are intermediate between covalent and ionic bonds. This means that the electrons are neither equally shared

Table A-2
Strength of Various Types of Chemical Bonds

Type of Bond	Force Required to Break (kcal/mole)
Covalent	50 to 100
Ionic	80 to 1
Hydrogen	3 to 6
Hydrophobic	0.5 to 3

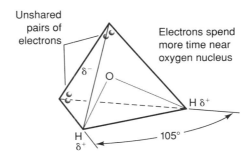

Figure A-7
Geometry and partial charge distribution of water.

nor completely transferred between atomic nuclei. For example, the hydrogen and oxygen in a water molecule are actually joined by polar covalent bonds; however, the shared electrons are held more tightly by the oxygen nucleus than by the smaller hydrogen nucleus. Since the bonds holding the two hydrogens to the oxygen form a 105° angle, there is a charge polarity to the water molecule (Figure A-7). The oxygen side of the molecule is more negatively charged than the hydrogen side, as indicated by the δ symbol, which stands for "partial."

Bonds between hydrogen, carbon, oxygen, nitrogen, and phosphorus are often polar covalent bonds. Hydrogen and nitrogen frequently are associated with a partial positive charge, while oxygen and phosphorus carry a partial negative charge. Carbon can carry either a partial positive or a partial negative charge (Figure A-8). When only carbon and hydrogen are present in a compound, the bonds are not polar, since carbon and hydrogen share electrons about equally. The four bonds formed by carbon show a tetrahedral symmetry (Figure A-9).

The existence of polar and nonpolar covalent bonds in biological compounds results in two additional classes of bonds that are very im-

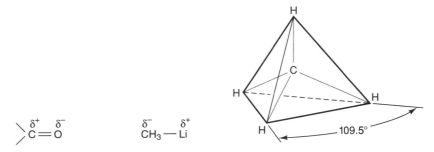

Figure A-8
Partial positive and negative charges on carbon.

Figure A-9
Tetrahedral carbon.

portant for the structure of macromolecules. These are the hydrogen bonds and hydrophobic bonds described in the next two sections.

HYDROGEN BONDS

When hydrogen is involved in a polar covalent bond with another atom, it carries a partial positive charge (Figure A-7). This partial positive charge will attract and associate with molecules (or parts of molecules) with a partial negative charge. For example, in water, the partially positively charged hydrogens will associate with the partially negatively charged oxygens of adjacent molecules. These hydrogen bonds are much weaker than a true covalent bond (Table 2), but when present in large numbers, they can contribute substantially to the stability of a mixture of molecules. As shown in Figure A-10, liquid water is actually a highly ordered structure with each water molecule involved in hydrogen bonds with three other molecules. This stable network gives water many of its unique properties, including its high boiling point and low freezing point. In addition, the ability to form hydrogen bonds with a variety of polar compounds makes water an extremely good solvent for most biological molecules.

Hydrogen bonds are also important for the structure of proteins and nucleic acids, as described in Chapters 3 and 4. Hydrogen bonds can be formed intramolecularly—that is, holding atoms in different parts of one molecule together, and intermolecularly—holding atoms in separate molecules together.

HYDROPHOBIC BONDS

Nonpolar bonds, such as those found in compounds made up only of carbon and hydrogen, also have unique properties. These bonds are

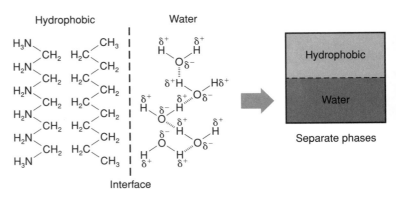

Figure A-10
Hydrogen bonded structure of water.

Figure A-11
Phase separation of hydrophobic molecules and water.

known as **hydrophobic** (literally "water-fearing"), because they do not readily associate with water. Remember that the structure of water is stabilized by the formation of hydrogen bonds. Nonpolar compounds cannot participate in hydrogen bond formation since there is no partial charge associated with their hydrogens. Mixing these compounds with water disrupts existing hydrogen bonds, a situation that is energetically unfavorable. Nonpolar compounds are therefore insoluble in water and tend to form separate phases or layers as shown in Figure A-11. In complex molecules, we will see examples where part of a molecule is hydrophobic and part is polar or **hydrophilic** (which means water-loving). The structure of the molecule is stabilized by the energetics of hiding the hydrophobic portions from water while exposing the hydrophilic portions for maximum interactions with water. These interactions are important for the structures of proteins, nucleic acids, and the macromolecules that make up cell membranes.

hydrophilic

RESONANCE BONDS

One final class of bonds is found in biological molecules. These are the resonance bonds, which are in between the single and double covalent bonds described above. Resonance bonds were first described in synthetic molecules, such as benzene (Figure A-12), where it was determined that electrons were actually shared between a network of atomic nuclei. Resonance bonds are extremely stable and are found in many biological molecules, including the nucleotide bases, which are the building blocks of nucleic acids, and some of the amino acids, which make up proteins. Although many resonance stabilized compounds are very hydrophobic (for example, benzene), sometimes resonance bonds are involved in stabilizing the components of an ionic bond as shown in Figure A-13).

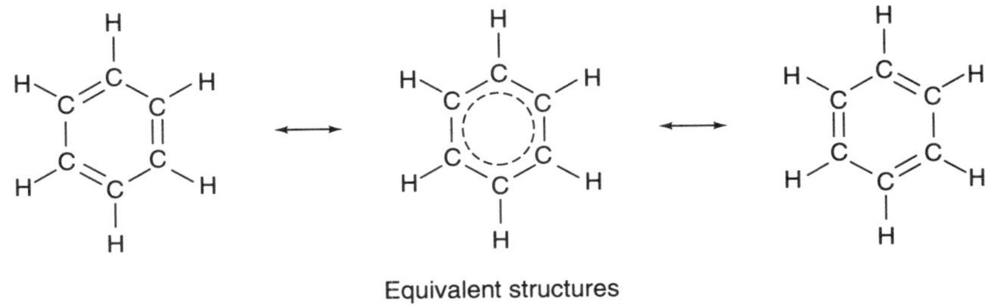

Equivalent structures

Figure A-12
Resonance forms of benzene.

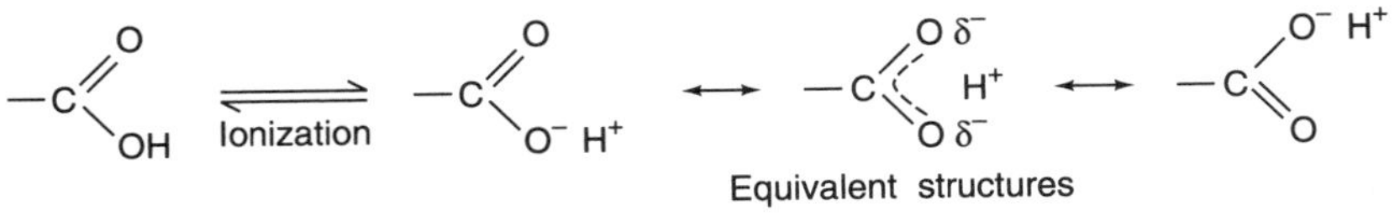

Figure A-13
Resonance forms of an ionic bond.

The Ionization of Water—the pH Scale

Now we will review the interaction between water and the biologically important small molecules and macromolecules. This interaction is critical to both the structure and function of virtually all the molecules in the living cell. Water is, of course, the main component in a cell. All macromolecules reside in it, and all chemical reactions take place in it.

Water molecules ionize in the following way

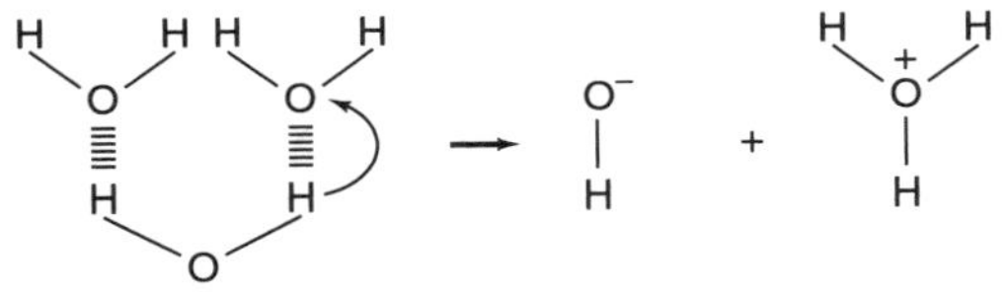

The polar covalent bonds between hydrogen and oxygen ionize forming hydroxide ion (OH^-) and hydronium ion (H_3O^+). The hydronium ion may be treated simply as a hydrogen ion or proton (H^+) making the dissociation reaction

$$H_2O \rightleftharpoons H^+ + OH^-$$

The frequency of this dissociation is described by the following equation

$$K_{eq} = \frac{[H^+][OH^-]}{[H_2O]} \text{ or } K_{eq}[H_2O] = [H^+][OH^-]$$

where K_{eq} is the equilibrium constant, and the terms in brackets are molar concentrations. The term, $K_{eq}[H_2O]$, is a constant such that in pure water

$$K_{eq}[H_2O] = K_w = [H^+][OH^-] = 1.0 \times 10^{-14}$$

where K_w is the dissociation constant or ion product of water. Thus, if the $[H^+]$ is known, the $[OH^-]$ can be calculated by solving the equation

$$[H^+][OH^-] = 1.0 \times 10^{-14}$$

In pure water, $[H^+] = [OH^-] = 1.0 \times 10^{-7}$ = neutral pH. Using such molar concentrations to express the $[H^+]$, however, would be difficult. Imagine if shampoo and soap makers had to say that their neutral products had a $[H^+]$ of 10^{-7}M. Instead, they say that the product is "pH balanced." The pH scale was devised to turn $[H^+]$ numbers into workable values. By definition

$$pH = -\log[H^+]$$

Therefore, neutral pH, where $[H^+] = [OH^-] = 1.0 \times 10^{-7} = -\log_{10}[1.0 \times 10^{-7}] = 7$. A pH scale is shown below. Note that it is a logarithmic scale such that the $[H^+]$ at pH 5 is 100 times greater than at pH 7; i.e., a change of two pH units causes a $10^2 = 100$ fold change in $[H^+]$.

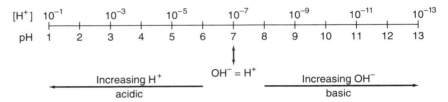

If a strong acid such as hydrochloric acid (HCl) is added to water, it will dissociate or ionize completely as follows

$$HCl \longrightarrow H^+ + Cl^-$$

and the increase in $[H^+]$ causes the pH of the solution to drop (move to the left on the pH scale). If HCl is added to a final concentration of 0.1M (ignoring the very minor contribution of the 10^{-7}M H^+ of the water), then the pH $= -\log[10^{-1}] = 1$. Conversely, if a strong base such as sodium hydroxide (NaOH) is added to a final concentration of 0.1M, after complete dissociation, the $[OH^-] = 0.1$M or 10^{-1}M. By solving the equation, $[H^+][OH^-] = 1.0 \times 10^{-14}$ for $[H^+]$, we get that $[H^+] = 10^{-13}$, since $(10^{-1})(10^{-13}) = 10^{-14}$, and the pH $= \log[10^{-13}] = 13$.

Most biological acids and bases are weak rather than strong, such as HCl and NaOH; therefore, they do not dissociate or ionize completely. An example is the carboxylic acid shown below.

The dissociation of this carboxylic acid is described by the equation

$$K_A = \frac{[H^+][A^-]}{[HA]}$$

where A^- represents the deprotonated acid (often referred to as the **salt** or **conjugate base** of HA), and HA represents the protonated acid. Note the similarity between this equation and the earlier equation for the dissociation of water. If we rearrange the terms and solve for $[H^+]$ we get

$$[H^+] = K_A \frac{[HA]}{[A^-]}$$

If we then take the $-\log$ of both sides of the equation we get

$$-\log[H^+] = -\log K_A - \log \frac{[HA]}{[A^-]}$$

This becomes

$$pH = pK_A + \log \frac{[A^-]}{[HA]}$$

by realizing that $-\log[H^+]$ is the definition of pH, and by analogy, $-\log K_A$ becomes pK_A. This is known as the Henderson–Hasselbach equation and permits us to calculate the pH of a solution if we know the ionization state of the constituents (the ratio of acid and conjugate base), as shown in Figure A-14.

When the pH $=$ pK_A,

$$\log \frac{[A^-]}{[HA]} = 0 \quad \text{and} \quad \frac{[A^-]}{[HA]} = 1$$

Therefore, the pK_A of an acid or base is the pH at which the ratio of protonated and deprotonated forms is 1:1 (ionization is 50%), and is a constant for each acid or base. If we are dealing with a **weak acid,** then at a pH above its pK_A, most of the molecules will have lost their H^+ and be negatively-charged. At a pH below its pK_A, most of the molecules will retain their H^+ and be neutral. If we have a **weak base** instead, then above its pK_A, most of the molecules will be deprotonated and hence neutral. Below its pK_A, almost every molecule will be protonated and hence carry a net positive charge as shown below.

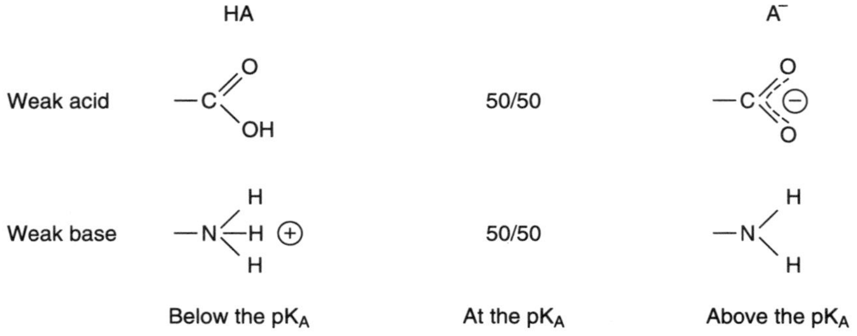

Remember that these relationships are logarithmic such that at a pH one unit below the pK_A, the ratio of $[A^-]:[HA]$ will be 1:10 while at two pH units below the pK_A, the ratio would be 1:100.

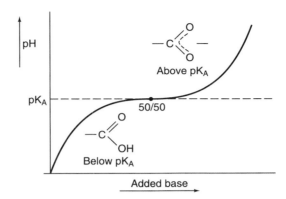

Figure A-14
Titration curve of a weak acid.

Organic Chemistry

Now, we turn our attention to a review of organic chemistry. Carbon is a special atom, because it needs to share four electrons to fill its outer shell. It is very versatile because it can form **strong** bonds with many other elements, in particular, other carbon atoms, oxygen, hydrogen, phosphorus, nitrogen, and sulfur. Carbon is also versatile in that it can form single, double, or triple bonds as shown below.

When four single bonds are formed, the shape of the molecule is a tetrahedron as shown in Figure A-9, but for simplicity it is not always drawn as such. Several characteristic ways of drawing the tetrahedral methane molecule are shown below.

Note that the Fisher Projection (on the left) ignores the three dimensional structure (it is merely assumed to be tetrahedral), while the other two drawings simulate the tetrahedral arrangement. Lines are in the plane of the page. Solid wedges project up out of the page, while dashed wedges project down into the page.

If there are four different substituents on one carbon atom, then this carbon is called a **chiral center,** and there are two possible **stereo-**

isomers. Stereoisomers are mirror images of each other and are not superimposable, rather like your left and right hands (see below).

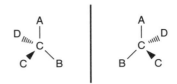

Any carbon that has four different groups attached to it is asymmetric. All living systems choose one version for each compound. Whereas ordinary chemical reactions cannot distinguish between stereoisomers, biological enzymes usually can act on only one stereoisomer, rather like a pair of gloves. One glove can fit the left hand, but not the right and vice versa.

Next, we will discuss several classes of organic molecules that are important in molecular biology.

ALKANES

Alkanes are one class of **hydrocarbons,** molecules consisting only of carbon and hydrogen. In addition they have only carbon–carbon single bonds, and have the general formula C_nH_{2n+2}. A series of alkanes containing from one to four carbons is shown in Figure A-15.

Methane, with one carbon, is the simplest alkane. For butane (four carbons) and all larger alkanes, **structural isomers** exist. Structural isomers have the same chemical formula but differ in the exact set of chemical bonds they contain. The different carbon chain lengths and different structural shapes give the various alkanes slightly different properties.

Alkanes are fully **reduced** (as opposed to oxidized) and are said to be **saturated,** meaning that they contain the maximum number of hydrogens possible (only single bonds).

Many of the shorter alkanes are used as amino acid side chains. The longer forms of alkanes are a major component of gasoline. These molecules, plus methane (natural gas) and propane, are very high energy compounds that can be burned or **oxidized** to produce CO_2 and H_2O (plus other by-products). Living cells obtain energy in a similar fashion. It is just a controlled burning of reduced carbon compounds to produce CO_2 and H_2O. However, alkanes do not work as a food source, since they are very hydrophobic and are therefore *not water soluble.* They are, in fact, toxic.

ALKENES

Another class of hydrocarbons are the alkenes, which, like alkanes, contain only carbon and hydrogen, but they differ from alkanes in that

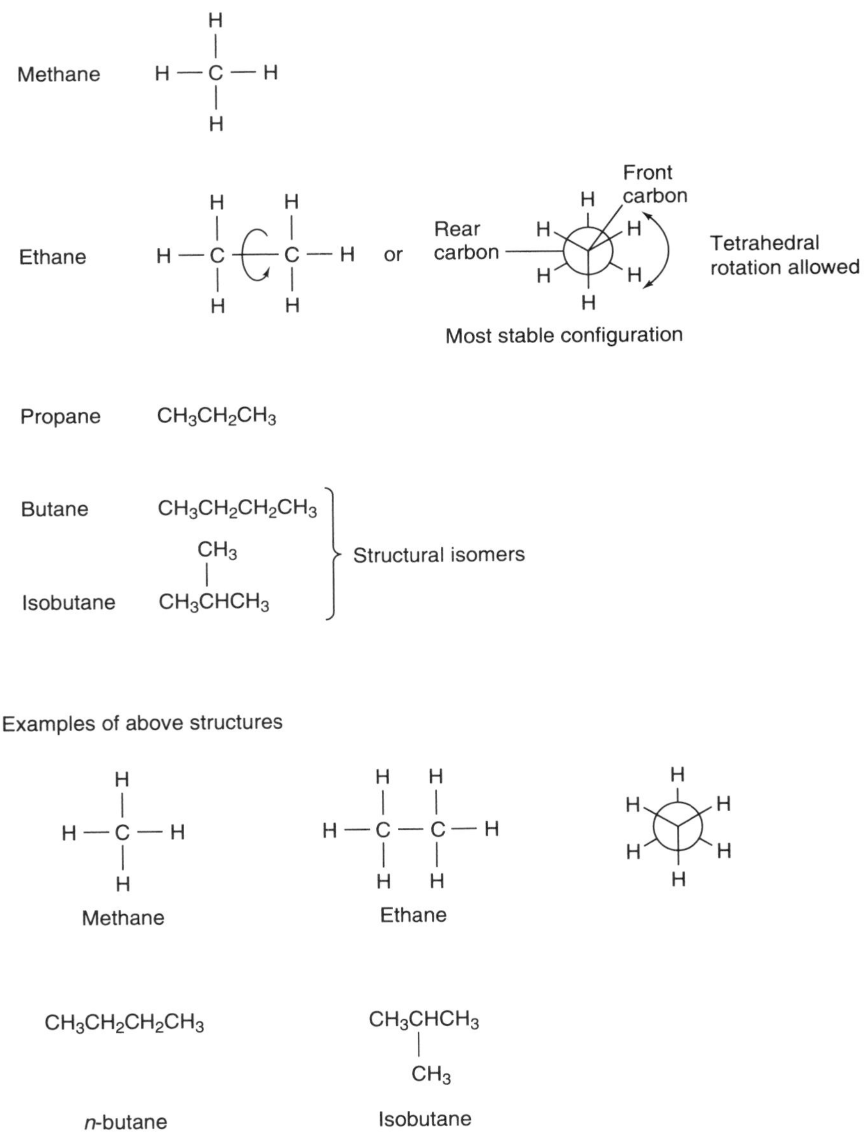

Methane

Ethane

Rear carbon

Front carbon

Tetrahedral rotation allowed

Most stable configuration

Propane $CH_3CH_2CH_3$

Butane $CH_3CH_2CH_2CH_3$

Structural isomers

Isobutane CH_3CHCH_3
with CH_3 branch

Examples of above structures

Methane   Ethane

$CH_3CH_2CH_2CH_3$ CH_3CHCH_3
with CH_3

n-butane   Isobutane

Figure A-15
Alkanes containing one to four carbons.

they have one or more double bonds between carbon atoms. A series of alkenes is shown in Figure A-16.

The double bonds in alkenes are somewhat strained and are, therefore, more reactive than single bonds. Alkenes are said to be **unsaturated,** since they do not contain the maximum amount of hydrogen possible, and hydrogen or other atoms can be added to the carbons of a double bond. These highly reactive bonds are used as the starting

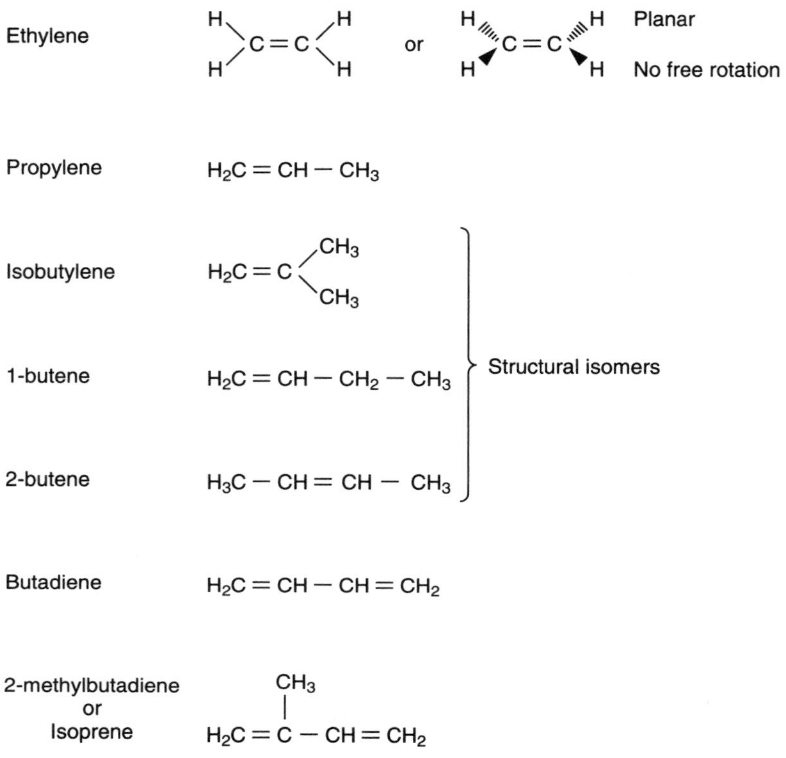

Ethylene	$\overset{H}{\underset{H}{>}}C=C\overset{H}{\underset{H}{<}}$ or $\overset{H}{\underset{H}{}}C=C\overset{H}{\underset{H}{}}$ Planar No free rotation
Propylene	$H_2C=CH-CH_3$
Isobutylene	$H_2C=C\overset{CH_3}{\underset{CH_3}{<}}$
1-butene	$H_2C=CH-CH_2-CH_3$
2-butene	$H_3C-CH=CH-CH_3$

Structural isomers

Butadiene	$H_2C=CH-CH=CH_2$
2-methylbutadiene or Isoprene	$H_2C=\overset{\overset{\textstyle CH_3}{\textstyle \vert}}{C}-CH=CH_2$

Figure A-16
Some short chain alkenes.

material for many industrial polymers, including plastics, as well as for complex cellular compounds. For example, isoprene (Figure A-16) is the starting material for the synthesis of cholesterol and steroid hormones.

The double bonds of alkenes also differ from the single bonds of alkanes in that the carbons on either side of the double bond cannot rotate freely around the double bond, causing the molecule to be planar rather than tetrahedral (compare ethane in Figure A-15 to ethylene in Figure A-16). Thus, in addition to the structural isomers that can form, **geometric isomers** are possible when different substituents are bonded to the carbons in an alkene as shown below.

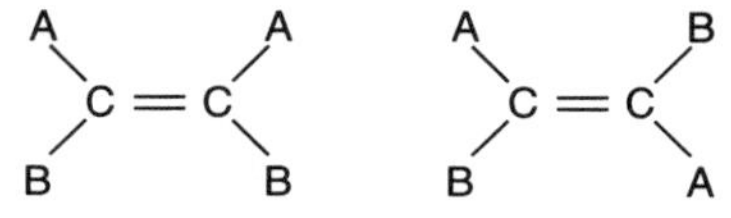

$$\overset{A}{\underset{B}{>}}C=C\overset{A}{\underset{B}{<}} \qquad \overset{A}{\underset{B}{>}}C=C\overset{B}{\underset{A}{<}}$$

Cis isomer   Trans isomer

Geometric isomers have the same set of chemical bonds (unlike structural isomers), but the groups are arranged differently in space. In the **cis** isomer, the identical groups are on the same side of the double bond, whereas in the **trans** isomer, they are on opposite sides.

ALKYNES

Alkynes are also hydrocarbons, but they contain one or more triple bonds between carbons. A simple alkyne, acetylene, which is the basis for the oxyacetylene torch, is shown below.

$$H - C \equiv C - H$$

Note that these molecules are linear around the carbon–carbon triple bond. The triple bonds of alkynes are very strained and, therefore, very reactive. Alkynes are even more unsaturated than alkenes. Carbon–carbon triple bonds are rare in biology.

CYCLOALKANES AND CYCLOALKENES

So far, we have mentioned only straight chain molecules plus a few examples of branched structures. There are also cyclic forms of alkanes and alkenes. Cycloalkanes still contain only single bonds, but they are now in a ring structure with a general formula of C_nH_{2n}. Cyclopropane is the smallest cycloalkane possible, but it is unstable due to highly strained bonds. Cyclohexane is stable. Note in the drawings below how the hydrogens and carbons may be omitted from the drawing of the structure for simplicity. They are assumed to be there.

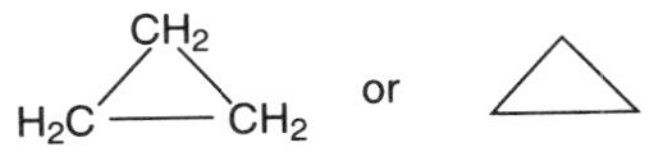

Cyclopropane 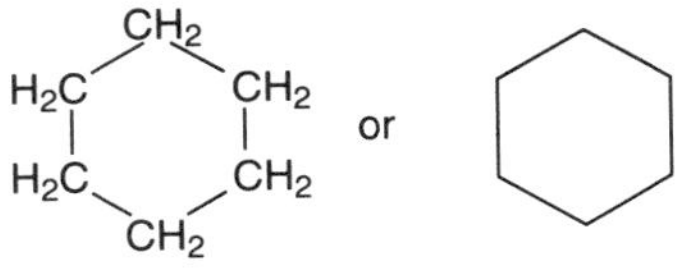   Cyclohexane

If a double bond is added, a cycloalkene is produced.

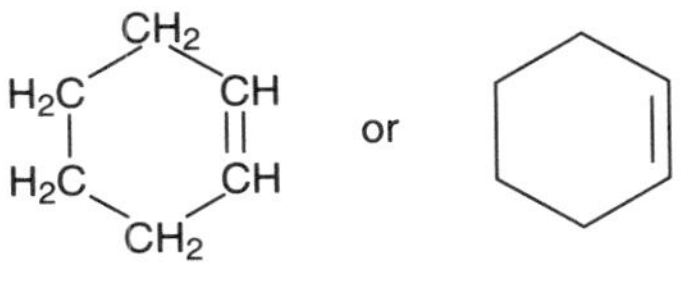

Cyclohexene

The alkanes, alkenes, alkynes, cycloalkanes, and cycloalkenes are all examples of **aliphatic** molecules, hydrocarbons that may be part of

an open–chain or cyclic configuration, may have a straight chain or
branched structure, and may contain single, double, or triple bonds.

Aromatic molecules are a special subclass of cycloalkenes that are
not aliphatic. These molecules have the atoms arranged in rings, and
the double and single bonds alternate in a resonance pattern that gives
these molecules a high degree of stability. Many aromatic molecules are
based on the structural backbone of benzene (shown below).

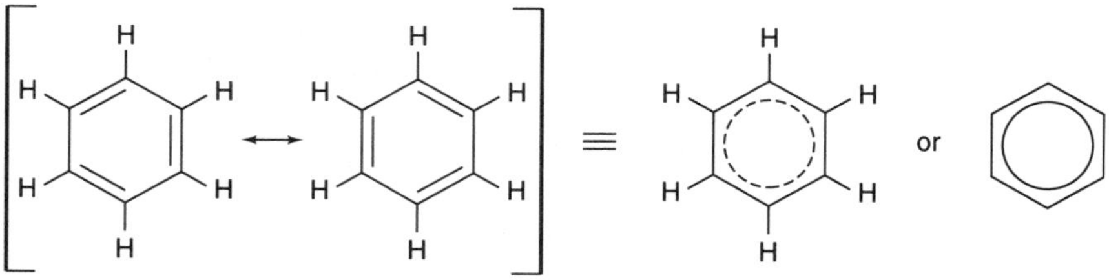

The actual structure of benzene (far right) is the instantaneous average
of the two resonance structures shown at the left. Several amino acids
have aromatic side chains. We will also see further examples of resonance
stabilization in the sections that follow.

So far, we have dealt only with carbon–carbon and carbon–hydrogen bonds. Now we will start adding other elements, such as oxygen,
sulfur, and nitrogen.

ALCOHOLS

Alcohols are compounds of the general formula ROH, where R is a
hydrocarbon chain in which one hydrogen has been replaced with a
hydroxyl group (OH). Alcohols are polar, hydrophilic molecules due
to the partial charges on the OH group, as shown below.

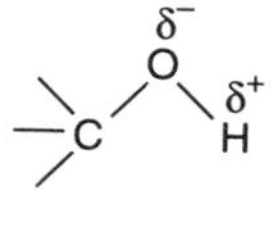

Many alcohols are very soluble in water due to their ability to form
hydrogen bonds with water molecules. A number of small alcohols are
shown in Figure A-17. Note that, as with hydrocarbons, structural
isomers of alcohols exist. However, since a functional group (the OH)
is present, this leads to a distinction between **primary, secondary,** and
tertiary alcohols. A primary alcohol is one in which only one carbon
chain (or only H in the case of methanol) is attached to the carbon with
the OH group, while secondary and tertiary alcohols have two and three
carbon chains attached, respectively (see Figure A-17). Thus, although
propanol and isopropanol are structural isomers, one is a primary alcohol,
while the other is a secondary alcohol. Also note that, since the water-
solubility of alcohols is due to the OH group, as the hydrocarbon chain

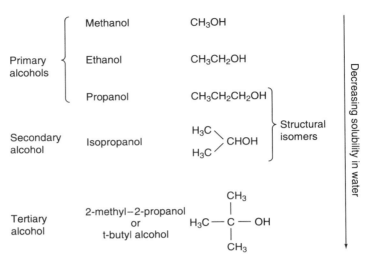

Figure A-17
Primary, secondary, and tertiary alcohols.

size increases, they become less soluble in water. Like other organic molecules, many alcohols are less dense than water.

The alcohols depicted in Figure A-17 are all examples of aliphatic alcohols. Below is an example of an aromatic alcohol, phenol.

Many biological molecules are alcohols. Sugars, for example, are polyalcohols. In fact, they are a special class of polyalcohols called **carbohydrates,** which have the general formula $(CH_2O)_n$. Glycerol is a simple three carbon sugar and is shown below.

$$H_2C — OH$$
$$HC — OH$$

Notice that an **–ol** suffix following each of the compounds just discussed indicates that they are related to an alcohol.

ETHERS

Ethers are another class of oxygen-containing compounds with the general formula R-O-R', where R and R' are identical or different hydrocarbon chains. Ethers may be thought of as the products of condensation reactions between two alcohols, specifically a **dehydration** reaction involving the loss of water as illustrated below.

$$ROH + R'OH \longrightarrow ROR' + H_2O$$

Ethers are less polar than water or an alcohol, but more polar than an alkane due to the partial charges on the oxygen and carbons, shown below for dimethyl ether.

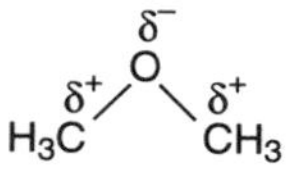

Ethers, in general, are not water soluble, but they are soluble in organic solvents and biological molecules, such as membrane lipids. Ethers are volatile, highly flammable, and frequently have strong odors. Some, such as diethyl ether (shown below), act as anesthetics at low doses by altering nerve transmission. Higher doses are toxic.

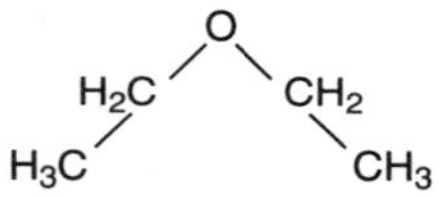

ALDEHYDES AND KETONES

Aldehydes and ketones are molecules that contain a carbon atom double bonded to an oxygen. This is called a carbonyl group. Aldehydes and ketones may be thought of as the products of the oxidation (or **dehydrogenation**) reactions of primary and secondary alcohols, respectively.

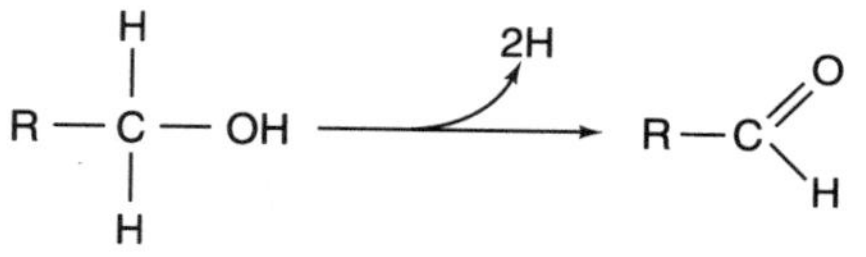

Primary alcohol   Aldehyde

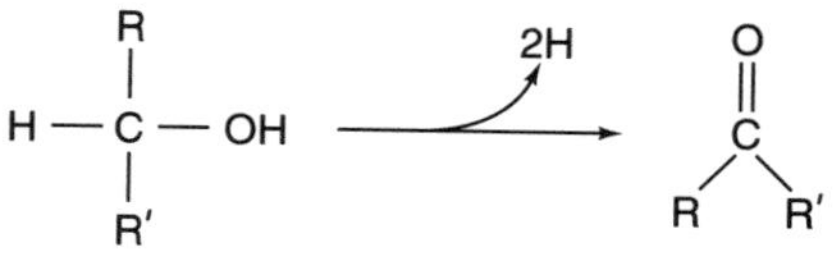

Secondary alcohol   Ketone

Thus aldehydes have the general formula RCHO, where one R chain is attached to the carbonyl, while ketones have the general formula RR'CO, where two R chains are attached to the carbonyl. As before, R or R' refers to aliphatic or aromatic side chains. A commonly used aldehyde and a ketone are shown on the next page.

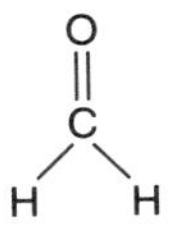

Formaldehyde 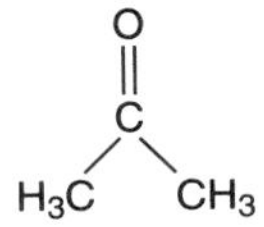   Acetone

Formaldehyde is used in biological laboratories as a fixative. Note that it has no carbon side chain, as it is derived from methanol (CH_3OH). Acetone is a laboratory solvent and also is used as nail polish remover.

Aldehydes and ketones are somewhat polar, so they are somewhat water soluble. The extent of their solubility depends on the rest of the molecule (as with alcohols). Aldehyde and ketone functional groups may be found in amino acids and nucleotides, but are particularly evident in carbohydrates. An **aldose** or aldehyde containing sugar and a **ketose** or ketone containing sugar are shown in Figure A-18. In solution, sugars can form ring structures in which the aldehyde or ketone group is

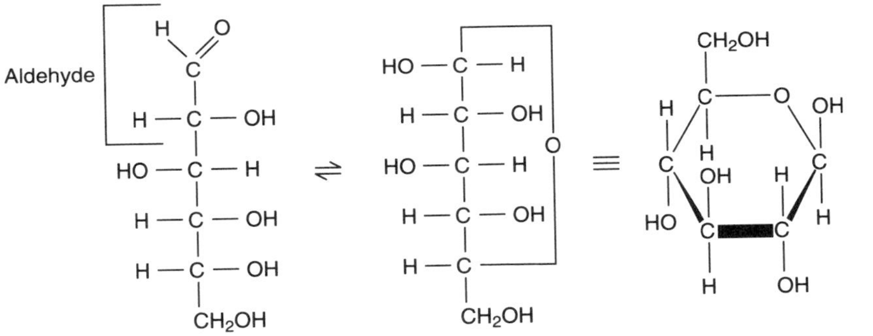

D-glucose, an aldose

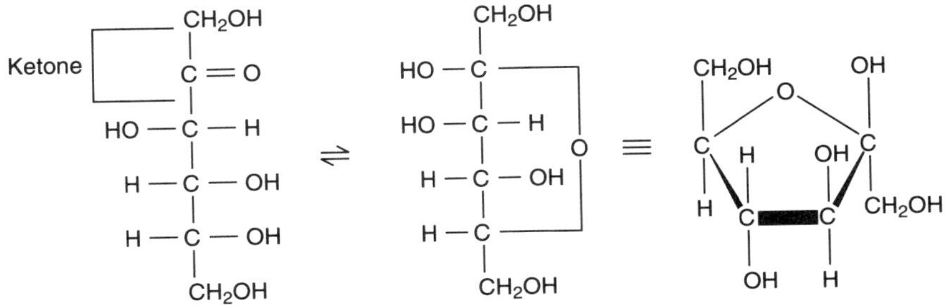

D-fructose, a ketose

Figure A-18
The sugars, glucose and fructose, in open and cyclic configurations.

reduced to a hydroxyl (alcohol) and an ether linkage is formed. Due to the presence of the hydroxyl on one of the carbons adjacent to the ether linkage, it is referred to as a **hemiacetal** (Figure A-18).

CARBOXYLIC ACIDS

Carboxylic acids are compounds that contain both a carbonyl and hydroxyl (alcohol) group bonded to the same carbon atom. They have the general structure shown below, where R, as usual, may be an aliphatic or aromatic side chain.

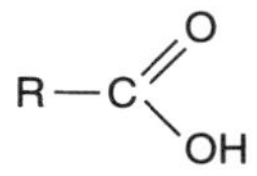

$$R-C\overset{O}{\underset{OH}{\diagup}}$$

They may be thought of as the oxidized products of aldehydes, as shown below.

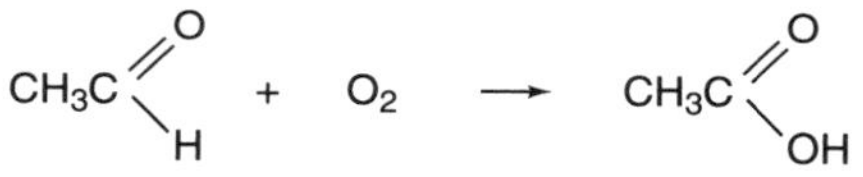

$$CH_3C\overset{O}{\underset{H}{\diagup}} + O_2 \longrightarrow CH_3C\overset{O}{\underset{OH}{\diagup}}$$

Acetaldehyde   Acetic acid, vinegar

For simplicity, the carbonyl functional group is often written as CO_2H or $COOH$.

Carboxylic acids are very important biological compounds. Many of the foods we eat derive their taste and smell from carboxylic acids. Carboxylic acids also give many animals their characteristic odors. Amino acids are carboxylic acids. In addition, carboxylic acids are involved in energy production (intermediates in the Krebs cycle, also referred to as the citric acid cycle or tricarboxylic acid cycle) and in energy storage (fatty acids). Some examples are given in Figure A-19.

Carboxylic acids tend to be water soluble, weak acids. As such, they can donate hydrogen ions (protons). This ionization is shown below.

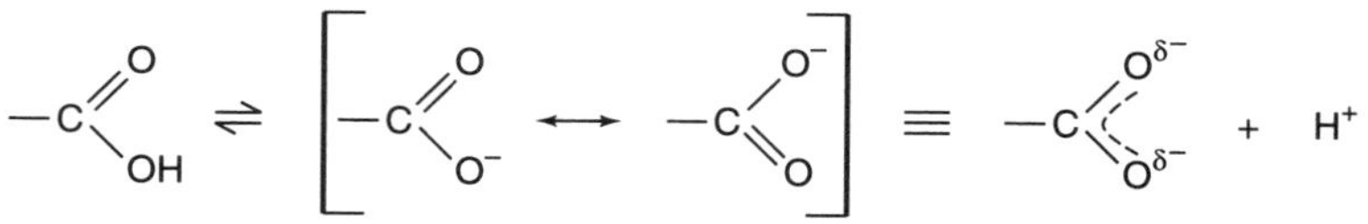

Resonance forms

For simplicity carboxylate ions are usually written as only one of the resonance forms. The pK_A values for this ionization differ from one carboxylic acid to another but are in the range of 2 to 5.

Formic acid (methanoic acid)	HCOOH	Ant pheromone
Butyric acid (butanoic acid)	$CH_3CH_2CH_2COOH$	Rancid butter
Caproic acid (hexanoic acid)	$CH_3(CH_2)_4COOH$	Odor of goats
Stearic acid (ocatadecanoic acid)	$CH_3(CH_2)_{16}COOH$	A saturated fatty acid

Lactic acid

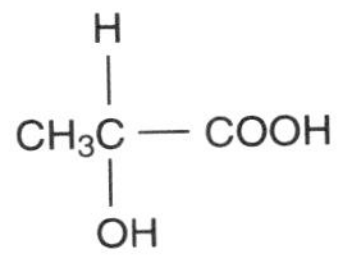

$$\underset{\underset{OH}{|}}{\overset{\overset{H}{|}}{CH_3C}} - COOH$$

Sour milk and overworked muscles

Citric acid

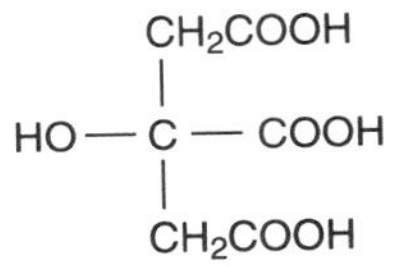

$$HO - \underset{\underset{CH_2COOH}{|}}{\overset{\overset{CH_2COOH}{|}}{C}} - COOH$$

Citrus fruits, Krebs cycle intermediate

Figure A-19
Several carboxylic acids and their biological significance.

ESTERS

A final class of oxygen containing compounds are esters, which are the product of a dehydration reaction between an alcohol and a carboxylic acid, as shown below. Note that an ester contains a carbonyl group and an ether linkage.

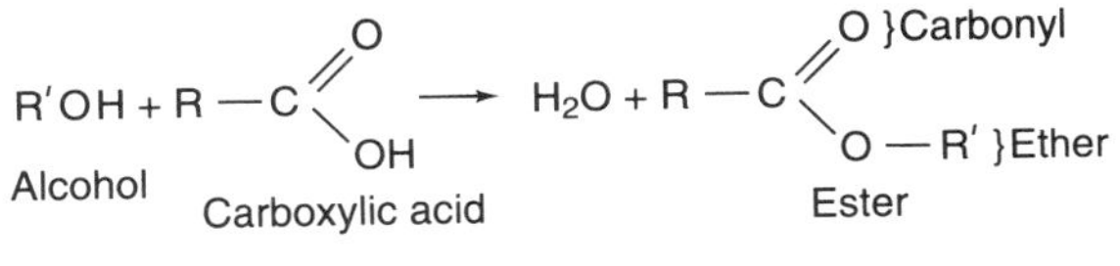

$$R'OH + R - C\overset{\displaystyle O}{\underset{\displaystyle OH}{\diagdown}} \longrightarrow H_2O + R - C\overset{\displaystyle O\ \}Carbonyl}{\underset{\displaystyle O - R'\ \}Ether}{\diagdown}}$$

Alcohol   Carboxylic acid   Ester

Esters tend to be volatile and reactive. They frequently give various fruits and flowers their pleasant odors and flavors. Figure A-20 shows several aliphatic and aromatic esters as well as the **triglyceride** stearin, containing three stearic acid molecules esterified to a glycerol backbone.

If one of the fatty acids in a triglyceride is replaced with a phosphate containing group, a phosphoester linkage is formed and a class of molecules called **phospholipids** is generated. They comprise a major constituent of cellular membranes. Another example of the phosphoester is seen in the DNA backbone, where in fact, a phosphodiester linkage occurs.

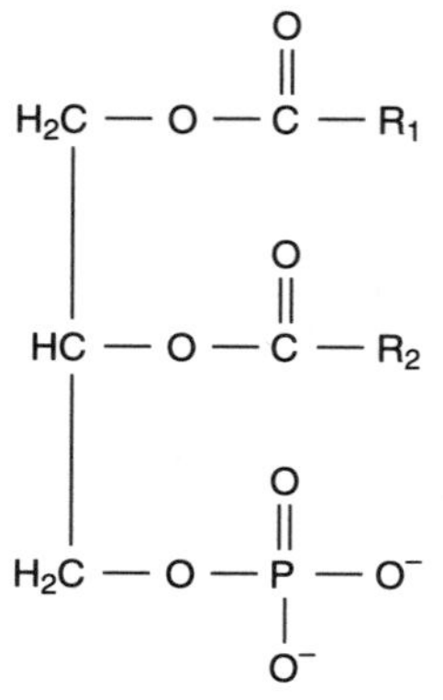

Phosphatidic acid
(precursor for phospholipids)

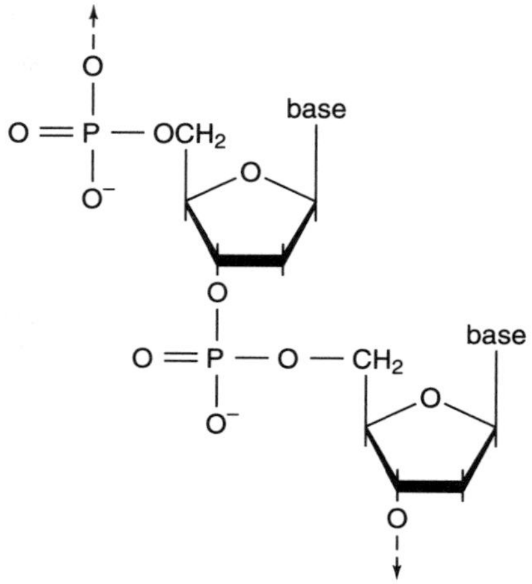

Strand of DNA

Note that phosphoesters are acidic and tend to be ionized at neutral pH.

THIOLS, THIOESTERS, AND DISULFIDES

Thiols (or mercaptans), thioethers, and disulfides may be thought of as the sulfur analogs of alcohols, ethers, and peroxides. They have the general formulas RSH, RSR′, and RSSR′, respectively. These sulfur containing compounds generally have very bad odors. Some thiols and a thioether are shown in Figure A-21.

Two thiols can react with one another to form a disulfide. This is a common occurrence in proteins where two cysteine side chains react to form a disulfide bond as shown below.

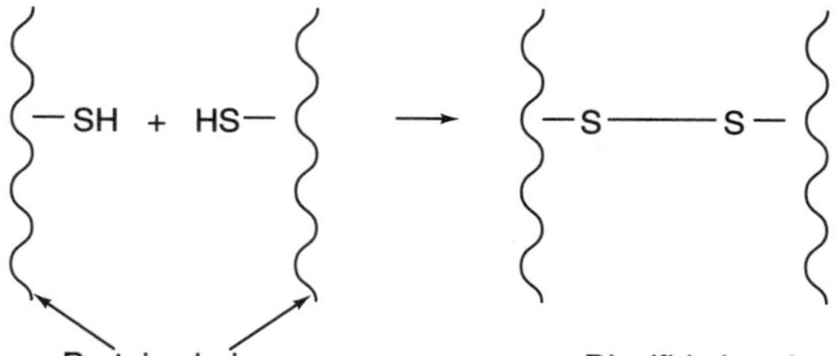

Protein chains
(reduced)

Disulfide bond
(oxidized)

These disulfide bonds, which can occur between different polypeptide chains or between different regions of the same chain, are involved in the structure and stability of proteins. At high concentrations, β-mercapto-ethanol can reduce (break) disulfide bonds and destroy a protein's structure.

Ethyl acetate — CH_3C (=O) OCH_2CH_3 — Odor of apples

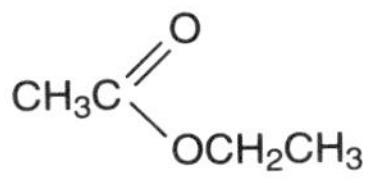

Amyl acetate — CH_3C (=O) $OCH_2(CH_2)_3CH_3$ — Odor of bananas

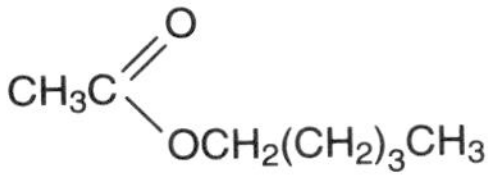

Methyl salicylate — (benzene ring with $C(=O)-OCH_3$ and OH) — Wintergreen oil

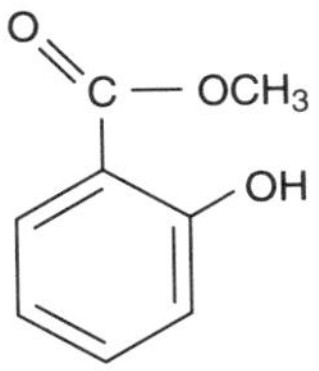

Acetylsalicylic acid — (benzene ring with $C(=O)-OH$ and $O-C(=O)-CH_3$) — Aspirin

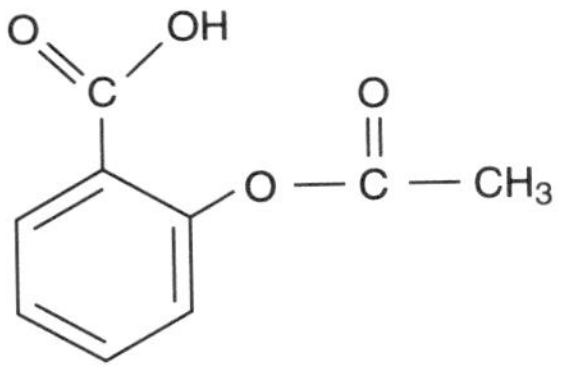

Stearin

$$H_2C-O-C(=O)-(CH_2)_{16}CH_3$$
$$HC-O-C(=O)-(CH_2)_{16}CH_3$$
$$H_2C-O-C(=O)-(CH_2)_{16}CH_3$$

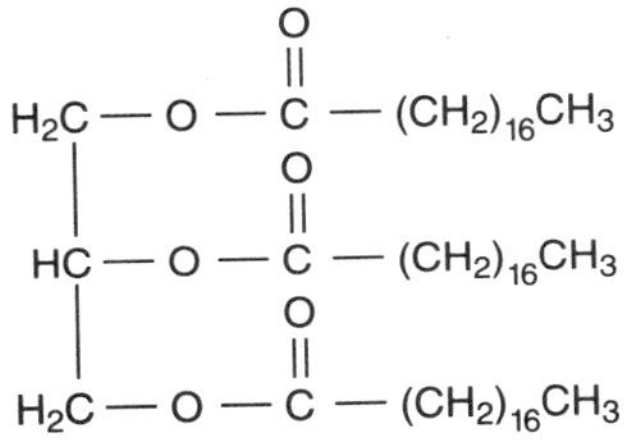

A triglyceride (fat)

Figure A-20
Several esters and their biological significance.

AMINES, IMINES, AND AMIDES

Several classes of compounds contain carbon bonded to nitrogen. Amines are derivatives of ammonia (NH_3) in which one or more hydrogen atoms are replaced with carbon chains. A series of amines is shown in Figure A-22. Note that the amines may be primary, secondary, or tertiary, depending on whether one, two, or three of the ammonia hydrogens are replaced with carbon chains.

In general, amines have fishy, ammonia-like odors. In contrast to carboxylic acids, which are weak acids, amines are weak bases and can be protonated as shown below.

$$R-N(H)(H) + H^+ \rightleftharpoons R-N^+(H)(H)-H$$

Methane thiol	CH_3SH	Rotten egg odor, cysteine side chain
Ethane thiol	CH_3CH_2SH	Sewer gas
Butane thiol	$CH_3CH_2CH_2CH_2SH$	Skunk secretion
β−mercaptoethanol	$HOCH_2CH_2SH$	Biochemical reagent, permanent wave solution
Thiophenol	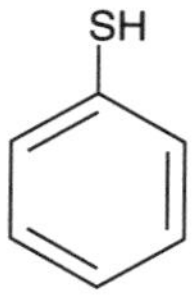	Odor of burnt rubber
Ethyl methyl thioether	$CH_3CH_2SCH_3$	Methionine side chain

Figure A-21
Thiols and a thioether with their characteristic odors and biological uses.

The pK_A values of amines range from 8 to 12. Therefore, at neutral pH, amines are protonated.

An imine is the nitrogen analog of a carbonyl and has the general form shown below.

$$>\!C=NR$$

Imines can also be protonated much like amines. They are often found in ring structures of biological importance as shown in Figure A-23.

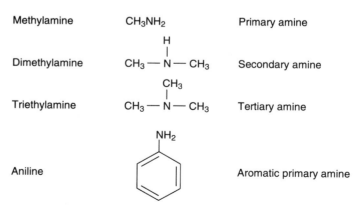

Methylamine	CH_3NH_2	Primary amine
Dimethylamine	$CH_3-\overset{\overset{\textstyle H}{\textstyle |}}{N}-CH_3$	Secondary amine
Triethylamine	$CH_3-\overset{\overset{\textstyle CH_3}{\textstyle |}}{N}-CH_3$	Tertiary amine
Aniline		Aromatic primary amine

Figure A-22
Some simple primary, secondary, and tertiary amines.

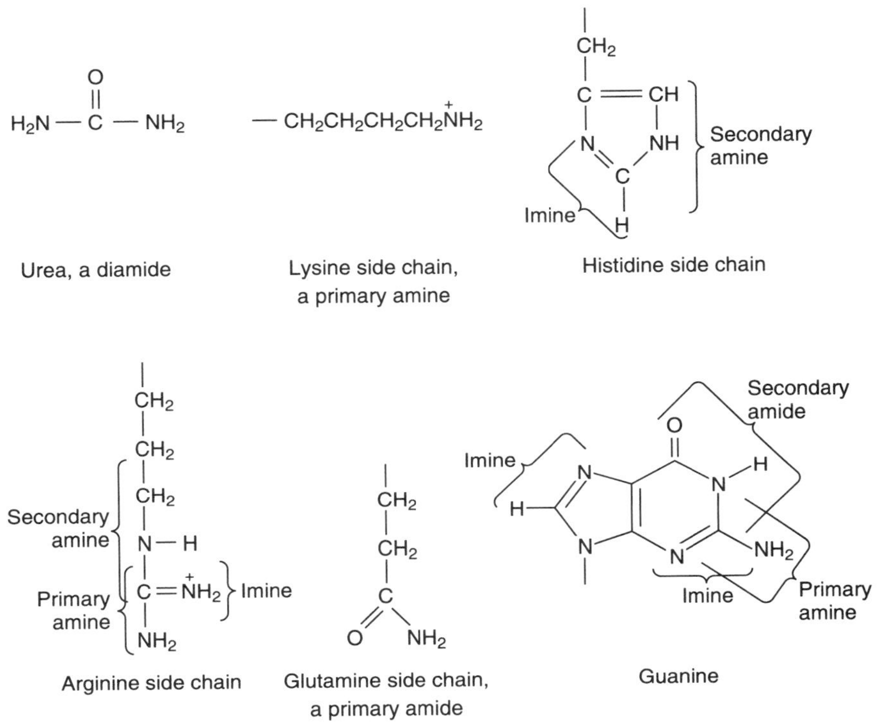

Figure A-23
Some important biological compounds containing carbon–nitrogen bonds.

Primary, secondary, or tertiary amides are a combination of a carbonyl group and a primary, secondary, or tertiary amine and have the general structures shown below.

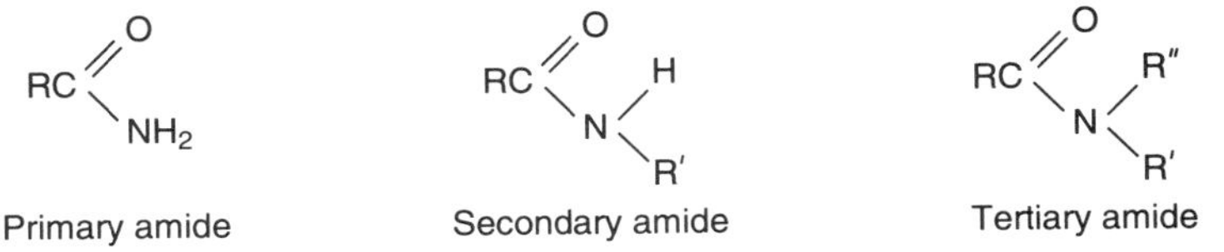

A number of biological compounds contain one or more of these carbon–nitrogen linkages. Several are shown in Figure A-23.

AMINO ACIDS

As we have seen, many molecules display more than one kind of functional group. One group of molecules that demonstrates a number of the functional groups and concepts previously discussed is the set of twenty, naturally occuring amino acids. All amino acids have a carbox-

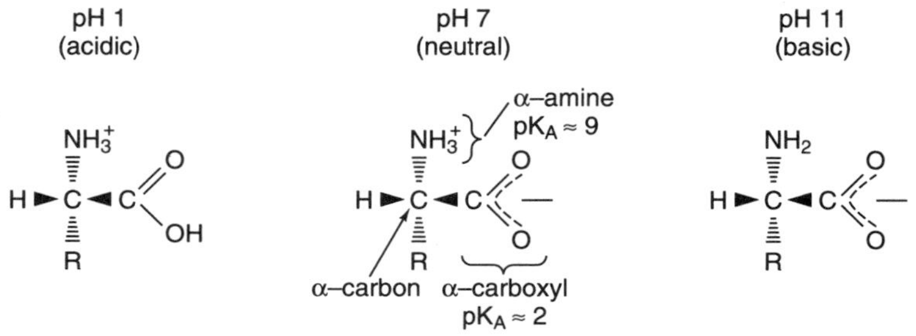

Figure A-24
A typical amino acid at acidic, neutral, and basic pH.

ylic acid and amine group (both of which are ionizable) attached to a central carbon atom called the α–carbon. One hydrogen atom and a variable side chain (R) are also attached to the α–carbon. With the exception of glycine, in which R is a hydrogen, there are four different groups attached to the α–carbon, so it is a chiral center. Thus all the amino acids except glycine have a specific stereochemistry. The side chain groups include aliphatic or aromatic carbon chains, primary or secondary alcohols, carboxylic acids, primary and secondary amines, primary amides, thiols, and thioethers.

Since amine and carboxylic acid groups have rather different pK_A values, amino acids have different charged states at different pH values. These charge states as well as the L–isomer configuration of amino acids are shown in Figure A-24. Although not shown, keep in mind that some of the amino acids have ionizable carboxylic acid or amine side chains that have pK_A values of about 3.9 to 4.2 and 10.5 to 12.5, respectively. The pK_A values for the α–carboxyl and α–amine are about 2 and 9 respectively, differing only slightly from one amino acid to another.

Note that at neutral pH, most amino acids are **zwitterionic,** from the German meaning both positive and negative charge in a molecule with a net charge of 0. All amino acids are **amphoteric,** from the Greek meaning both acid and base. However, in proteins, the α–amine and α–carboxyl are involved in peptide bonds linking one amino acid to the next and can no longer be ionized. Only the terminal carboxyl and amine groups as well as the side chain carboxyls and amines can be ionized and lead to the net charge of the protein. The structures of all twenty naturally occurring amino acids may be found in Chapter 2.

Concluding Note

The chemistry of the complex macromolecules found in biological compounds is a simple extension of the rules described in this appendix.

They contain a variety of the functional groups discussed previously. Remember that the types of bonds formed (covalent, polar covalent, ionic, hydrophobic, or hydrogen bonding) will determine how a molecule interacts with water and with other macromolecules. Also, many of the reactions we have seen (dehydration, dehydrogenation, reduction, oxidation, esterification, etc.) are common reactions during biochemical synthesis and degradation.

Supplement to Appendix

The Molecular Biology of Plants

Virtually all of the features of macromolecules and their functions, which we learned about in Essentials of Molecular Biology primarily from the study of bacterial/viral prokaryotes and animal eukaryotes, pertain to plants as well. The major differences between plants and animals emerge when they are compared at the level of organelles (e.g., plants have chloroplasts), cell boundaries (e.g., plants have rigid cell walls), and tissues (e.g., plant tissues often display unending [indeterminate] growth).

PLANT HORMONES

One major difference between animals and plants does, however, apparently reside at the molecular level. Plants appear to employ relatively small, and often remarkably simple, molecules as hormones or signal molecules. Those molecules initiate series of physiological changes, including modifications in gene expression patterns. Included in the roster of plant hormones are auxins, cytokinins, gibberellins, abscisic acid, and ethylene. Most of the hormones discovered so far in animals are usually more complex than plant hormones and include steroids and proteins. Plant hormones include molecules as simple as ethylene (C_2H_4). The manner in which plant hormones act—via initiating transcription—may, however, not be very different from similar processes in animal cells.

One other difference between plants and animals is, of course, the plant's response to light. Proteins in the plant cell contain a pigment that absorbs light. As those proteins, called **phytochrome** proteins, absorb light they become activated. The activated protein then triggers several metabolic processes, including calcium influx, activation of other proteins, and, eventually, activation of gene expression (transcription).

PLANT GENETICS

As we learned throughout *Essentials of Molecular Biology,* studies on genetics often provide the initial insight into our understanding of a molecular process. In the early days of plant molecular biology, the impact that genetics had was limited by the cumbersome nature of plant breeding protocols. Multiple crosses, which are usually necessary to formulate a hypothesis and test it adequately, require several months or longer due to the prolonged generation time of the typical plant.

Genetic engineering procedures are now short-circuiting many of the traditional breeding routines. The production of transgenic plants has, in many ways, revolutionized the discipline of plant molecular biology. New insights into metabolic processes have been achieved, and applications for agriculture have been developed.

One plant, *Arabidopsis,* is being developed as a model system for plant molecular biology studies. It is fast becoming to plant scientists what the fruit fly (*Drosophila*) has been to animal scientists. It has a relatively short generation time (six weeks), yields large numbers of seeds, and can be self-crossed (i.e., fertilizes itself). It is small in size, and its seeds can be mutagenized without difficulty. Unlike many plants, its genome is relatively small. Several laboratories are now engaged in an effort to sequence its entire genome! That sequence data will accelerate a detailed understanding of the regulation of gene expression in plants.

GENETIC ENGINEERING

The production of transgenic plants is similar, in principle, to the production of transgenic animals (see pages 285–289). Several properties of plants can be modified by the introduction of foreign genes into the host plant's genome. Those properties include nutrient content, insect resistance, herbicide resistance, disease resistance, and flower color, among others. Introduction of a foreign gene is usually facilitated by using the Ti plasmid vector, as mentioned on page 293. Ti plasmid vector replicates in the bacterium *Agrobacterium* independently of the bacterial DNA. Those bacteria can infect many plants. Occasionally, part of the plasmid is transferred from the bacterial cell into the plant cell. The plasmid sometimes integrates into the plant's genome and is expressed through the normal transcription/translation pathway. Here are a few examples of successful applications of genetic engineering methods.

Insect resistance. Essentially all plants are susceptible to damage caused by one or another insect. Historically, the main treatment for combating insect damage has been chemical insecticides. With the environment already overloaded with substantial amounts of some pesticides (e.g., DDT), the development of alternative strategies is highly desirable. One approach is biological control—the carefully monitored introduction of predator insects or natural bacterial parasites.

Another, more modern approach involves generating transgenic plants that express bacterial toxins active against insect pests. That is, the bacterial genes that code for toxins harmful to pest insects are isolated and inserted into an appropriate vector such as the Ti plasmid system mentioned. Those genes are then introduced into the plant where, if all goes as expected, they will be expressed along with the plant's own genes.

Insects that attack such a plant are in for trouble! In the insect's digestive system, the toxin is chemically modified by digestive enzymes such as proteases to yield highly active toxin. That toxin inhibits the action of the cells that line the insect's digestive tract. Eventually, the insect dies of starvation. Since such toxins are typically highly specific for pest insects and have no effect on harmless insects or humans, they show great promise as alternatives to chemical pesticides.

Herbicide resistance. Weed growth, like insect pests, substantially diminishes agricultural yields. Weed killer chemicals, however, present the same problems associated with chemical pesticides. They are not highly selective and thereby kill useful species. They also accumulate in the environment to levels that pose human health risks.

Genetic engineers are therefore attempting to use plasmid vectors to introduce herbicide-resistant genes into crop plants. For each herbicide a unique strategy is required. In those cases where the herbicide inactivates an enzyme crucial for crop yield, genes that cause the plant to overexpress that enzyme are being engineered into the plant genome. In other cases a herbicide-resistant version of the key enzyme is introduced. In a few instances genes that code for herbicide-destroying enzymes are being introduced into crop plants. Because the herbicide and the plant need to be closely matched, much of the research is carried out by large commercial firms that produce and market both herbicides and seeds.

Novel flowers. Because ornamental flowers represent a growing segment of agribusiness, seed companies are very interested in generating new varieties of flowers. Transferring color genes from one species to another—using genetic engineering methods—represents a potentially useful (and profitable) avenue of research. With such technology the traditional color schemes for many common flowers will probably be greatly expanded in the next decade.

BRIGHT FUTURE

These examples should serve to convince the beginning student of molecular biology that plant molecular biology, especially its commercial applications, represents a discipline with many opportunities for the newcomer.

Supplement to Appendix

The Molecular Biology of Mammalian Viruses

MAMMALIAN VIRUSES AS MODEL SYSTEMS

As was the case for bacteriophage (Chapter 15), mammalian viruses also represent relatively simple experimental systems for studies in molecular biology. Substantial amounts of information about basic features of DNA replication, transcription, and translation can be accumulated with simple systems such as virus infected cells.

Mammalian viruses offer many of the advantageous features of their bacterial counterparts. They can be grown in cultured (mammalian) cells; they incorporate radioactive isotopes efficiently; and they can be harvested in sufficient quantities for biochemical analyses. In addition, some mammalian-cell viruses cause tumors, so they can be studied to gain insight into the causes of cancer.

GENERAL PROPERTIES

Virtually all mammalian viruses contain a nucleic acid genome (DNA or RNA) encapsulated by a protein coat. Among the hundreds of viruses that have been described, substantial diversity exists. Some viruses contain as few as four or five genes, while others encode a few hundred.

Each virus is specialized in one way or another for the absorption, entry, replication, and lysis stages of the lytic cycle. In some cases, as will be detailed on the next page, the virus persists in the host cell rather than complete its life cycle, which would destroy its host.

DNA TUMOR VIRUSES

DNA tumor viruses employ DNA as genetic material and can replicate within some types of cells without killing the cell. The replication of the DNA tumor virus within those kinds of cells does occasionally interfere with the normal mechanisms that control growth rate. Thus, these "transformed" cells proliferate out of control and eventually form a tumor.

Two examples of such DNA tumor viruses commonly studied by molecular biologists are simian virus-40 (SV 40), isolated from monkey cells, and polyoma, isolated from mice. Those viruses are remarkably simple. Their DNA is small and organized as a circle. It consists of only approximately 5,200 base pairs and codes for fewer than a dozen gene products. The virus replicates within the nucleus of the host cell, since host enzymes (which are localized in the nucleus) are required for viral replication and transcription. A complete infection cycle, which results in the release of thousands of free-virus particles, instead of transformation of the host cell to a wildly growing tumor-forming cell, takes about three days.

RNA TUMOR VIRUSES

RNA tumor viruses use RNA as their genetic material. Like other mammalian-cell viruses they are simple in structure and consist largely of genetic material packaged within a protein envelope. During their replication they occasionally alter the host cell's genome so that, as is the case with DNA tumor viruses, growth control is lost, and these host cells proliferate out of control to form a tumor.

The modification by RNA of the host genome is accomplished indirectly. A viral encoded enzyme, reverse transcriptase, transcribes the virus genomic RNA into complementary DNA (cDNA) molecules. Some of those cDNA molecules become integrated into the host genome, where they are expressed and thereby alter the host cell's properties.

The first virus to be assigned a role in tumor formation, the Rous Sarcoma virus, as well as the now famous AIDS virus (which causes Acquired Immune Deficiency Syndrome) are examples of so-called "retroviruses."

AIDS VIRUS REPLICATION

The replication of the AIDS virus in a host cell is mediated by the reverse transcriptase encoded in the RNA genome of the virus. Upon infection that enzyme makes a DNA copy of the RNA genome, forming an RNA/DNA hybrid. The enzyme next makes a DNA/DNA double helix copy using the RNA/DNA hybrid as a template. That DNA/DNA copy of the virus genome then integrates into the host's chromosome, using recombination enzymes encoded in the genome of the virus. From its location in the host's chromosome, the viral genome is repeatedly transcribed. Some of the transcripts are translated to form coat proteins for the virus. Eventually, complete length transcripts are assembled with the coat proteins into mature virus particles.

DIFFICULTIES IN COMBATTING AIDS VIRUS

Another name for the AIDS virus is "human immunodeficiency virus" (HIV). It kills one specific type of cell—the so-called "helper T cell"—

which is important in the early stages of the immune response. The virus inactivates the immune system by destroying one of the first steps in the production of antibodies. Lacking a functional immune system, AIDS patients usually die of infections, such as pneumonia, which in normal healthy individuals are routinely eliminated by natural antibodies.

Several factors combine to frustrate attempts of molecular biologists to develop drugs, vaccines, or other treatments for combatting the AIDS virus. These include the following: the virus inserts itself into the host genome and is therefore not easily removed from host cells; a long latency period permits the spread of the virus before infected individuals are identified; frequent mutations alter the virus, so that its coat protein, the usual target for vaccines, changes often; the target of the virus—the immune system—represents the body's main line of defense against viral infections.

VIRUSES AS GENE CARRIERS

The introduction of foreign genes into mammalian cells is, in a few cases, a routine procedure. Naked DNA can be employed to transfect susceptible cells using experimental protocols that are, in principle, remarkably similar to the original bacterial transformation procedures described in Chapter 6 (Figures 6-1 and 6-2).

In most instances, however, mammalian cells are either incapable of being transformed with those direct methods, or the transferred gene is expressed at very low levels. Thus, the use of so-called "transfected cells" for routine use in research projects is somewhat limited. This is unfortunate for in the future research in molecular medicine will be increasingly directed towards gene therapy (see Chapter 13, p. 292). Suitable means for delivering correct copies of disease-causing genes to target cells will be required.

Hence, molecular biologists are exploring using viruses to transport foreign genes into mammalian cells. One strategy involves using recombinant DNA technology (Chapter 13) to substitute some of the nonessential genes of a DNA virus with a foreign gene. Another strategy employs RNA viruses. The double stranded DNA replication version of the RNA virus genome is manipulated by inserting a foreign gene. That modified DNA is then used to infect a culture of mammalian cells that has the capacity to transcribe the viral DNA into RNA and package the RNA into infectious RNA virus particles. Those viruses are subsequently used to infect target cells that, for instance, display a disease syndrome. The enzyme reverse transcriptase integrates the entire viral genome, including the foreign gene, into the host chromosome. In this manner the virus serves to deliver the foreign gene to the host chromosome. Those genes are often expressed and can cancel the effects of disease-causing viruses.

During the genetic engineering of the virus, its genome is altered so that the virus is incapable of undergoing a complete replicative cycle in its host cell. The virus therefore does not kill the cell it infects. Rather, the virus serves as a "shuttle" or "vector" for transporting a foreign gene into the host cell.

Potential problems with viral vectors exist. First, the virus's genes are integrated into the host chromosome at random locations, and their expression is not guaranteed. Second, although foreign genes can be added to cells, disease-causing genes cannot be removed. For many metabolic diseases, additional medical strategies will be required.

Glossary

Active site – region of an enzyme to which the substrate binds.

Adenine (A) – purine base that pairs with thymine (T) in DNA.

Amino acid – small molecule building block of protein.

Aminoacyl-tRNA synthetases – enzymes that covalently link amino acids to the 2′ or 3′-OH position of tRNA.

Annealing (of DNA) – reassociation to form a hybrid molecule of two complementary single-stranded DNA molecules.

Antibody – protein (immunoglobulin) that recognizes a specific foreign "antigen."

Antigen – molecule that provokes synthesis of an antibody (immunoglobulin).

Anticodon – three base sequence in a tRNA that bases pairs with a specific triplet codon in mRNA.

Antiparallel – opposite polarities of two strands of a DNA double helix; one strand is 5′-3′ top to bottom and the other strand is 3′-5′.

Attenuation – regulatory mechanism for controlling frequency of transcription of a gene that employs a nucleotide sequence that leads to premature transcription.

Autoradiography – method for detecting radioactively labeled molecules by the image they produce on photographic film.

Auxotroph – organism that requires one or more growth supplements in the culture medium.

Bacteriophages – virus that infects and reproduces in bacteria; often denoted as phage.

Base-pair – pair of bases (either A-T or G-C) that are opposite one another in a double-stranded DNA molecule.

B-galactosidase – enzyme that hydrolyzes lactose into two sugar molecules.

Bidirectional replication – DNA replication process that involves two replication forks that move in opposite directions.

B lymphocytes (or B cells) – cells in which antibodies are synthesized.

bp – abbreviation for base-pair.

CAAT box – a conserved nucleotide sequence located upstream of the transcription startpoints of most eukaryotic genes.

Cap – a G nucleotide (which is sometimes methylated) that is added to the 5′ end of eukaryotic mRNA after transcription.

CAP – regulatory protein that is activated by cyclic AMP and is involved in transcription of the lac operon of *E. coli.*

Carboxyl terminus – end of a polypeptide that contains a free carboxyl group.

Catabolite repression – reduced expression of bacterial genes, which results from growth of excess glucose.

cDNA – single-stranded DNA with a nucleotide sequence, which is complementary to an RNA.

cDNA clone – double-stranded DNA sequence that is complementary to a specific RNA and inserted in a cloning vector such as a plasmid.

Charging – enzymatic attachment of an amino acid to its appropriate tRNA.

Chromosome walk – method for sequentially isolating overlapping segments of DNA, so that very large genes can be studied.

Clone – large number of either molecules (e.g., DNA) or cells (e.g., *E. coli*) derived from a single progenitor.

Cloning vector – plasmid of phage into which a foreign DNA is inserted for replication.

Coding strand – DNA strand that is transcribed into mRNA.

Codon – triplet of nucleotides that codes for specific amino acid, or translation start-stop signal.

Colony hybridization – method for detecting bacteria that carry a vector with desired inserted sequence.

Complementation – non-allelic gene functions in a double mutant organism to produce the necessary product for restoration of the wild type phenotype.

Conditional lethal mutation – kills the cell virus that harbors it, but only under certain (nonpermissive) conditions, such as elevated temperature.

Conjugation – "mating" between two bacteria, which involves movement of genetic material from an F+ to an F− cell.

Consensus sequence – composite nucleotide sequence commonly believed to serve the same function in different genes (e.g., promoter sequences).

Core particle – enzyme digestion product of nucleosome that contains histone octamer and approximately 150 bp of DNA.

Cytoplasm – collection of cell components that exist within the cell, but does not include nucleus.

Cytosine (c) – pyrimidine base that pairs with guanosine in DNA.

Deletion – removal of gene region from chromosome.

Denaturation – for DNA or RNA, describes separation of double-stranded molecule to a single-stranded state, usually by heating; for protein, describes change in physical shape, which usually renders it inactive.

Deoxyribose – five carbon sugar found in DNA.

Diploid – the normal number of chromosomes (two copies of each −2N) in virtually all eukaryotes.

Direct repeats – multiple identical (or closely related) nucleotide sequences in the same orientation in a DNA molecule.

Discontinuous replication – synthesis of DNA in many short (Okazaki) fragments that are later joined to form a single, continuous strand.

DNA (deoxyribonucleic acid) – polynucleotide that contains deoxyribose sugar.

DNAase – enzyme that cleaves phosphodiester bonds in DNA, to break the molecule into pieces.

DNA ligase – enzyme that joins two double-stranded DNAs together, end to end.

Double-stranded helix – three-dimensional shape exhibited by two complementary base-paired DNA strands.

Downstream – direction away from transcription start site in DNA and towards the last section of the coding region to be transcribed.

Editing – system of enzymes that monitors each nucleotide to ensure that it is correctly base paired during DNA replication.

Electrophoresis – separation technique that involves placing charged macromolecules in a voltage field and permitting them to migrate according to their individual net charges.

Elongation factors - regulatory proteins that associate with ribosomes to promote either the tRNA binding or the translocation step in the elongation phase of translation.

Endonuclease - enzyme that cleaves phosphodiester bonds within a nucleic acid.

Endoplasmic reticulum (ER) - folded membrane sheets to which polysomes (sites of protein synthesis) are anchored.

Enhancer - nucleotide sequence that increases the utilization of eukaryotic promoters; it can function in either an upstream or downstream location relative to the promoter.

Escherichia coli (*E. coli*) **-** intestinal bacterium widely used because it can be conveniently cultured for procaryotic genetics studies.

Excision-repair - an enzyme system that removes a short, single-stranded sequence of double-stranded DNA, containing mispaired or damaged bases, and replaces it by synthesizing a sequence complementary to the remaining strand.

Exon - nucleotide sequence of a eukaryotic gene that is represented in the mRNA.

Exonuclease - enzyme that cleaves nucleotides one at a time off the end of a DNA or RNA molecule.

F factor - bacterial plasmid that functions to generate sex or fertility characteristics.

F plasmid - fertility plasmid in *E. coli,* which permits a donor bacterial cell to conjugate (mate) with a recipient bacterial cell.

F1 generation - progeny of a cross between two parental types that differ in one or more genes.

Filter hybridization - method for nucleic acid hybridization performed by soaking a denatured DNA preparation, immobilized on a nitrocellulose filter, in a solution of radioactively labeled RNA or DNA.

Fingerprint (DNA) - specific pattern of fragments of DNA on gel electrophoresis, generated by first treating DNA sample with endonucleases.

Frameshift mutation - insertion or deletion of either one or two bases in the coding region of a gene, which results in an altered downstream amino acid sequence.

Gene - basic unit of heredity consisting of the nucleotide sequences of DNA involved in synthesis of a protein.

Gene family – set of very similar genes derived by duplication of an ancestral gene and subsequent minor alteration in each gene in the family.

Genetic code – set of 64 codons (triplets) in DNA and the amino acids they represent.

Genetic marker – mutant gene that is useful in genetic mapping studies for locating sites of other genes.

Genome – complete set of genetic information in a specific organism.

Genomic library – collection of cloning vectors (plasmids or phages) which contains fragments of chromosomal DNA.

Genotype – genetic constitution of an individual organism.

Guanine (G) – purine base that pairs with cytosine in DNA.

Gyrase – topoisomerase (enzyme) that can introduce negative super-coils into DNA.

Haploid – chromosome number in the gametes of a species, symbolized by "N."

Helicase – enzyme that unwinds DNA double helix during DNA replication at the replication fork.

Heterogeneous nuclear (hn) RNA – transcripts of varying size synthesized by RNA polymerase II.

Heterozygote – diploid genotype with different alleles for a specific gene.

Hfr – strain of *E. coli* that exhibits an unusually high frequency of recombinations.

Histones – class of approximately five small, basic DNA binding proteins of eukaryotics that together with DNA and RNA comprise chromatin.

Homeobox – nucleotide sequence, found in many genes, that codes for a polypeptide with DNA-binding properties.

Hotspot – region in DNA where mutations occur at exceptionally high frequency.

Hybridization – formation of a double-stranded nucleic acid molecule from complementary single-stranded molecules.

Hydrophilic – describes molecules (e.g., some amino acids) that prefer to be in contact with water.

Hydrophobic – describes molecules that prefer to avoid direct contact with water (e.g., many lipids).

Inducer – small molecule (e.g., galactose) that causes the operon to be expressed.

Induction – expression of specific genes in response to the presence of substrates for the proteins (enzymes), coded by the induced genes.

Initiation factors – regulatory proteins that associate with the small subunit of the ribosome to facilitate initiation of translation of the mRNA.

Inosine (I) – nucleotide containing the base hyporanthine, which base pairs with cytosine (C).

Insertion – placement of additional nucleotide pairs in a specific site in DNA.

Insertion sequence (IS) – DNA sequence in bacteria capable of relocating itself to a new location in the genome, using transposase included in its nucleotide sequence.

Intervening sequence – intron.

Intron – nucleotide sequences in a gene that is transcribed, but eventually removed from the mRNA by specific splicing enzymes.

Inverted repeats – symmetrical nucleotide sequence of DNA that is repeated in opposite orientations on same molecule.

Isozymes – multiple forms of enzymes that differ from one another in either catalytic or physical properties.

Joining region (J) – short segment of an antibody gene that is joined to a larger segment to constitute a complete variable region of an immunoglobulin gene.

kb – an abbreviation for 1,000 nucleotide base-pairs of DNA or RNA.

Lac **operon** – collection of genes in some bacteria that produce enzymes for metabolizing the milk sugar lactose.

Lagging strand (DNA) – replication of DNA employing short segments (Okazaki fragments).

Leader (sequence) – nontranslated sequence of mRNA, upstream of the initiation codon.

Leading strand (DNA) – replication of DNA as a continuous strand.

Library – collection of cloned fragments in a vector (plasmid or phage); usually "cDNA" or "genomic."

Ligation – enzymatically catalyzed formation of a phosphodiester bond that links two DNA molecules.

Linker (DNA) – DNA contained in a nucleosome that is not directly complexed with histenes.

Lipid bilayer – model for cell membranes in which hydrophobic portions of two layers of lipid molecules face inward and hydrophilic portions face outward.

Long-terminal repeat (LTR) – nucleotide sequence that is repeated at the end of a DNA molecule.

Looping (of DNA) - double-stranded DNA bends back on itself, usually when in contact with proteins.

Lysis – the last step in the bacteriophage life cycle, which involves bursting of the bacterial cell membrane and release of multiple phage progeny.

Melting temperature (Tm) – midpoint of the heat denaturation curve for double-stranded DNA.

Membrane – for eukaryotic cell, a lipid bilayer that contains various proteins and surrounds the cytoplasm.

Messenger RNA (mRNA) – single-stranded template RNA that contains information for amino acid sequence of the protein.

Missense mutation – change in DNA nucleotide sequence that leads to substitution of one amino acid for another in the protein.

Mutagen – an agent that causes changes in the nucleotide sequence of DNA.

Mutant – organism that carries a modified inherited gene.

Mutation – change in the nucleotide sequence of DNA that is inherited.

Negative supercoiling – twisting of double-stranded DNA in the opposite direction of the natural double helix twist.

Nick – gap in one strand of DNA of a double helix, caused by breakage of a phosphodiester bond.

Nick translation – method for radioactive labeling double-stranded DNA; DNA polymerase begins at a nick and replaces the original strand with a new one.

Nuclease – enzyme that cleaves phosphodiester bonds in nucleic acids.

Nucleosome – chromatin subunit that consists of DNA and a set of eight histone proteins.

Okazaki fragments – short run of nucleotide's synthesized lagging strand during DNA replication.

Oncogenes – genes that when expressed transform cells into a tumor-like state, characterized by rapid growth.

Operator – nucleotide sequence that binds repressor protein and prevents transcription of adjacent gene.

Operon – set of genes in bacteria that is coordinately regulated and contributes to a single cellular function.

Ori – origin of replication in DNA.

Palindrome – DNA sequence consisting of adjacent inverted repeats that read the same in left direction on one strand as in right direction on the other strand.

Peptide – polymer synthesized by forming a covalent bond between the X-amino group of one amino acid and the X-carboxyl group of another amino acid.

Permissive condition – circumstance, usually an environmental condition, that allows a conditional, lethal mutant organism to live.

Phage – bacterial virus (bacteriophage).

Phenotype – the characteristics or traits of an organism that the genes produce.

Plasmid DNA – circular DNA that replicates inside a bacterial host.

Point mutation – alteration in a single base in DNA.

Polyadenylation (poly A) – addition to the 3' end of a eukaryotic RNA sequence of adenylic acids.

Polycistronic mRNA – mRNA that contains coding sequences of more than one gene.

Polymerase chain reaction (PCR) – method for producing multiple copies of a nucleotide sequence, using primers and DNA replication enzymes that function by repeatedly priming, copying, priming, copying, etc.

Polyribosome (polysome) – mRNA and ribosome complex that is engaged in synthesis of a protein.

Positive supercoiling – twisting of double-stranded DNA in the same direction as the natural double helix twist.

Primary structure – the sequence of subunits, such as amino acids or nucleotides, in a polymer.

Primary transcript – the full length, unmodified RNA synthesized from a single gene.

Primer – a structure that serves as the site for initiation of the synthesis of a polymer, such as DNA.

Prokaryote – organism, such as bacterium or virus, that lacks an organized nucleus.

Promoter – the DNA sequence that binds RNA polymerase and serves as a starting point for transcription.

Pulse chase – experiment in which cells are briefly exposed to radioactive precursors, flushed with nonradioactive precursors, and analyzed to determine fate of initial radioactive precursor.

Purine – a heterocyclic ring molecule with various side chains. (Adenine and quanine are purines found in DNA and RNA.)

Pyrimidine – a heterocyclic ring molecule with six carbons and various side chains. (Cytosine and uraci are pyrimidines found in RNA, while cytosine and thymine are found in DNA.)

Reading frame – the nucleotide sequence of an mRNA as it is translated into a polypeptide, beginning with the AUG initiation codon.

Rec A – *E. coli* DNA repair gene capable of exchanging single-stranded DNA segments.

Recessive allele – form of a gene that normally does not contribute to the phenotype unless present in two copies in homozygous recessive individuals.

Regulation – control of the activity of a gene or enzyme by mechanisms which increase or decrease that activity.

Regulatory gene – gene that codes for an RNA or protein used to control the expression of other separate genes.

Renaturation – process of a macromolecule returning to its native three-dimensional structure. For DNA this involves the two strands base pairing; for protein, it involves folding to its active configuration.

Replication fork – the Y-shaped region of DNA, which serves as the site of synthesis of new strands.

Repressor protein – a regulatory protein that binds to the operator locus on DNA and blocks transcription of the adjacent gene.

Resolvase – enzyme that carries out recombination between two repeated transposons.

Restriction enzymes – enzymes that recognize specific, short nucleotide sequence DNA in double-stranded DNA and cleaves at that site.

Restriction fragment length polymorphism (RFLP) – individual

DNA nucleotide sequence differences that lead to variation in lengths of fragments when DNA is digested with specific restriction endonuclease.

Restriction map – linear representation of DNA sequence that illustrates sites cut by specific restriction endonucleases.

Reverse transcriptase – enzyme that synthesizes DNA from an RNA template.

Reversion – change in nucleotide sequence of DNA that reverses the effects of the original mutation.

Rho factor – *E. coli* protein that recognizes specific nucleotide sequences and causes RNA polymerase to terminate transcription.

Ribose – five carbon sugar found in RNA.

Ribosomal RNA – RNA component of a ribosome.

Ribosome – component of protein synthesis machinery comprised of RNA and protein; it provides structural support for mRNA.

RNA (ribonucleic acid) – polynucleotide that contains ribose sugar.

RNA polymerase – enzyme that synthesizes RNA from ribonucleotide triphosphates; it requires a DNA template.

RNA splicing – process that removes segments of RNA (usually "introns") from a long RNA molecule and joins remaining segments (usually "exons").

Rolling circle replication – process for replicating double-stranded circular DNA.

Satellite DNA – short nucleotide sequence repeated several times (in tandem) that, upon centrifugation, separates away from the main DNA fraction.

Secondary structure – the folding pattern of proteins or nucleic acids, which results from interaction of subunits.

Sex plasmid – plasmid which is capable of initiating conjugation (mating) between two bacteria.

Shine–Dalgarno sequence – the AGGAGG purine sequence on bacterial mRNA, which facilitates binding of the mRNA to a ribosome.

Sigma factor – subunit of RNA polymerase that facilitates binding to promoter sites on DNA.

Splicing – process which removes introns from RNA and joins exons (coding sequences) to form "translatable" mRNA.

SSB protein – protein that binds to single-stranded DNA during replication.

Stacking (of bases) – the tight packing of purine and pyrimidine base-pairs in double-stranded DNA, favored by the energy of interaction.

Supercoiling – coiling of a large, circular, double-stranded DNA molecule back around itself.

Suppressor – a mutation which acts to restore the original nucleotide sequence.

Tandem repeats – nucleotide sequence of DNA that is serially repeated many times.

TATA box – an A-T-rich sequence, usually consisting of 7 or 8 nucleotides, located approximately 25 base-pairs upstream of the transcription start site. It serves as a binding site for RNA polymerase.

Tertiary structure – the native three-dimensional configuration of a macromolecule, which results after its subunit interactions are complete.

Thymine – pyrimidine base found in DNA.

Thymine dimer – adjacent thymine residues in DNA that have been chemically linked, usually by the action of ultraviolet irradiation.

Topoisomerase – enzyme that alters amount of supercoiling in large, double-stranded DNA molecules.

Trailer (sequence) – nucleotide sequence at the 3′ end of a mRNA that is not translated into part of the protein.

Transcription – synthesis of RNA from a complementary DNA template.

Transduction – transfer of bacterial gene from one bacterium to another by a bacteriophage.

Transfer RNA (tRNA) – small RNA that transfers amino acids to the mRNA template for protein synthesis.

Translation – synthesis of a protein from an mRNA template.

Transposase – enzyme that functions to insert a transposon into a new site in DNA.

Transposon – nucleotide sequence that can insert itself, with the aid of an enzyme (transposase), into new locations in the genome.

Upstream – direction ahead of the start site for transcription or translation.

Uracil – pyrimidine base found in RNA.

van der Waals forces – weak attractive forces that result from transitory fluctuations in electron-charged densities in neighboring atoms.

Wobble hypothesis – explanation for ability of certain tRNAs to exhibit unusual base-pairing with more than one triplet codon.

References

Chapter One: Systems of and Analytical Approaches to Molecular Biology

Lerner, R. A., and F. J. Dixon. "The human lymphocyte as an experimental animal." *Scientific American:* (June, 1973): 82–91.

Phaff, H. J. "Industrial microorganisms." *Scientific American:* (September, 1981): 76–89.

Stahl, F. W. "Genetic recombination." *Scientific American:* (February, 1987): 90–101.

Weintraub, H. M. "Antisense RNA and DNA." *Scientific American:* (January, 1990): 40–46.

White, R., and J-M Lalovel. "Chromosome mapping with DNA markers." *Scientific American:* (February, 1988): 40–48.

Youvan, D. C., and B. L. Marrs. "Molecular mechanisms of photosynthesis." *Scientific American:* (June, 1987): 42–48.

Chapter Two: Macromolecules

Sharon, N. "Carbohydrates." *Scientific American:* (November, 1980): 90–116.

Weinberg, R. A. "The Molecules of Life." *Scientific American:* (October, 1985): 48–57.

Chapter Three: Nucleic Acids

Bauer, W. R., F. H. C. Crick, and J. H. White. "Supercoiled DNA." *Scientific American:* (July, 1980): 118–133.

Britten, R. J., and D. E. Kohne. "Repeated segments of DNA." *Scientific American:* (April, 1970): 24–31.

Cech, T. R. "RNA as an enzyme." *Scientific American:* (November, 1986): 64–75.

Darnell, J. E. Jr. "RNA." *Scientific American:* (October, 1985): 68–78.

Dickerson, R. E. "The DNA helix and how it is read." *Scientific American:* (December, 1983): 94–111.

Felsenfeld, G. "DNA." *Scientific American:* (October, 1985): 58–67.

Chapter Four: The Physical Structure of Protein Molecules

Capra, J. D., and A. B. Edmundson. "The antibody combining site." *Scientific American:* (January 1977): 50–59.

Doolittle, R. F. "Proteins." *Scientific American:* (October, 1985): 88–99.

Grey, H. M., A. Sette, and S. Buus. "How T cells see antigen." *Scientific American:* (November, 1989): 56–64.

Karplus, M., and J. A. McCammon. "The dynamics of proteins." *Scientific American:* (April, 1986): 42–51.

Koshland, D. E. Jr. "Protein shape and biological control." *Scientific American:* (October, 1973): 53–64.

Lerner, R. A., and A. Tramontano. "Catalytic antibodies." *Scientific American:* (March, 1988): 58–70.

Perutz, M. "The hemoglobin molecule." *Scientific American:* (November, 1964): 64–71.

Perutz, M. "Hemoglobin structure and respiratory control." *Scientific American:* (December, 1978): 92–125.

Phillips, D. C. "The three–dimensional structure of an enzyme molecule." *Scientific American:* (November, 1966): 78–84.

Unwin, N., and R. Henderson. "The structure of proteins in biological membranes." *Scientific American:* (February, 1984): 78–94.

Chapter Five: Macromolecular Interactions and the Structure of Complex Aggregates

Bauer, W. R., F. H. C. Crick, and J. H. White. "Supercoiled DNA." *Scientific American:* (July, 1980): 118–133.

Bretscher, M. S. "The molecules of the cell membrane." *Scientific American:* (October, 1985): 100–108.

Capaldi, R. A. "A dynamic model of cell membranes." *Scientific American:* (March, 1974): 26–33.

Fox, C. F. "The structure of cell membranes." *Scientific American:* (February, 1972): 30–38.

Kornberg, R. D., and A. Klug. "The nuclosome." *Scientific American:* (February, 1981): 52–64.

Lodish, H. F., and J. E. Rothman. "The assembly of cell membranes." *Scientific American:* (January, 1979): 48–63.

Murray, A. W., and J. W. Szostak. "Artificial chromosomes." *Scientific American:* (November, 1987): 62–68.

Chapter Six: The Genetic Material

Eigen, M., W. Gardiner, P. Shuster, and R. Winkler-Oswatitisch. "The origin of genetic information." *Scientific American:* (April, 1981): 88–118.

Mirsky, A. E. "The discovery of DNA." *Scientific American:* (June, 1968): 78–92.

Patterson, D. "The causes of Down's syndrome." *Scientific American:* (August, 1987): 52–57.

Stahl, F. W. "Genetic recombination." *Scientific American:* (February, 1987): 90–101.

Varmus, H. "Reverse transcription." *Scientific American:* (September, 1987): 56–64.

Chapter Seven: DNA Replication

Mullis, K. B. "The unusual origin of the polymerase chain reaction." *Scientific American:* (April, 1990): 56–65.

Radman, M., and R. Wagner. "The high fidelity of DNA replication." *Scientific American:* (August, 1988): 40–46.

Chapter Eight: DNA Repair

Howard-Flanders, P. "Inducible repair of DNA." *Scientific American:* (November, 1981): 72–80.

Radman, M., and R. Wagner. "The high fidelity of DNA replication." *Scientific American:* (August, 1988): 40–46.

Wang, J. C. "DNA topoisomerases." *Scientific American:* (July, 1982): 94–109.

Chapter Nine: Transcription

Chambon, P. "Split genes." *Scientific American:* (May, 1981): 60–71.

Darnell, J. E. Jr. "The processing of RNA." *Scientific American:* (October, 1983): 90–100.

Darnell, J. E. Jr. "RNA." *Scientific American:* (October, 1985): 68–78.

Miller, O. L. Jr. "The visualization of genes in action." *Scientific American:* (March, 1973): 34–42.

Spiegelman, S. "Hybrid nucleic acids." *Scientific American:* (May, 1964): 48–56.

Steitz, J. A. "Snurps." *Scientific American:* (June, 1988): 56–63.

Chapter Ten: Translation

Abraham, E. P. "The beta–lactam antibiotics." *Scientific American:* (June, 1981): 76–86.

Darnell, J. E. Jr. "RNA." *Scientific American:* (October, 1985): 68–78.

Eigen, M., W. Gardiner, P. Shuster, and R. Winkler–Oswatitisch. "The origin of genetic information." *Scientific American:* (April, 1981): 88–118.

Lake, J. A. "The ribosome." *Scientific American:* (August, 1981): 84–97.

Miller, O. L. Jr. "The visualization of genes in action." *Scientific American:* (March, 1973): 34–42.

Nirenberg, M. "The genetic code." *Scientific American:* (March, 1963): 80–94.

Rich, A., and S. H. Kim. "The three–dimensional structure of transfer RNA." *Scientific American:* (January, 1978): 52–62.

Chapter Eleven: Mutagenesis, Mutations, and Mutants

Devoret, R. "Bacterial tests for potential carcinogens." *Scientific American:* (August, 1979): 40–49.

Sigurbjörnsson, B. "Induced mutations in plants." *Scientific American:* (January, 1971): 86–95.

Upton, A. C. 1982. "The biological effects of low–level ionizing radiation." *Scientific American:* (February, 1971): 41–49.

Wills, C. "Genetic load." *Scientific American:* (March, 1970): 98–107.

Chapter Twelve: Plasmids and Transposons

Chilton, M-D. "A vector for introducing genes into plants." *Scientific American:* (June, 1983): 50–59.

Clowes, R. C. "The molecule of infectious drug resistance." *Scientific American:* (April, 1973): 19–27.

Cohen, S. N., and J. A. Shapiro. "Transposable genetic elements." *Scientific American:* (February, 1980): 40–49.

Federoff, N. V. "Transposable genetic elements in maize." *Scientific American:* (June, 1984): 84–98.

Novik, R. P. "Plasmids." *Scientific American:* (December, 1980): 102–127.

Chapter Thirteen: Recombinant DNA and Genetic Engineering: Molecular Tailoring of Genes

Anderson, W. F., and E. G. Diacumakos. "Genetic engineering in mammalian cells." *Scientific American:* (July, 1981): 106–121.

Brown, D. D. "The isolation of genes." *Scientific American:* (August, 1973): 21–29.

Cohen, S. "The manipulation of genes." *Scientific American:* (July, 1975): 25–33.

Gilbert, W., and L. Villa-Komaroff. "Useful proteins from recombinant bacteria." *Scientific American:* (April, 1980): 74–94.

Hopwood, D. A. "The genetic programming of industrial microorganisms." *Scientific American:* (September, 1981): 91–102.

Lawn, R. M., and G. A. Vehar. "The molecular genetics of hemophilia." *Scientific American:* (March, 1986): 48–54.

Milstein, C. "Monoclonal antibodies." *Scientific American:* (October, 1980): 66–74.

Mullis, K. B. "The unusual origin of the polymerase chain reaction." *Scientific American:* (April, 1990): 56–65.

Pestka, S. "The purification and manufacture of human interferons." *Scientific American:* (August, 1983): 36–43.

Weinberg, R. A. "Finding the anti-oncogene." *Scientific American:* (September, 1988): 44–51.

Chapter Fourteen: Regulation of Gene Activity in Prokaryotes

Maniatis, T., and M. Ptashne. "A DNA operator-repressor system." *Scientific American:* (January, 1976): 65–76.

Nomura, M. "The control of ribosome synthesis." *Scientific American:* (January, 1984): 102–114.

Ptashne, M. "How gene activators work." *Scientific American:* (January, 1989): 40–47.

Ptashne, M., and W. Gilbert. "Genetic repressors." *Scientific American:* (June, 1970): 36–44.

Chapter Fifteen: Bacteriophage

Campbell, A. M. "How viruses insert their DNA into the DNA of the host cell." *Scientific American:* (December, 1976): 102–113.

Fiddes, J. C. "The nucleotide sequence of a viral DNA." *Scientific American:* (December, 1977): 56–67.

Ptashne, M., A. D. Johnson, and C. O. Pabo. "A genetic switch in a bacterial virus." *Scientific American:* (November, 1982): 128–140.

Chapter Sixteen: Regulation of Gene Activity in Eukaryotes

Britten, R. J., and D. Kohn. "Repeated segments of DNA." *Scientific American:* (April, 1968): 24–31.

Holliday, R. "A different kind of inheritance." *Scientific American:* (June, 1989): 60–70.

Hunter, T. "The proteins of oncogenes." *Scientific American:* (August, 1984): 70–79.

Leder, P. "The genetics of antibody diversity." *Scientific American:* (May, 1982): 102–115.

Moses, P. B., and N-H. Chua. "Light switches for plant genes." *Scientific American:* (April, 1988): 88–93.

O'Malley, B. W., and W. T. Schrader. "The receptors of steroid hormones." *Scientific American:* (February, 1976): 32–43.

Ptashne, M. "How gene activators work." *Scientific American:* (January, 1989): 40–47.

Scheller, R. H., and R. Axel. "How genes control an innate behavior." *Scientific American:* (March, 1984): 54–62.

Stein, G. S., J. S. Stein, and L. J. Kleinsmith. "Chromosomal proteins and gene regulation." *Scientific American:* (February, 1975): 46–57.

Tonegawa, S. "The molecules of the immune system." *Scientific American:* (October, 1983): 122–131.

Answers

Answers to Drill Questions (Ch. 1)

1. a. Eukaryote.
   b. Only bacteria are prokaryotes.

2. 1, D; 2, A; 3, B; 4, C.

3. Haploid—having only one copy of each chromosome. Diploid—having two copies of each chromosome.

4. A partial diploid is a cell that contains one full set of genes and duplicates of a subset of these genes. It is used in haploid organisms such as bacteria to allow genetic recombination and hence genetic mapping and to study complementation between different mutations.

5. Transfection is a technique by which bacterial cell walls are made permeable such that free DNA may get inside the cell without the aid of a phage particle of other infective agent.

6. Cells in culture have been separated from other cells with which they may have been in contact and have lost their normal route for receiving nutrients and so forth. Furthermore, cells in culture grow and divide, while most cells in a living organism are quiescent. Last, cells in culture usually die within a few weeks. Cells from an established cell line, however, do not die, but continue to grow indefinately and are said to be immortalized.

7. In general, at least four. However, intragenic complementation may be possible. That is, if the protein molecule is folded in such a way that it consists of two interacting regions, a deleterious change in one region (which destroys the interaction) could be compensated for by a change (which, by itself, is also deleterious) in the other region. In other words, the two incorrectly folded regions could interact correctly to yield a functional protein.

8. A model is a conceptual explanation of a complex phenomenon, often in the form of a diagram or cartoon. A good model is one that is consistent with all the known data and can be experimentally tested.

Answers to Problems (Ch. 1)

1. Growth is for five generations. Thus the initial number of bacteria increases by a factor of 2^5 or 32, and the final concentration is $32 \times 10^5 = 3.2 \times 10^6$ cells per ml.

2. a. *kyuQ*.
   b. A regulatory mutant that prevents synthesis of each gene product.

3. That the *g*-gene and *h*-gene products probably join together to form an active macromolecular complex. The regions in which mutations occur (that can be compensated or are compensatory) are probably in the binding sites.

4. a.

a	c		b	d
	2	13	4	

   b. Production of *ABD* progeny from this cross requires a double recombination: one between *a* and *b* and another between *b* and *d*. The probability of this is the probability or recombination between *a* and *b* times the probability of recombination between *b* and *d* or $0.15 \times 0.04 = 0.006$ or 0.6%.

5. a. Six genes.
   b. Some of the closely linked mutations might lie in closely linked genes, that is, they might complement one another. Moreover, other genes in the pathway for which no mutants have yet been isolated might exist. Ten mutations distributed over seven or eight genes according to the Poisson distribution would stand a good chance of leaving one gene untouched.

6. Four genes. The groups are (1,7), (2,4), (3,6), and (5).

Answers to Drill Questions (Ch. 2)

1. proteins—amino acids; nucleic acids—nucleotides; polysaccharides—sugars (most often glucose).

2. An amino group and a carboxyl group are linked to form a peptide linkage.

3. The termini of a protein molecule consist of an amino group and a carboxyl group. These groups are also contained in the side chains of some of the amino acids forming proteins. For example, both aspartic acid and glutamic acid have a free carboxyl group, and both asparagine and glutamine have a free amino group.

4. DNA—thymine; RNA—uracil

5. The 5'-phosphate of one nucleotide and the 3'-OH of the adjacent nucleotide; a phosphodiester group.

6. A nucleotide is a nucleoside phosphate.

7. Two molecules that are poorly soluble, that is, interact poorly with water, can reduce contact with water by clustering, because clustering minimizes the ratio of surface to volume.

8. strongest—ionic bond; weakest—van der Waals attraction

9. All DNA molecules have a negative charge and, therefore, move in the same direction when electrophoresed. This is not true for protein molecules, however, because they may have a net positive or negative charge.

10. One molecule of SDS binds per amino acid, giving each protein the same charge per amino acid, and, thus, the same net charge. When SDS is bound and there are no disulfide bonds, all proteins have nearly the same shape.

Answers to Problems (Ch. 2)

1. a) No, since differences in shape can affect mobility—that is, two proteins differing in both molecular weight and shape might have the same electrophoretic mobility. b) Yes. All linear DNA fragments have the same basic shape and same charge–to–mass ratio. Therefore, if all bands seen were of equal intensity, the conclusion would be that five different DNA fragments were present.

2. cysteine—disulfide bonds; lysine—ionic bonds and hydrogen bonds; isoleucine—hydrogen bonds; glutamic acid—ionic bonds and hydrogen bonds. All could participate in van der Waals attractions.

3. a) The compactness probably results from an

ionic bond(s) between unlike charged groups. b) The molecule probably contains many groups having like charges, and in the lower ionic strength of 0.01 M NaCl these groups repel one another.

4. Three. Other possible interactions involving amino acid side chains such as hydrophobic interactions, ionic bonds, and van der Waals attractions would determine the naturally occurring structure of the protein.

5. Electrophoretic mobility increases with charge but is reduced by friction between the moving molecule and the solvent molecules. Since valine has a longer side chain than alanine, it generates greater friction and moves more slowly.

Answers to Drill Questions (Ch. 3)

1. The fraction that is A is 0.2 (10%).

2. The virus must have as its genome a single-stranded DNA molecule (at least one), because $[A]+[G]$ is not equal to $[T]+[C]$.

3. In a double-stranded molecule, $[A]+[G] = [T]+[C]$. The relationship between $[A]+[T]$ and $[G]+[C]$ varies from molecule to molecule.

4. The B helix is considered to be the common form of DNA, but DNA forms an A helix under conditions of dehydration. A Z helix is favored by segments with alternating Gs and Cs.

5. a) No—there would be nothing to prevent the circle from untwisting.
   b) Yes—this would form a supercoil.
   c)G No—this would merely form a relaxed circle.

6. Hydrophobic and hydrophilic properties of the bases and phosphate groups, respectively, and hydrogen bonding between the bases of a base pair.

7. The salt concentration must be high enough to neutralize the mutually repulsive negative charges of the phosphate groups, and the temperature must be sufficient to avoid extensive intrastrand base–pairing.

8. With no cations to complex with the phosphate groups, the repulsion among them would be great enough to denature the DNA.

9. 480 turns of the helix are present (4800 bases at 10 bases per turn). The molecule would be 1.632 μm long (480 turns × 34 Å/turn).

10. No, there would be no free ends to the molecule.

Answers to Problems (Ch. 3)

1. a) would be least likely to re-form the original structure, as it could have extensive intrastrand base pairing.

2. b) would have the highest T_m, as it has the highest $G+C$ content. a) would require a higher temperature for reannealing (relative to its T_m) in order to overcome the intrastrand base–pairing.

3. $5'-ATCAAGGTA-3'$

4. Central 10% deleted:

   Central 10% replaced with nonhomologous DNA:

5. 30%.

6. The solution contained a mixture of copies of two different DNA molecules. One type had a considerably higher $G+C$ content than the other.

Answers to Drill Questions (Ch. 4)

1. (a) The polar amino acids are arginine, asparagine, aspartic acid, cysteine, glutamic acid, glutamine, histidine, lysine, serine, threonine, and tyrosine. The nonpolar amino acids are alanine, glycine, isoleucine, leucine, methionine, phenylalanine, proline, tryptophan, and valine.
   (b) Isoleucine is more nonpolar than alanine, because it has a long, nonpolar side chain.

2. The peptide bond.

3. Cysteine—disulfide; arginine—ionic; valine—hydrophobic; aspartic acid—ionic and hydrogen bonds. All of course could be involved in van der Waals attractions.

4. Both will probably be near an amino acid with the opposite charge.

5. Set (c), in a hydrophobic cluster.

6. They would probably be scattered throughout the primary sequence, though there would probably also be a few in small clusters.

7. Primary structure is simply the linear sequence of amino acids. Secondary structure involves hydrogen bonding between peptide groups. Examples are the α helix and β structure. Tertiary structure is the folding of a polypeptide chain and is formed from secondary structure elements plus several other interactions between amino acid side chains including ionic bonds between oppositely charged side chains, various hydrogen bonds, hydrophobic clustering of nonpolar side chains, and the coordination of metal ions.

8. Since the α helix and β structure are both rigid, the former protein would be long, and thin or fibrous; the latter would be more flexible, and spherical or globular in shape.

9. Both the enzyme active site and the substrate change shape slightly upon binding and the strain placed on the substrate by the shape-change aids in the catalysis.

Answers to Problems (Ch. 4)

1. (a) Yes. A van der Waals attraction would favor aggregation. This would be aided by a hydrophobic interaction if the side chains on the surface were very nonpolar.
(b) No. For geometric reasons, the hydrophobic regions probably cannot come into contact.
(c) No. Charge repulsion by the lysines will effectively counteract any tendency to form a hydrophobic cluster.
(d) Yes. If the geometry is appropriate, the alternation of unlike charges will allow unlike charges in the two molecules to attract one another.

2. No. In general, enzymes must be slightly flex-ible in order to adapt their shape to the substrate and to carry out the required chemical reaction.

3. The enzyme could be a multisubunit protein in which one defective subunit is sufficient to eliminate the enzymatic activity.

4. When there is a single binding site, it is probably located at or near the junction of all the subunits. This location is certainly unlikely if there are several identical binding sites. If the number of binding sites equals the number of subunits, one might guess that the binding sites are far from all regions of contact between the subunits. If the number of binding sites is half the number of subunits, each site probably includes the contact region of two subunits.

5. If you looked at the protein structures with and without substrate bonds, there should be no change in the structure for a lock–and–key mechanism, but there will be a change in the structure for an induced–fit mechanism.

Answers to Drill Questions (Ch. 5)

1. An octameric disk is an aggregate of two each of histones H2A, H2B, H3, and H4. A core–particle consists of an octameric disk plus the 140 base–pairs of DNA that encircle it. A nucleosome includes the core–particle, linker DNA, and the histone H1.

2. A eukaryotic cell has more DNA than a prokaryotic cell, and that DNA must be contained within a nucleus.

3. During cell division, the DNA must be organized into a compact unit in order to be moved from the metaphase plate toward the poles of the cell. During other stages of the cell cycle (when replication and transcription occur) proteins must interact with specific base sequences of DNA, so the DNA must be less compact to permit these interactions.

4. The DNA would be most compact during mitosis, to facilitate movement of the chromosome as a unit. During S phase, DNA is being replicated, and so must be less compact.

5. 3.4 hm, the length of one turn of the helix.

6. The polar regions are on the "ends" of the protein that are outside of the membrane,

while the non–polar region is the part of the protein that is actually within the membrane.

7. Van der Waals attractions and hydrophobic interactions are the main factors stabilizing a lipid bilayer.

Answers to Problems (Ch. 5)

1. In this case, there are 200 base pairs per nucleosome. A DNA molecule 2.4 cm long contains about 7×10^7 base–pairs (3.4 nm per base–pair). Therefore, there would be about 3.5×10^5 nucleosomes.

2. Electrostatic interactions.

3. The unknown component is a covalently linked nucleotide–amino acid complex; the protein–DNA binding is therefore through a covalent bond.

4. The phospate groups of the backbone of the helix.

5. Since the Cro protein binds stereospecifically, it is likely that the Cro–DNA complex would not form. Also, specific contact points might be blocked.

Answers to Drill Questions (Ch. 6)

1. The conversion of the genotype of a bacterium to the genotype of a second bacterium by exposure of the first bacterium to DNA from the second.

2. Many plant and animal viruses and some bacteriophages.

3. A codon.

4. ^{32}P for DNA and ^{35}S for proteins.

5. Mutation.

6. Uracil.

7. Thymine is 5–methyluracil.

8. Chemical alterations and replication errors.

Answers to Problems (Ch. 6)

1. The RNA in a virus is protected from the environment by a protein coat.

2. Progeny of a transformed cell all had the new character; in genetic terms, the cells bred true.

3. The chemical mechanism for hydrolysis of a phosphodiester bond requires participation of a free OH group, which is present on the 2′ carbon in RNA, but not in DNA.

4. It is widely believed that DNA was a simple tetranucleotide incapable of carrying the information required of a genetic material.

5. Transforming activity was not lost by reaction with either proteolytic enzymes or ribonucleases. Treatment with DNases inactivated the transforming principle.

6. During replication, an incorrect base or an extra base is inserted or deleted in the daughter molecule.

Answers to Drill Questions (Ch. 7)

1. Semiconservative

2. Template

3. All, 1/2, 1/4, respectively

4. Theta replication

5. Rolling circle replication

6. Positive supercoiling

7. Negative supercoiling

8. Polymerizing activity, 5′—3′ exonuclease, 3′—5′ exonuclease

9. A 3′–OH group

10. Polymerization, and the 5′—3′ exonuclease

Answers to Problems (Ch. 7)

1. No. After one round of replication it has hybrid density, but an ^{15}N strand always remains in the circle.

2. The 5′—3′ activity removes ribonucleotides from the 5′ termini of precursor fragments, and the 3′—5′ exonuclease removes a base that has been incorrectly added to the growing end of a DNA strand.

3. Pol III requires a helicase to unwind the DNA; pol I can unwind a helix without an accessory protein.

4. DNA polymerase joins a 5'–triphosphate to a 3'–OH group and in so doing removes two phosphates, so that the phosphodiester bond contains one phosphate; a ligase joins a 5'–monophosphate to a 3'–OH group.

5. One strand is copied from the 3' end to the 5' end, and the other strand is copied in the direction opposite to the movement of the replication fork by synthesis in short pieces (see Figures 7-16 and 7-17).

6. The RNA primer must be removed from the precursor fragment that was made first, because DNA ligase cannot join DNA to RNA.

Answers to Drill Questions (Ch. 8)

1. Depurination and deamination of cytosine

2. a. the system for repair of deaminated cytosine (uracil N–glycosylase)
   b. photolyase
   c. the systems for repair of depurination and deamination of cytosine (endonuclease)

3. It is inducible and it increases the frequency of replication errors.

4. One explanation is that the T4 possesses its own repair system.

5. d. DNA polymerase I

Answers to Problems (Ch. 8)

1. The result of having an incorrect base in DNA is much more drastic than in RNA. Many mRNA molecules (and thus proteins) are made from a single DNA molecule, while relatively few proteins are made from a single RNA molecule before it is degraded. Therefore, a mutation in DNA will be amplified, and if uncorrected, passed on to other generations.

2. a. X is not inducible, because the enzymes were present before protein synthesis was blocked by chloramphenicol.
   b. X would probably be considered to be inducible. The residual 5% could be due to either a second, noninducible system, or to a small amount of synthesis of X proteins when chloramphenicol is present.

3. If the dimer is in the strand being copied by the leading strand, only the leading strand would have a gap. If the dimer were able to block helicase however, then a dimer in the template for the lagging strand would block the advance of the leading strand, and gaps would be produced in both the leading and lagging strands.

Answers to Drill Questions (Ch. 9)

1. a. Ribonucleoside 5'–triphosphates
   b. Double–stranded DNA
   c. RNA polymerase
   d. No

2. The reactions are identical—reaction of a nucleoside 5'–triphosphate with a 3'–OH terminus of a nucleotide to form a 5' to 3'–diester bond. The substrates differ in that DNA polymerase joins deoxynucleotides while RNA polymerase joins ribonucleotides.

3. A 5'–triphosphate and a 3'–OH

4. a. An RNA molecule that is translated into one or more proteins.
   b. A primary transcript is a complementary copy of a DNA strand. It could be a precursor to mRNA, tRNA, or rRNA and may be processed to form a functional RNA.
   c. A cistron is a DNA segment between and including translation start and stop signals that contains the base–sequence corresponding to one polypeptide chain. A polycistronic mRNA molecule contains sequences encoding two or more polypeptide chains.
   d. Leaders, spacers, and the unnamed region following the last stop codon of a mRNA are not translated.

5. Five subunits, the sigma subunit is responsible for positioning.

6. a. $^-10$ and $^-35$
   b. TATAAAA

7. a. A terminal structure in which a methylated guanosine is in a 5' to 5'-triphosphate linkage at the 5' terminus of mRNA.
   b. The 3'–OH end
   c. All mRNA molecules (except those of sev-

eral viruses) are capped. Some mRNA molecules lack the poly(A) tail.

8. a. Untranslated sequences that interrupt the coding sequence of a transcript which are removed before translation begins.
   b. Removal of introns and linking together of exon sequences.

Answers to Problems (Ch. 9)

1. a. It is probably a sequence that existed at very early times and from which others were derived by mutation. Furthermore, it indicates that the biochemical system that used the sequence existed very long ago.
   b. It is essential for some stage or promotion—in particular, binding of RNA polymerase or initiating polymerization.
   c. The rates of initiation may differ slightly. It is difficult to be certain of this point though because the rate of initiation is mainly determined by the -35 region.

2. Since the triphosphate of the primary transcript is absent, the molecule has been processed. The processing could be extensive, or as simple as triphosphate hydrolysis; but there is no way of knowing from the information given.

3. Eukaryotic mRNA molecules contain a terminal poly(A) which can hydrogen–bond to the dT in the column under renaturing conditions. Thus, the mRNA will be retained by the column, and other RNA species will pass through. The mRNA can be eluted from the column simply by heating.

4. The number of times each of the four bases appears at each position is noted. The maximum value (as a percentage) is indicated as a subscript in the consensus sequence that follows:
   $A_{80}T_{70}G_{80}C_{90}A_{80}C_{70}$.

5. 5′—AGCUGCAAUG—3′ and 5′—CAUUGCAGU—3′

6. *E. coli* is a prokaryote and its RNA polymerase does not recognize all eukaryotic promoters. The eukaryote yeast, on the other hand, provides not only a suitable eukaryotic RNA poly-

merase, but also various transcription factors that act to facilitate transcription.

Answers to Drill Questions (Ch. 10)

1. a., b., and c. are true. d. Amino acyl tRNA synthetases are required (see question 2). e. Rho is involved in the termination of transcription, not translation. However if the anticodon of a tRNA has been altered, it can sometimes read a termination codon.

2. a., d., and e.

3. Prokaryotic: 70S with 30S and 50S subunits containing 5S, 16S, and 23S RNA. Eukaryotic: 80S with 40S and 60S subunits containing 5S, 5.8S, 18S, and 28S RNA.

4. 1. Prokaryotes initiate translation at an AUG codon just downstream of the Shine–Dalgarno sequence (ribosome binding site) AGGAGGU, which base–pairs with a section of the 16S rRNA near its 3′ end. The eukaryotic ribosomal small subunit binds at the 5′ cap of mRNA, and the first AUG encountered is used.
   2. Prokaryotes initiate translation with formylmethionine, whereas eukaryotes initiate with an unmodified methionine.
   3. Prokaryotic mRNAs may be polycistronic, since the ribosome may reinitiate translation at a second AUG following termination of translation of the first reading frame. Eukaryotic ribosomes do not reinitiate translation without first dissociating.

5. Steps a., b., and d. are. Step c. is false. Amino acids are coupled to tRNA molecules by aminoacyl tRNA synthetases.

6. A reading frame is a string of sequential, nonoverlapping codons beginning with an initiation codon (AUG), ending with a termination codon (UAA, UAG, or UGA), and containing codons for amino acids in between. In addition to a reading frame, most mRNAs have 5′ leader and 3′ tail sequences. A 5′ cap and poly A tail is present in eukaryotic mRNAs. Prokaryotic mRNAs can have several reading frames separated by short spacer regions.

7. Translation occurs along the mRNA in the 5′ to 3′ direction, the same direction as the syn-

thesis of the mRNA itself. Thus protein synthesis can occur while the mRNA is being copied from the DNA. In the reverse polarity, protein synthesis would have to await the completion of the molecule of mRNA. Thus, in the existing system, protein synthesis can start earlier than would be possible with reverse polarity, and the mRNA is relatively resistant to nuclease attack. This is only true in prokaryotes, of course.

8. b.

9. Formation of the 70S initiation complex and translocation.

Answers to Problems (Ch. 10)

1. Met Pro Leu Ile Ser Ala Ser

2. There are two families of codons for arginine, the CGX family and the AG_G^A family. Single base changes in the first two bases of the CGX family of in any of the bases of the AG_G^A family could yield the following amino acid replacements: histidine, glutamine, cysteine, tryptophan, serine, glycine, leucine, proline, isoleucine, threonine, lysine, or methionine.

3. For UAG, the amino acids are Tyr, Leu, Trp, Ser, Lys, Glu, and Gln. For UAA they are Tyr, Lys, Glu, Gln, Leu, and Ser.

4. The Arg–2 codon must have been AGG (the only arginine codon which can become AUG in one step), and Arg–2 could therefore also be replaced by Gly, Trp, Lys, Thr, or Ser. The Arg–3 codon must have been AGA, and Arg–3 could also be replaced by Ser, Lys, Thr, and Gly. The Arg–1 codon could have been any one of the six arginine codons, and cannot be unambiguously identified.

5. Val-Cys-Val-Cys-Val-Cys . . ., and peptides of various sizes starting with Val or Cys.

6. a. Threonine, i.e. 5′–ACX–3′. Remember that the codon–anticodon pairing is antiparallel.
b. ACU, ACC, and ACA since I in the wobble position can pair with U, C, or A.
c. CGU, since only ACG remains to be read, and C in the wobble position will only pair with G. U in the wobble position would read G, but also A, so that both ACG and ACA codons would be read, and ACA is already read by the first tRNA discussed.

Answers to Drill Questions (Ch. 11)

1. A nonsense mutation stops chain growth at the mutational site; a missense mutation causes an amino acid substitution at the mutational site.

2. BU is a base–analogue mutagen. It is a thymine analogue and is incorporated into the DNA during replication by base pairing with adenine. Due to the bromine atom, a shift in the keto–enol equilibrium occurs, such that the enol form is more prevalent than it would be with thymine. The enol form may base–pair with guanine in future rounds of replication, leading to A.T → G.C or T.A → C.G transitions after one more round of replication. BU also inhibits the synthesis of dCTP without inhibiting the synthesis of TTP, so the level of TTP relative to dCTP becomes very high, and T is then often incorporated into DNA by pairing with G due to the shortage of C. This causes G.C → A.T or C.G → T.A transitions.

3. Many amino acid changes will not yield a mutant phenotype, and many mutations will be chain–termination mutants.

4. The ratio of polymerizing activity to exo–nuclease function will decrease in the antimutator, since errors will be removed more often.

5. A silent mutation is a change in the base–sequence of the gene which either does not change the corresponding amino acid in the protein (due to the redundancy of the genetic code), or does change the amino acid but does not cause any noticeable effect on the phenotype (because some amino acids may be substituted for others without harmful effects). In either case wild–type activity is maintained. A leaky mutation is one which changes an amino acid, but does not completely abolish activity, though activity is altered in some way.

6. A transition is a base–pair change in which the

purine–pyrimidine orientation is not altered; a transversion is a base–pair change in which the orientation is changed.

7. There are two possibilities: There are two genes for the original tRNA species, and only one of these has been mutated, or natural chain–termination sequences might usually consist of two or more different termination codons. The nonsense suppressor would suppress only one, and chain termination would still occur. Both possibilities occur.

8. They clearly interact. Since a charge sign–change yields a mutant and a second sign–change in another amino acid yields a revertant, amino acids 28 and 76 are probably held together by an ionic bond.

Answers to Problems (Ch. 11)

1. This mutant could not be isolated because there is no temperature at which it could grow.

2. a., c., g., and h., as they involve either changes in polarity, charge sign, or chemical properties.

3. No. If the original rate of mutation is one in 10^5 for a sequence 1000 base pairs long, this gives the approximate probability of hitting any base in the sequence and causing a detectable mutation. Since the reversion rate is also one in 10^5, this probably represents other random changes in the sequence. The frequency of an exact replacement should be the probability of a change × the number of sites that can be changed. For this problem the answer would be one in (10^5 × 1000), which is one in 10^8, about 1000 times less than the initial mutation rate. However since not all changes at the original site would regenerate the original amino acid, the actual rate of exact reversal would still be somewhat lower.

4. The original mutation is probably a frameshift mutation. The mutagenes listed cause a variety of transitions and transversions, some of which would almost certainly suppress a missense or nonsense mutation. To suppress a frameshift mutation, would require an intercalating agent such as proflavine or acridine orange; muta-

gens which themselves cause frameshifts. The original mutation could also be caused by a transposable element, which would not be suppressed by the mutagens available.

5. Chemical mutagens alter (damage) bases in DNA giving them novel base–pairing properties. For mutation to occur, however, incorrect base–pairing during replication must occur. Repair systems exist and can often repair or replace damaged bases correctly. Under slow growth conditions, the repair systems have more time to repair the damage before replication occurs. Thus, fewer mutants would be recovered from the slower growing culture.

Answers to Drill Questions (Ch. 12)

1. The plasmid would have an origin of replication similar enough to the host origin of replication so that it would be recognized by the host replication machinery without the aid of a plasmid–encoded Rep protein. The plasmid might also contain a partition, or *par* sequence.

2. a. Replication begins at the *ori* sequence of the plasmid.
   b. The Rep protein binds the *ori* sequence and then directs the assembly of the replication machinery.
   c. DNA polymerase III
   d. A replisome
   e. Production of the Rep protein is regulated by repressing its synthesis by a protein repressor. CopB. binding to the *rep* promoter and preventing transcription and by the CopA RNA. An anti–sense RNA which binds to the *rep* mRNA and prevents its translation.

3. Synthesis in the donor provides the single–strand that is copied. Synthesis in the recipient converts the transferred strand to double–stranded DNA.

4. a. The various genes of the *tra* operon
   b. A pilus, or conjugational bridge
   c. Rolling circle replication
   d. An endonucleolytic nick at *ori*T producing $5'-P_4$ and a $3'-OH$ ends.

5. a. An Hfr or high frequency of recombination strain

b. One of its several IS elements

c. Conjugation is usually interrupted before the entire chromosome can be transferred, though it is theoretically possible.

d. An excised F plasmid which contains some chromosomal DNA

6. Direct repeat: ABCD . . . . ABCD in which the dots represent nonrepeated bases. Inverted repeat: ABCD . . . . D'C'B'A' in which X' is the complement of X.

7. Conservative transposition involves no replication of the original element, only its movement to a new location. Replicative transposition involves the duplication of the transposon such that one copy remains in the original location, and a new copy is produced in a new location. In either case, the target sequence is duplicated.

8. An insertion element is essentially a transposase gene flanked by terminal inverted repeats of 16–41 base–pairs depending on the IS element. A simple transposon is a single IS element. One type of complex transposon contains an antibiotic–resistance gene flanked by two identical (or nearly identical) IS elements, which may be in either direct repeat or inverted repeat orientation. Two copies of the transposase gene are present, one in each IS element. The second type of complex transposon has only one copy of the transposase gene. It is not flanked by IS elements, but still has inverted repeats at its termini. In addition, an antibiotic–resistance gene and resolvase gene are present between the inverted repeats.

9. Most IS element insertion points are essentially random. However, some regions are inserted into more often than others, and these are called hot–spots.

Answers to Problems (Ch. 12)

1. a. Lac$^-$

   b. The F' (Ts) lac^+ has integrated into the bacterial chromosome.

   c. Integration has occurred inside a *gal* gene.

2. a. Amp$^-$r Tet$^-$s, plasmid A only; Amp$^-$s Tet$^-$r, plasmid B only; Amp-r Tet-r, both plasmid A and plasmid B.

   b. The numbers for the colonies which are Amp$^-$r Tet$^-$s or Amp$^-$s Tet$^-$r are very close to 2/9 of the total colonies (22.2) which would be the expected numbers for the loss of one or the other of two incompatible plasmids from the cell. Using Figure 12–7, this would be (1/2) (2/3) (1/2) + (1/2) (1/3) (1/3) (1/2) + (1/2) (1/3) (1/3) (1/2) = 2/9.

   c. The plasmids would be compatible in this case.

3. a. The RNase H mutant may be leaky. Alternatively, ColE1 may have an RNase H independent method of replication.

   b. Such mutations would destabilize the RNA I–primer RNA interaction such that the primer RNA–plasmid DNA hybrid would form more often, leading to more functional primer complexes, and replication initiation would occur more often. These mutations however, must not seriously affect the secondary structure of the primer RNA.

   c. The first mutation probably disrupts an important secondary structure interaction within the primer RNA and decreases the probability of the primer RNA–plasmid DNA hybrid forming, thus leading to a decreased copy number. The suppressing change probably restores his interaction, and the necessary secondary structure, though in a novel way.

4. a.

```
                                    transposon
    5'-----GATCCAGCAATGGCA--------------TGCCATTGCGATCCA-----3'
    3'-----CTAGGTCGTTACCGT--------------ACGGTAACGCTAGGT-----5'
b.  5'-----TTAGCA-3'            5'-TTAGCA-----3'
    3'-----AATCGT-5'            3'-AATCGT-----5'
    double-stranded break in donor molecules
c.
                                    transposon
    5'-----TTAGCAGCAATGGCA--------------TGCCATTGCTTAGCA-----3'
    3'-----AATCGTCGTTACCGT--------------ACGGTAACGAATCGT-----5'
```

5. a. If the mutation affects only the transposase repressor function, transposition frequency would go up, and the products would be an unaltered chromosome and a plasmid with the Tn3 element. If the mutation affects only the recombination function, transposition frequency would remain the same but the product would be a cointegrate, i.e. the plasmid would be integrated into the chromosome. If the mutation affects both functions, transposition frequency would go up, but only cointegrates would form.

b. No resolvase is made, so transposition frequency goes up but only cointegrates form, since transposase is not repressed, and resolution of the cointegrate cannot take place.

c. Transposition would be abolished either because the transposon excision or target DNA cutting functions are impaired. Either no products would be formed, or the donor or recipient molecules might be lost due to cleavage of only one of these molecules.

d. Transposition would be abolished since no transposase is made, and no products or intermediates would be detected.

e. There would be no effect on transposition or resolution, but screening for transposition becomes impossible since the selectable marker has been knocked out.

f. Transposition frequency would remain the same, but only cointegrates would form since resolvase cannot recognize the *res* sites.

6. The number of copies of the transposon in the new plasmid would answer the question. The number would be two copies if fusion were mediated by a transposon.

Answers to Drill Questions (Ch. 13)

1. a. A restriction enzyme is an endonuclease that makes cuts in DNA at one particular base sequence.

b. The biological function is to destroy foreign DNA.

c. Scientists use restriction enzymes to make defined, reproducible DNA fragments from DNA, and for cloning *in vitro*.

d. Each sequence is a palindrome. That is, it has rotational (dyad) symmetry. In other words, it reads the same in both directions.

2. The single–strand breaks may be directly opposite one another at the center of symmetry of the recognition site, or they may be staggered (several nucleotides apart, but symmetric around the line of symmetry of the recognition site). Breaks directly opposite one another produce blunt ends, while staggered cuts produce either 5′ or 3′ overhangs. Overhanging breaks are called cohesive ends, because they can stick to similar overhangs by base–pairing.

3. They must both recognize the same base–sequence.

4. An ideal cloning vector must have an origin of replication. Two selectable marker genes are helpful: one conferring antibiotic resistance or complementing a genetic mutation in the host such that only transformed cells live on a selecting medium, and another which distinguishes recombinant molecules from nonrecombinant ones by being inactivated when a fragment is cloned into it. A polylinker containing numerous, unique cloning sites would make the vector more versatile. If the fragment is to be expressed, a promoter sequence is necessary.

5. Insertional inactivation is the process of cloning a DNA fragment into a site within a vector gene such that the gene is no longer active. Usually the gene encodes an antibiotic resistance such that colonies transformed with a recombinant plasmid will be sensitive to the antibiotic. With α–complementation, only a portion of a selectable gene is present on the plasmid, while the host strain contains a different portion of the same gene. Neither is active alone, but together they function in a wild–type manner. Insertion into the vector–encoded portion of the gene prevents complementation.

6. If the DNA fragment is available, it may be radiolabeled and used for colony hybridization or DNA hybridization of membrane lifts taken from plates containing isolated plasmids or phage. If the fragment has a known restriction map, then a number of plasmid or phage DNA fragments can be isolated, cut with one or more restriction enzymes, and

analyzed by gel electrophoresis. The resulting patterns can be compared to the predicted ones. Ultimately, DNA from the plasmids or phage can be sequenced to verify the insert.

7. a. Kanamycin
    b. Kan-r Amp-r, Kan-r Amp-s
    c. Kan-r Amp-s; Colonies resistant to kanamycin can be replicated on agar containing ampicillin. Those that do not grow are Amp-s. Sensitive colonies can be obtained from a master plate.

8. Since DNA ligase requires a 5′–phosphate and a 3′–hydroxyl for ligation, removal by a phosphatase of the 5′–phosphate from the cut vector molecules will prevent religation of the vector alone. Another method is to use directional cloning which involves cutting the vector and using an insert with two different restriction enzymes. As long as the ends generated are not compatible, the plasmid can not religate. The small vector DNA fragment between the cut sites must, of course, be removed to prevent its favored religation into the plasmid over the desired insert.

9. A genomic library is made by cloning pieces of genomic (chromosomal) DNA into an appropriate vector. A cDNA library contains DNA fragments derived from cellular mRNA by reverse transcriptase. A genomic clone could contain promoter and regulatory elements as well as introns. The advantage here is that the gene may be studied as it exists in the organism with control elements intact. A cDNA clone would have no introns or control elements, but if supplied with a bacterial promoter, it could have been removed.

10. Site–specific mutagenesis is the alteration of a cloned gene *in vitro* either by synthesizing the entire gene using short, overlapping, single–stranded oligomers containing one or more specific changes from the wild-type sequence or by replacing a wild-type region of a cloned gene with a mutant region. The advantages over conventional mutagenesis are two–fold. First, the changes made are *specific;* they are the exact ones that the experimenter wishes to make. Second, the changes occur only in the gene of interest, and the time-consuming search for mutants in the gene and the complication of second–site mutations are eliminated. Several disadvantages are that the gene of interest often must be properly re-introduced into the organism, and since the experimenter chooses the changes that are made, some interesting mutations may be missed. Information gained from using site–specific mutagenesis includes understanding the role of specific amino acid residues in enzyme structure and function, the interaction of multiple proteins, and the location within an enzyme of its active site. Likewise, mutagenesis within promoter and regulatory regions can lead to information on the regulation of gene expression.

Answers to Problems (Ch. 13)

1. a. The EcoRI digest will generate eleven fragments from the linear molecule (n + 1 fragments, where n is the number of sites) and ten fragments (n fragments) from the circular molecule.
    b. Only nine of the fragments (n − 1) from the linear molecule will circularize, since the two end fragments will only have one EcoRI cohesive end. All of the fragments from the circular molecule can circularize. This assumes that all fragments are long enough to circularize.

2. a. BamHl and Bg1II
    plasmid:

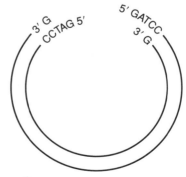

gene fragment:

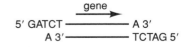

Yes, they are compatible as follows:

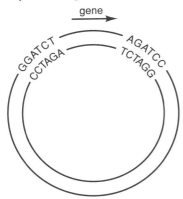

Neither site is regenerated. Note that the opposite cloning orientation is also possible.
b. BamHl and Sau3A
plasmid:

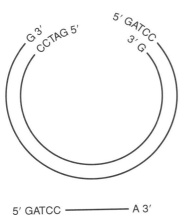

5′ GATCC ———————— A 3′
3′ G ———————— TCTAG 5′

These ends are also compatible as follows:

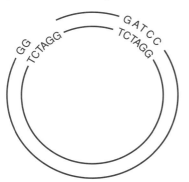

Again both cloning orientations are possible. In all cases, the Sau3A sites are regenerated. In this orientation, the left BamHl site is regenerated, but the right one is not.
c. BamHl and Dpnl
plasmid: same as in (a) and (b)
fragment:

GATCC ———————— AGA3′

TCTAGG ———————— TCT5′

These are not compatible. Although Dpnl recognizes the same sequence as Sau3A, the cut position is different. Dpnl produces blunt ends which are not compatible with the BamHl cohesive ends.

3. The following circles would be produced. In addition, double T5 and double T7 circles would probably form.

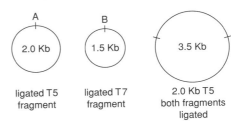

Since low concentrations of DNA ligase allow the double fragment circle to form, the enzymes probably produce cohesive ends with either 5′ or 3′ overhangs. Blunt end ligations require high concentrations of DNA ligase. Since both enzymes can cut the double circle, the enzymes probably recognize identical sequences. It is less likely that compatible cuts would regenerate sites cleavable by both enzymes.

4.

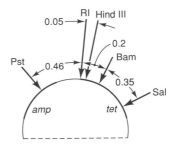

5.

(a)

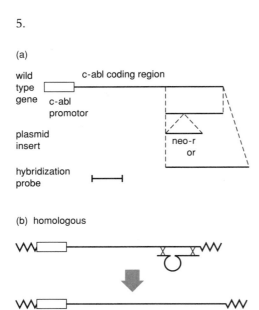

wild type gene

c-abl coding region

c-abl promotor

plasmid insert

neo-r or

hybridization probe

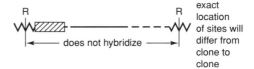

R   R
exact location of sites will differ from clone to clone

|← does not hybridize →|

All symbols are as in (a) and (b) above. R represents the restriction sites used for Southern analysis.

(b) homologous

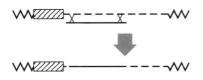

nonhomologous

Note that the open box again represents the c-*abl* promotor, while the hatched box represents a different cellular promotor. Sawtooth lines represent flanking chromosomes, and the dashed line represents the coding region of the gene downstream of the hatched promotor.

(c) homologous

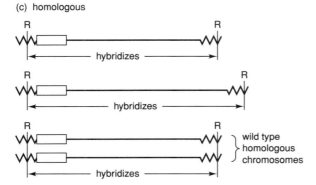

R   R
|← hybridizes →|

R   R
|← hybridizes →|

R   R
|← hybridizes →|

} wild type homologous chromosomes

6. Chromosome walking using a human genomic clone bank is required. The first step is to create a genomic bank. Human genomic DNA can be partially digested and the fragments cloned into a phage vector. A partial digest is necessary to produce the overlapping fragments required for chromosome walking.

An initial genomic clone can be isolated by probing the library with a radiolabeled probe from one of the cDNA clones by plaque hybridization (similar to colony hybridization but on phage plaques). Radioactive probes corresponding to both ends of this genomic clone should be prepared by radiolabeling short restriction fragments or synthesized oligomers. These may be used to screen the genomic library for clones on either side of the first fragment. The walk must proceed in both directions, since the orientation of the two genes on the chromosome is not known. The new fragments can be properly oriented by restriction mapping and/or nucleotide sequencing. These new fragments should be probed with a radioactive fragment from the second gene by Southern or phage plaque hybridization to see if the walk has proceeded into the second gene. If the second gene is not found within a reasonable distance in both directions from the first gene, they are probably not linked.

Answers to Drill Questions (Ch. 14)

1. In negative regulation an inhibitor bound to the DNA, must be removed before transcription can occur. In positive regulation an activator molecule must bind to the DNA. Negative and positive autoregulation is the same, except the gene product is its own inhibitor or activator.

2. A repressor protein binds to operator on DNA to prevent transcription. A corepressor is a small molecule that binds to a regulatory protein and an aporessor to prevent transcription.

3. Derepressed state is the same as induced when describing the normal state of a gene; it has the same meaning as constitutive in describing the effect of mutation.

4. F' O^c $lacZ^+/O^+$ $lacZ^-$; constitutive expression of lacZ.
F' O^c $lacZ^-/O^+$ $lacZ^+$; inducible expression of lacZ.

5. In the absence of arabinose, AraC binds to operator and prevents transcription. In the presence of arabinose, arabinose–AraC complex activates transcription from the *ara* promoter.

6. In eukaryotes transcription and translation are not coupled.

7. Frequency of transcription initiation
Frequency of transcription termination (attenuation)
mRNA stability

8. mRNA stability
Frequency of translation initiation
Protein stability

Answers to Problems (Ch. 14)

1. b. and probably also c.

2. In all cases *lacZ* will not be expressed.

3. Attenuation is controlled by the levels of charged tRNA^{trp} inside the cell. If there is no tryptophan present, the concentration of charged tRNA^{trp} becomes inadequate, and a translating ribosome stalls at the tryptophan codons of the leader peptide, preventing the terminator from forming, and hence preventing premature transcription termination. However, other mutations (RNase P) that decrease the levels of charged tRNAs in the cell will not induce transcription termination.

4. There might be a mutation in the gene for adenylate cyclase, or for the cyclic AMP receptor protein. Another possibility is a membrane transport mutant, if these sugars were to utilize the same transport system.

Answers to Drill Questions (Ch. 15)

1. One

2. Tailed icosahedra, Nontailed icosahedra, filaments

3. Single- and double-stranded DNA (linear or circular); single-stranded RNA and segmented double-stranded RNA (linear).

4. T4, T7, λ: linear, double-stranded DNA. M13: circular, single-stranded DNA. MS-2: linear, single-stranded RNA.

5. In general, the bacterium will lack an absorption site.

6. Proteins

7. Complementary single-stranded ends.

8. Virulent

9. 5-hydroxymethylcytosine (HMC)

10. Transducing

Answers to Problems (Ch. 15)

1. No, because DNase cannot penetrate a phage head.

2. Adsorption, entry of nucleic acid, transcription, production of phage proteins, production of phage nucleic acid, assembly of particles (including encapsidation of nucleic acid), release of progeny particles.

3. A protein coat to enclose the nucleic acid, a component capable of adsorbing to a bacterium, an element designed to enable the nucleic acid to penetrate the cell wall. The latter could be an injection system or a component that stimulates the bacterium to take in the particles.

4. A phage gene is considered nonessential if its product is not needed for growing the phage in the laboratory. Those genes that duplicate host genes may be truly nonessential, even in nature.

5. Phage-host specificity refers to the fact that a particular phage can grow on only one or a small number of bacterial species or strains. The most frequent cause is the inability of a phage to adsorb, except to certain bacteria.

6. No. Phage can only develop in a metabolizing bacterium.

7. It must be a host enzyme.

8. After the first transcript is made by the host RNA polymerase, T7 inactivates *E. coli* RNA polymerase and synthesizes a T7-specific polymerase. λ uses the bacterial polymerase throughout the life cycle.

9. A double-stranded circular molecule.

10. The linear molecule spontaneously circularizes via the single-stranded termini, and the single-stranded breaks are sealed by DNA ligase.

Answers to Drill Questions (Ch. 16)

1. RNA polymerase I transcribes genes encoding ribosomal 18S, 5.8S, and 28S RNA. RNA polymerase III transcribes genes encoding transfer RNA, ribosomal 5S RNA, and U6 small nuclear RNA (U6 snRNA). RNA polymerase II transcribes protein-encoding genes, and most of the sn RNA types.

2. The position of eukaryotic regulatory sequences with regard to the genes they regulate is often of great importance. However, in contrast to such regulatory sequences in prokaryotic genomes, far more difference in type and exact position of these regulatory sequences is seen in eukaryotes. In fact, some of these sequences (known as enhancers) can be experimentally moved, often hundreds of nucleotide pairs upstream or downstream, without affecting regulatory activity.

3. Variable region; A V segment and a J segment are joined to the C region of the gene. The C region encodes the constant region.

4. Increase in the stability, and thus, lifetime, of the mRNA. Activation of a strong promoter by a transcription factor. Activation of transcription through binding of a hormone to its receptor.

5. Active chromatin is at least partially devoid of histones and hypomethylated. To define such active regions chromatin can be treated with DNase to locate hypersensitive sites.

6. Both cis- and trans-splicing can occur in the processing of eukaryotic mRNA. With cis-splicing, exons from the same transcript are joined. In trans-splicing, an exon from one transcript is spliced to an exon located on a different mRNA.

7. The gene encoding the protein could have a unique promoter. An effector molecule could cause synthesis of two new proteins, an inhibitor of the cellular RNA polymerase, and a new RNA polymerase that could recognize only that promoter. There are also other possibilities.

8. An embryonic gene contains a large number of different base-sequences that constitute the coding sequences for all antibodies. These sequences are contiguous in the DNA. In the course of development, a genetic recombinational event removes large blocks of DNA that include many adjacent sequences. There are many blocks that can be removed, so that many different coding sequences can remain after this recombinational event occurs. One event occurs in a particular cell leaving that cell with a unique coding sequence that enables it to make a particular antibody.

Answers to Problems (Ch. 16)

1. a. Increase the number of J genes.

   b. $(150)(12)(3)(5000) = 2.7 \times 10^7$.

2. Very likely, unless gene B was somehow separated from some element needed for its activity.

3. Some upstream sites at which regulatory proteins seem to act are too far from the start site for a polymerase to bind both to a regulatory protein and the start site. Furthermore, many upstream sites can be moved somewhat without significantly affecting the rate of transcription, and enhancers can even be moved downstream of the genes with which they are associated.

4. E. somehow regulates excision of a single, 30-nucleotide intron near the 5' terminus of the

RNA. The 5′ initiation AUG is located either slightly upstream of the intron, or is contained within the intron. With the intron present, there is also an in–frame stop codon that may or may not be in the intron. Thus, without splicing, only a short product can be made due to the early stop caused by the in–frame stop codon. Once the intron is excised, and the downstream exon placed in the correct frame, the Q product can be translated utilizing either the upstream exon's AUG if present, or the first AUG contained in the downstream exon. Translation continues until the proper in–frame stop codon in the downstream exon is encountered.

5. a. The promoters for both transcription units probably have a common sequence acted on by either the effector itself, another factor which requires the effector for activity, or a negative regulator which is inactivated by the effector.

   b. Both primary transcripts have a common sequence acted on by an element that prevents some stage of processing. This element is inactivated by the effector.

   c. Both processed mRNA molecules have a common sequence involved in ribosome binding. The effector may remove a protein bound to this sequence, or it may denature a double–stranded region containing the ribosome binding site.

6. Probably the simplest explanation would be that the two mRNAs had different lifetimes. More complicated explanations include post-translational modification of one of the proteins decreasing its stability, or some inhibition of translation of one of the mRNAs.

7. No. Bent DNA can yield DNase hypersensitive sites. Furthermore, altered histone–phasing, no matter what its purpose, will show these sites. Not all lengths of DNA exposed by such phasing are regulatory binding sites such as promoters and enhancers.

Index